# Universality in Nonequilibrium Lattice Systems

Theoretical Foundations

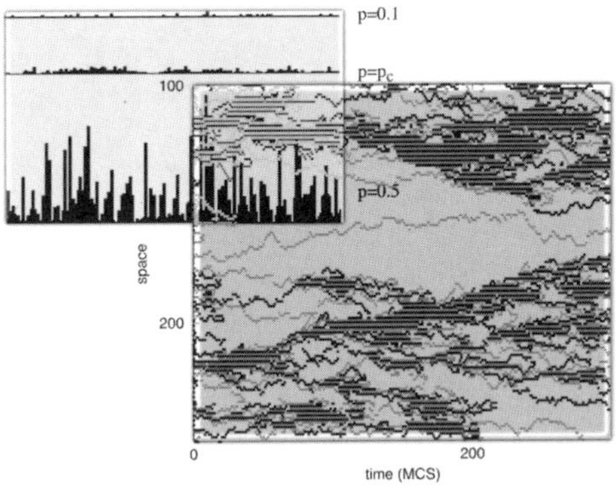

# Universality in Nonequilibrium Lattice Systems

Theoretical Foundations

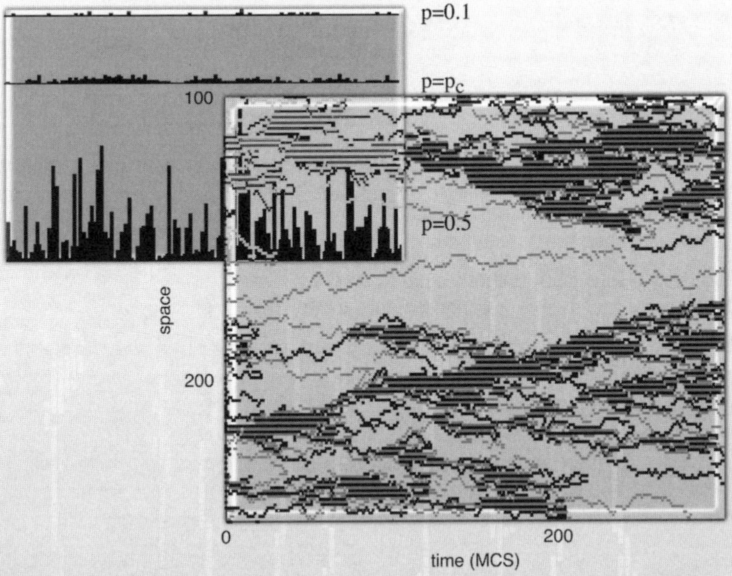


Géza Ódor

Research Institute for Technical Physics
and Materials Science, Hungary

World Scientific

NEW JERSEY • LONDON • SINGAPORE • BEIJING • SHANGHAI • HONG KONG • TAIPEI • CHENNAI

*Published by*

World Scientific Publishing Co. Pte. Ltd.
5 Toh Tuck Link, Singapore 596224
*USA office:* 27 Warren Street, Suite 401-402, Hackensack, NJ 07601
*UK office:* 57 Shelton Street, Covent Garden, London WC2H 9HE

**British Library Cataloguing-in-Publication Data**
A catalogue record for this book is available from the British Library.

**UNIVERSALITY IN NONEQUILIBRIUM LATTICE SYSTEMS**
**Theoretical Foundations**



ISBN-13 978-981-281-227-8
ISBN-10 981-281-227-X

Printed in Singapore by World Scientific Printers

# Preface

Universal scaling behavior is an attractive feature in statistical physics because a wide range of models can be classified purely in terms of their collective behavior due to a diverging correlation length. Scaling phenomenon has been observed in many branches of physics, chemistry, biology, economy etc., most frequently by critical phase transitions and surface growth. Nonequilibrium phase transitions appear, for example, in models of

- origin of life [Cardozo and Fontanari (2006b)]
- biological control systems [Kiyono *et al.* (2005)]
- enzyme biology [Berry (2003)]
- brain [Werner (2007)]
- population [Albano (1994)]
- spatiotemporal intermittency [Jabeen and Gupte (2005)]
- socio-physics [Baronchelli *et al.* (2007)]
- epidemics [Ligget (1985); Mollison (1977)]
- catalysis [Ziff *et al.* (1986); yin Hua (2004)]
- itinerant electron systems [Feldman (2005)]
- cooperative transport [Havlin and ben Avraham (1987); Chowdhury *et al.* (2000)]
- plasma physics [Knapek *et al.* (2007)]
- stock-prize fluctuations and markets [Bouchaud and Georges (1990); Kiyono *et al.* (2006)]
- collision of solids [Kun and Hermann (1999)]

The concept of self-organized critical (SOC) phenomena has been introduced some time ago [Bak *et al.* (1987)] to explain the frequent occurrence of scaling laws experienced in nature. The term SOC usually refers to a mechanism of slow energy accumulation and fast energy redistribution, driving a system toward a critical state. The prototype of SOC systems is the sand-pile model in which particles are randomly dropped onto a two dimensional lattice and the sand is redistributed by fast avalanches. Therefore, in SOC models, instead of tuning the parameters, an inherent mechanism is responsible for driving it to criticality. SOC mechanism has been proposed to model

- earthquakes [Bak and Tang (1989); Sornette (1989)]
- the evolution of biological systems [Bak and Sneppen (1993)]
- solar flare occurrence [Lu and Hamilton (1991)]
- fluctuations in confined plasma [Politzer (2000)]
- snow avalanches [Faillettaz *et al.* (2004)]
- rainfall [Peters *et al.* (2002)]

However, SOC critical classes can be shown to be equivalent to those of the ordinary critical ones by identifying the control parameters and the boundary conditions properly [Vespignani and Zapperi (1997); Dickman *et al.* (1998); Dickman (2002a)], therefore I shall discuss the critical behavior of SOC models briefly in Sect. 6.10.

Diverging correlation length — necessary to change the global symmetry at a second order phase transition point — and scaling may also occur away from the critical point. Naturally, in a fully ordered state (below the ordering temperature) the correlation length is infinite. If the interactions of the system is such that reaching this state requires diverging time, one finds dynamical scaling near that point. This happens usually in case of multi-particle, reaction-diffusion systems in the ordered phase (experimentally observed in [Kroon *et al.* (1993)]). In quantum matter, near absolute zero temperature thermal equilibration can be obstructed in the case of topological ordered ground states, where only the slow dynamical relaxation of defects pairs — via annihilation-diffusion — can occur [Chamon (2005)]. By quenching magnets to zero, temperate domain coarsening occurs by power laws since topological defects such as interfaces or vortices slow down the dynamics [Bray (1994)].

Rough surfaces and interfaces can also exhibit temporal and spatial scaling if the correlation length and time diverges. They are ubiquitous in nature and from a technological point of view the control of their roughness is becoming critical for applications in fields such as micro-electronics, image formation, surface coating or thin film growth [Chow (2000)]. Understanding the fundamental laws driving tumor development is one of the biggest challenges of contemporary science. Internal dynamics of a tumor reveals itself in a number of phenomena, one of the most obvious ones being the growth [Brutovsky *et al.* (2006); Martins *et al.* (2007)].

Earlier most of the models were investigated on regular lattice type of systems (approximating a smooth field theory by continuum limit). It is well known that for systems with translational symmetry (like a lattice) the influence of underlying structure becomes negligible at the critical point (i.e. when the correlation length is much larger than the cell spacing). Note that in systems with non-integer (fractal) dimensions the translational symmetry is replaced by discrete scale invariance and the topological details of the generating cell are present at any scale. As the consequence, log-periodic corrections to scaling — described by complex exponents — occur [Sornette (1998)], which I shall not discuss here. The advantage of lattice realizations is that they are simpler to handle than models in continuum space, e.g., they sometimes allow for exact results and are easier to be implemented in a computer. Furthermore, a bunch of emerging techniques may now be applied to lattice systems, including nonequilibrium statistical field theory. A general amazing result from these studies is that lattice models often capture the essentials of

- social organisms [Antal *et al.* (2001); Washenberger *et al.* (2007)]
- epidemics [Hinrichsen (2007a)]
- glasses [Chamon (2005)]
- electrical circuits
- transport [Ez-Zahraouy *et al.* (2006)]
- hydrodynamics [Díez-Minguito *et al.* (2005)]
- colloids, computational neuro-science [Furtado and Copelli (2006)]
- botany [K. A and Peak (2006)]

In the past few years, the interest focused on the research of complex scale-free networks [Albert and Barabási (2002); Barabási (2002); Dorogovtsev and Mendes (2003)]. Recently the dynamics and the phase transitions of network systems came under study [Aldana and Larralde (2004)]. Contrary to the regular lattices universality in network models is not so well defined, usually it depends on the underlying topology, therefore I shall not discuss it in this book. A very recent analytical study has shown that the asymptotics of random walks on uncorrelated random networks is essentially the same as that for regular Bethe lattices [Samukhin *et al.* (2007)].

Dynamical extensions of static universality classes — established in equilibrium — are the simplest nonequilibrium models systems, but beyond that critical phenomena, with classes have been explored so far [Marro and

Dickman (1999); Grassberger (1996); Hinrichsen (2000a)]. While the theory of phase transitions is quite well understood in thermodynamic equilibrium, its research in nonequilibrium is rather new. In general phase transitions, scaling and universality retain most of the fundamental concepts of equilibrium models. The basic ingredients affecting universality classes are again the collective behavior of systems, the symmetries, the conservation laws and the spatial dimensions as described by renormalization group (RG) theory. Besides these, several new factors have also been identified recently. Low dimensional systems are of primary interest because the fluctuation effects are relevant, hence the mean-field type of description is not valid. In the past decades, this field of research grew very quickly and now we are faced with a zoo of models, universality classes, strange notations and abbreviations. This book aims to help newcomers as well as researchers to navigate in the literature by systematically reviewing most of the explored universality classes. I define models by their field theory (when it is available), show their symmetries or other important features and list the critical exponents and scaling relations.

Nonequilibrium systems can be classified into two categories:

(a) Systems which do have a hermitian Hamiltonian and whose stationary states are given by the proper Gibbs-Boltzmann distribution. However, they are prepared in an initial condition which is far from the stationary state and sometimes, in the thermodynamic limit, the system may never reach true equilibrium. These nonequilibrium systems include, for example, phase ordering systems, spin glasses, glasses etc. I show the scaling behavior of the prototypes of such systems in the second chapter (Out of Equilibrium Classes). These are defined by the addition of simple dynamics to static models. The initial condition can be regarded as a boundary condition in the time direction, hence these models exhibit strong resemblance to static models near surfaces. They are discussed in Chapter 2.

(b) Systems without a hermitian Hamiltonian defined by transition rates, which do not satisfy the detailed balance condition (the local time reversal symmetry is broken). They may or may not have a steady state and even if they have one, it is not a Gibbs one. Such models can be created by combining different dynamics or by generating currents in them externally. In some cases, their critical behavior is insensitive to such changes and these are discussed in Chapter 2. There are also systems, which are not related to equilibrium models, in the simplest case these are lattice Markov processes of interacting particle systems [Ligget (1985)]. These are referred here as "genuinely nonequilibrium systems" and are discussed in the rest of the work.

The discussion of latter type of classes is split into five parts: In Chapter 3, transition classes of models possessing fluctuating ordered states are provided. There are not too many of them known yet, but the exploration of the phase transitions of current driven systems attracts much interest nowadays. In Chapter 4, phase transition classes of models with absorbing states are presented. These are usually reaction-diffusion (RD) type of models, but sometimes they are defined by spin systems or via a coarse grained Langevin equation. In Chapter 5, I briefly touch on the point of discontinuous nonequilibrium phase transitions and tricritical phenomena, especially because dynamical scaling may occur in such nonequilibrium cases. In Chapter 6, I list known classes, which occur by combinations of basic genuine class processes. These models are coupled, multi-component RD systems. While the former three chapters are related to critical phenomena near to extinction, in Chapter 7, I discuss universality classes in systems where site variables are non-vanishing in surface growth models. The bosonic field theoretical description is applicable for them. I point out mapping between growth and RD systems when it is known.

I define a critical universality class by the complete set of exponents at the phase transition. Therefore, different dynamics split up the basic static classes of homogeneous systems. I emphasize the role of symmetries and boundary conditions which affect these classes. I also point out very recent evidence, according to which in low-dimensional systems, symmetries are not necessarily the most relevant factors of universality classes. Although the systems covered here might prove to be artificial to experimentalists or to application-oriented people they constitute the fundamental blocks of understanding of nonequilibrium critical phenomena. Note that even the understanding of models so simple runs into tremendous difficulties very often.

I shall not discuss the critical behavior of quantum systems [Rácz (2002)] and just briefly mention experimental realizations. However, it is well known that a quantum phase transition (occurring at $T = 0$ due to quantum fluctuations) in $d$ space dimensions can be mapped onto a classical (finite temperature) transition at $d + Z$ dimensions (where $Z$ (see Sect. 1.3) is the dynamical exponent, and $Z = 1$ in space-time isotropic systems). The effect of boundary conditions in static models is reviewed elsewhere. The detailed discussion of the applied methods is also omitted due to the lack of space, although in Sect. 1.6 I give a brief introduction to the field theoretical approach in the first chapter. This section shows the formalism for defining nonequilibrium models. This is necessary to express the symmetry relations affecting critical behavior. The detailed discussion of renormalization group solutions is not provided in this book, and only mean-field theoretical derivations and scaling arguments are shown in some important cases.

Researchers from other branches of science are provided with a kind of catalog of classes in which they can identify their models and find corresponding theories. The classification of phase transition classes of one-component, bosonic reaction-diffusion models has been attempted very recently [Elgart and Kamenev (2006)]. This is based on topological portraits of "zero energy" lines of the reaction Hamiltonians in the phase-space, similarly to the Ginzburg-Landau potential minima in the case of equilibrium systems. Although the predictions of this scheme is not always in agreement with the results of other methods, especially in case of non-bosonic models (with topological site restrictions) it gives a constructive, organizational view for the zoo of nonequilibrium models and classes. I discuss this new method in Sect. 1.6.1 and the corresponding phase portraits will be shown in the field theoretical introduction of the reaction-diffusion classes.

Another very recent advance in this field is the recognition of more general scale transformations than mere rescaling. Similar to the conformal invariance in equilibrium systems, the concept of local scale invariance (LSI) has been introduced [Henkel (2002)]. I provide LSI scaling exponents and forms determined recently for some basic models and discuss their limitations.

Besides scaling exponents and scaling relations, there are many other interesting features of universality classes like scaling functions, extremal statistics, finite size effects, fluctuation-dissipation theory etc., which I do not discuss in this review. Still, I believe the material shown provides a useful frame for orientation in this huge field. There is no general theory of nonequilibrium phase transitions, hence a widespread overview of known classes can help theorists deduce the relevant factors determining universality classes.

There exists some recent, similar reviews in the literature. One of them is by [Marro and Dickman (1999)], which gives a pedagogical introduction to driven lattice gas systems and to fundamental particle systems with absorbing states. The other one [Hinrichsen (2000a)] focuses more on basic absorbing state phase transitions, methods and experimental realizations. A more technically detailed review on universal scaling of basic models exhibiting absorbing phase transitions supplements the second one [Lübeck

(2004)], focusing on scaling functions, external fields, crossovers and upper-critical behavior.

This book is based on a former review article by the author, which considered nonequilibrium universality classes systematically [Ódor (2004a)]. That overview aimed to give a comprehensive overview of known nonequilibrium dynamical classes, incorporating surface growth classes, classes of spin models, percolation and multi-component system classes and damage spreading behavior. Relations and mappings of the corresponding models were pointed out. Now effects of anisotropy, boundary conditions, long-range interactions, external fields and disorder are shown more systematically for each class. The crossover between classes is emphasized by boldface letters. This book provides an updated and extended review (as a result of lack of the size limitation of an article). The extensions includes local scale invariance, phase space topologies, non-perturbative renormalization group etc.

To help navigating in the text and in the literature I provided a list of the most common abbreviations in the appendix. Naturally this review cannot be complete and I apologize for any references I have inadvertently omitted.

*Géza Ódor*

# Acknowledgments

Support from the Hungarian Research Fund OTKA (Grant No. T-046129) during the preparation of this book is gratefully acknowledged.

# Contents

# Chapter 1

# Introduction

## 1.1 Critical exponents of equilibrium (thermal) systems

In this section I briefly summarize the definition of well known critical exponents of homogeneous equilibrium systems and show some scaling relations [Fisher (1967); Kadanoff *et al.* (1967); Stanley (1971); Ma (1976); Amit (1984)]. The basic thermal exponents (denoted by subscript '$H$' to avoid confusion with the some nonequilibrium ones defined later) are defined via the scaling laws:

$$c_H \propto \alpha_H^{-1} \left( (|T - T_c|/T_c)^{-\alpha_H} - 1 \right) , \tag{1.1}$$

$$m \propto (T_c - T)^{\beta} , \tag{1.2}$$

$$\chi \propto |T - T_c|^{-\gamma} , \tag{1.3}$$

$$m \propto H^{1/\delta_H} , \tag{1.4}$$

$$G_c^{(2)}(r) \propto r^{2-d-\eta_\perp} , \tag{1.5}$$

$$\xi \propto |T - T_c|^{-\nu_\perp} . \tag{1.6}$$

Here $c_H$ denotes the specific heat, $m$ the order parameter, $\chi$ the susceptibility and $\xi$ the correlation length. Note that the anomalous dimension exponent in the spatial two-point correlation scaling-law is denoted by $\eta_\perp$ in nonequilibrium systems — where $\perp$ corresponds to perpendicular to the time direction — therefore in the time being I shall use this notation. The presence of another degree of freedom besides the temperature $T$, like a (small) external field (labeled by $H$), leads to other interesting power laws when $H \to 0$. The $d$ present in the expression of two-point correlation function $G_c^{(2)}(r)$ is the space dimension of the system.

Some laws are valid both to the right and to the left of the critical point; the values of the relative proportionality constants, or *amplitudes*, are in general different for the two branches of the functions, whereas the exponent is the same. However there are universal amplitude relations among them. We can see that there are altogether six basic exponents Nevertheless they are not independent of each other, but related by some simple scaling relations:

$$\alpha_H + 2\beta + \gamma = 2, \;\; \alpha_H + \beta(\delta_H + 1) = 2, \tag{1.7}$$
$$(2 - \eta_\perp)\nu_\perp = \gamma_H, \;\; \nu_\perp d = 2 - \alpha_H \; .$$

The last relation is a so-called **hyper-scaling law**, which depends on the spatial dimension $d$ and is **not valid above the upper critical dimension** $d_c$, for example by the Gaussian theory. Therefore below $d_c$ there are only two independent exponents in equilibrium.

Below the **lower critical dimension** ($d_c^-$) the fluctuations are so strong that they completely destroy the ordered state, hence no order-disorder phase transition can exist. In equilibrium models finite-range interactions cannot maintain long-range order below $d_c^- = 2$ [Landau and Lifshitz (1981)].

One of the most interesting aspects of second order phase transitions is the so-called **universality**, i.e., the fact that systems which can be very different from each other share the same set of critical indices (exponents and some amplitude ratios). One can therefore hope to assign all systems to **classes** each of them being identified by a set of critical indices.

## 1.2 Static percolation cluster exponents

Universal behavior may occur at percolation [Stauffer and Aharony (1994); Grimmett (1999)], which can be considered a **purely geometrical phenomenon** describing the occurrence of infinitely large connected clusters on lattices. On the other hand such clusters emerge at critical phase transition of lattice models indeed. The definition of connected clusters is not unambiguous. It may mean the set of sites or bonds with variables in the same state, or connected by bonds with probability $b = 1 - \exp(-2J/kT)$. The differences in correlated cluster behavior of critical domains will be discussed in Chapter 2. following the definition of basic models.

By changing the system control parameters ($p \to p_p$) (that usually is the temperature in equilibrium systems) the coherence length between sites may diverge as

$$\xi(p) \propto |p - p_p|^{-\nu_\perp} \;\; , \tag{1.8}$$

hence percolation at $p_p$ like standard critical phenomena exhibits renormalizibility and universality of critical exponents. At $p_p$ the cluster size ($s$) distribution (the number of $s$-clusters divided by the total number of lattice sites) follows the scaling law:

$$n_s \propto s^{-\tau} f(|p - p_p|s^\sigma) \ . \tag{1.9}$$

Formally we may define a free energy $F$ as a generating function of a ghost field (in analogy with equilibrium statistical physics)

$$F(h) = \sum_s n_s e^{-hs} \ , \tag{1.10}$$

such that its $h$ derivatives, the moments of 1.9 exhibit the leading singular behavior, characterized by the exponents:

$$\sum_s n_s(p) \propto |p - p_p|^{2-\alpha_p} \ , \tag{1.11}$$

$$\sum_s s n_s(p) \propto |p - p_p|^{\beta} \ , \tag{1.12}$$

$$\sum_s s^2 n_s(p) \propto |p - p_p|^{-\gamma} \ , \tag{1.13}$$

$$\sum_s s n_s(p) e^{-hs} \propto h^{1/\delta_p} \quad . \tag{1.14}$$

These critical exponents are formally related by the same scaling laws as the thermal ones (1.7). The cluster exponents $\tau$, $\sigma$ are connected to them via

$$\tau = 2 + 1/\delta_p, \quad 1/\sigma = \beta\delta_p \ , \tag{1.15}$$

hence only two of them are independent. Note that for exponents $\alpha$ and $\delta$ I introduced the subscript '$p$' here, referring to percolation (contrary to the subscript '$H$' of analogous exponents characterizing thermal properties in the previous section) and there are also plain $\alpha$ and $\delta$ (for historical reasons) to be introduced in Sect. 1.37, which describe a different scaling law.

In case of completely random placement of (sites, bonds, etc) variables (with probability $p$) on lattices we find **random isotropic (ordinary) percolation** (see Sect. 4.4.1). Percolating clusters may arise at critical, thermal transitions or by nonequilibrium processes. If the critical point ($p_c$) of the order parameter does not coincide with $p_p$ than at the percolation transition the order parameter coherence length is finite and does not influence the percolation properties. We observe random percolation in that case. In contrast if $p_p = p_c$ percolation is influenced by the order parameter behavior and we find different, **correlated percolation** universality [Fortuin and Kasteleyn (1972); Coniglio and Klein (1980); Stauffer and Aharony (1994)] whose exponents may coincide with those of the order parameter.

According to the Fortuin-Kasteleyn construction of clusters [Fortuin and Kasteleyn (1972)] two nearest-neighbor spins of the same state belong to the same cluster with probability $b = 1 - \exp(-2J/kT)$. It was shown that using this prescription for $Z_n$ and $O(n)$ and symmetric models [Coniglio and Klein (1980); Bialas *et al.* (2000); Fortunato and Satz (2001); Blanchard *et al.* (2000)] the **thermal phase transition** point coincides with the percolation limits of such clusters. On the other hand in case of "pure-site clusters" ($b = 1$) different, universal cluster exponents are reported in two dimensional models [Fortunato and Satz (2001); Fortunato (2002)] (see Sects. 2.3.1, 2.4.1, 2.6.1).

## 1.3 Dynamical critical exponents

Nonequilibrium systems were first introduced to study relaxation to equilibrium states [Hohenberg and Halperin (1977)] and phase ordering kinetics [Binder and Stauffer (1974); Marro *et al.* (1979)] (see Fig. 1.1). Since at the critical point ($T_c$) and at the ordered $T = 0$ situation the the spatial correlation length and time are divergent (or they are much larger than the microscopic length and time scales) universal dynamical scaling is natural. Power-law time dependences were investigated away from the critical point as well, example by the domain growth in a quench to low temperature $0 < T < T_c$. Later the combination of different heat-baths, different dynamics, external currents became popular investigation tools of fully nonequilibrium models.

To describe the dynamical behavior of a critical system scaling hypothesis was introduced [Ferrell *et al.* (1967); Halperin and Hohenberg (1967)] and additional exponents were introduced. For example the relation of the divergences of the relaxation time $t_R$ and correlation length $\xi$ is described by the **dynamical exponent** $Z$

$$t_R \propto \xi^Z \quad . \tag{1.16}$$

Systems out of equilibrium may show anisotropic scaling of two (and n) point functions

$$G(b\mathbf{r}, b^{\varsigma} t) = b^{-2x} G(\mathbf{r}, t) \tag{1.17}$$

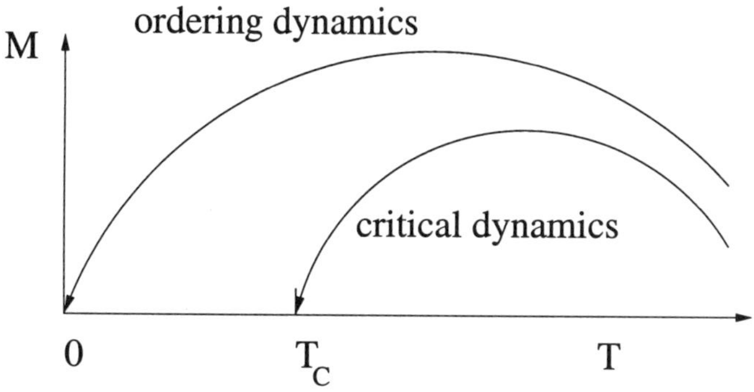


Fig. 1.1 Different dynamics considered by a quench from high temperature to $T = 0$ (ordering) or to $T_c$ (critical).

where $\mathbf{r}$ and $t$ denote spatial and temporal coordinates, $x$ is the scaling dimension and $\zeta$ is the anisotropy exponent. As a consequence the temporal ($\nu_{||}$) and spatial ($\nu_{\perp}$) correlation length exponents may be different, described by $\zeta = Z$.

$$Z = \zeta = \frac{\nu_{||}}{\nu_{\perp}} \quad . \tag{1.18}$$

For some years it was believed that dynamical critical phenomena are characterized by a set of three critical exponents, comprising two independent static exponents (other static exponents being related to these by scaling laws) and the dynamical exponent $Z$. However, the large correlation time induces a memory effect. In a magnetic system exhibiting a small initial magnetization $m_0$ a quench to $T_c$ (Fig. 1.1) a short time scaling emerges (see Fig. 1.2). First the magnetization increases as $\propto t^{\eta}$ until the ordered domains begin to feel each other ($t <\sim m_o^{Z/x_0}$), than this crosses over to the long-time power-law decay ($\propto t^{-\beta/\nu_{||}}$), finally exponential cutoff may occur due to finite size effects ($t >\sim L^Z$). Note that in many papers $\eta$ is denoted by $\theta$. This is very similar to the magnetization profile near a surface with different bulk and surface magnetizations (caused by an external magnetic field), because a $d$ dimensional dynamical system is like a $d+1$ dimensional static one. In this sense the initial state corresponds to the surface state of a $d+1$ static model. This kind of so-called 'initial slip' behavior occurs at spreading from an initial active site in reaction-diffusion models as well (see. [Grassberger and de la Torre (1979)], Sect. 1.5).

A universal dynamic scaling form, which sets in right after a time scale $t_{mic}$ which is enough large in *microscopic* sense but still very small in *macro-*

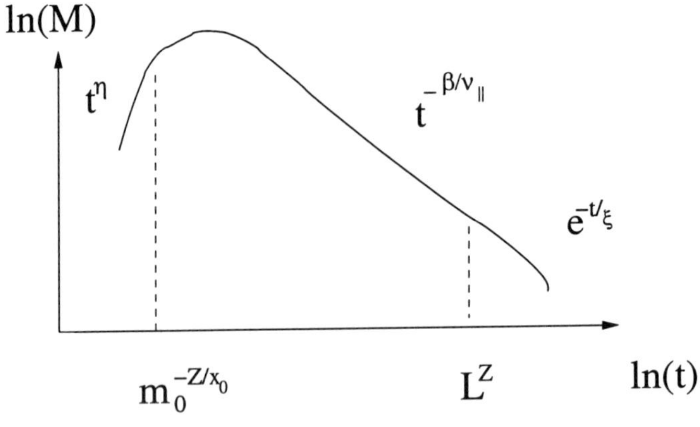


Fig. 1.2 Sketch of the evolution of magnetization in a quench to $T_c$ from a high temperature state with initial magnetization $m_0$.

*scopic* sense, has been derived with an $\epsilon$-expansion [Janssen *et al.* (1989)]. Important is that extra critical exponents should be introduced to describe the dependence of the scaling behavior on the initial conditions. For example, for the $k$-th moment of the magnetization, the finite size scaling form is written as [Janssen *et al.* (1989)]

$$M^{(k)}(t,\tau,L,m_0) = b^{-k\beta/\nu_\perp} M^{(k)}(b^{-Z}t, b^{1/\nu_\perp}\tau, b^{-1}L, b^{x_0}m_0) \qquad (1.19)$$

here the **new independent** exponent $x_0$ is the scaling dimension of the initial magnetization $m_0$. For $\tau = 0$ in a large lattice ($L \to \infty$) with small $t$ and $m_0$ we can derive

$$M(t) \sim m_0\, t^{\eta}, \qquad \eta = (x_0 - \beta/\nu_\perp)/Z. \qquad (1.20)$$

For almost all statistical systems studied up to now, the exponent $\eta$ is *positive*, i.e. the magnetization undergoes surprisingly a **critical initial increase**. The time scale of this increase is $t_0 \sim m_0^{-Z/x_0}$. As we shall see later similar initial slip may occur in genuinely nonequilibrium systems as well. The slip exponents may be related to others, depending on the symmetries of the system.

By studying phase ordering kinetics of magnets it was discovered that in case of a quench to critical or zero temperature magnets exhibit slow, non-exponential relaxation, dynamical scaling, and time dependent observable depend on the **age** of the system (for a review see [Bray (1994)]). **Physical ageing** was first identified in glassy systems [Struik (1978)] (for a recent review of ageing see [Cugliandolo (2003)]). For characterizing ageing the introduction of two-point functions are need, where in general **translational invariance in time is not satisfied.**

First it was discovered that there is another dynamical exponent, the 'nonequilibrium' or **short-time exponent** $\lambda_C$, needed to describe **two-time correlations** in a spin system ($\{s_i\}$) of size $L$ relaxing to the critical state from a random initial condition ($m_0 = 0$) [Janssen *et al.* (1989); Huse (1989)].

$$A(t,0) = \frac{1}{L^d} < \sum_i s_i(0)s_i(t) > - < s_i(0) >< s_i(t) > \propto t^{-\lambda_C/Z} . \quad (1.21)$$

Generally the $\lambda_C$ autocorrelation exponent may or may not be independent from other dynamical exponents (see for example Sects. 2.3.2, 2.6, 4.2.1). But even for the same model it depends on the initial conditions. For example in case of $O(N)$ models with $m_0 \neq 0$ magnetized initial state it is not independent [Calabrese and Gambassi (2007)].

If the initial time of the auto-correlator $A(t,s)$ is not zero ($s > 0$) in the asymptotic (ageing) limit: $t, s \to \infty$ and $y = t/s > 1$ one expects the scaling form

$$A(t,s) = s^{-b} f_C(t/s) , \quad (1.22)$$

where $b$ is the "ageing" exponent and $f_C$ is the corresponding scaling function such that

$$f_C(y) \sim y^{-\lambda_C/Z} \quad (1.23)$$

for $y \gg 1$.

Another two-point function characterizing ageing is the linear **autoresponse function**, which describes the system response for an applied conjugated magnetic field at time $s$ and location $\tau$. In the limits $t, s \to \infty$ and $y = t/s > 1$ (ageing regime) it exhibits the scaling

$$R(t,s) = s^{-1-a} f_R(t/s) \quad (1.24)$$

where $a$ is the autoresponse ageing exponent and $f_R$ is a scaling function such that

$$f_R(y) \sim y^{-\lambda_R/Z} \quad (1.25)$$

for $y \gg 1$. For **non-disordered magnets**, with short-ranged initial correlations

$$\lambda_C = \lambda_R \quad (1.26)$$

usually (for more discussion see Sect. 2.1).

For **quenches to** $T = T_c$ equilibrium state scaling arguments [Godrèche and Luck (2000a)] and renormalization approach [Janssen *et al.* (1989); Janssen (1992)] predict that the relevant length-scale is set by the time dependent correlation length

$$L(t) \sim \xi(t) \sim t^{1/Z} , \tag{1.27}$$

hence the ageing exponents are related to the equilibrium ones via the spatial two-point correlator $G(\mathbf{r}, t)$

$$a = b = (d - 2 - \eta_\perp)/Z , \tag{1.28}$$

and the autoresponse function follows the form **for every** $t$ **and** $s$

$$R(t, s) = A_R s^{-1-a}(t/s)^{\eta}(t/s - 1)^{-1-a} f_R(t/s) , \tag{1.29}$$

where $A_R$ is a non-universal constant, and $f_R(t/s)$ is a universal scaling function such that $f_R(t/s \gg 1) = 1$.

On the other hand for **quenches to** $T < T_c$ one usually observes simple scaling of $A(t, s)$ with $b = 0$. The value of the response function ageing exponent $a$ depends on whether the equilibrium correlator is short (S) or long (L) (see Table 1.1) and expression (1.29) **holds in the ageing regime only**.

Table 1.1 Nonequilibrium exponents for non-conserved ferromagnets with $T_c > 0$ and $m_0 \neq 0$

| | $Z$ | $a$ | $b$ | Eq. corr. range |
|---|---|---|---|---|
| $T = T_c$ | dynamics dep. | $(d-2+\eta_\perp)/Z$ | $(d-2+\eta_\perp)/Z$ | L |
| $T < T_c$ | 2 | $(d-2+\eta_\perp)/Z$ | 0 | L |
| $T < T_c$ | 2 | $1/Z$ | 0 | S |

More recently the **persistence exponents** $\theta_l$ and $\theta_g$ were introduced by [Derrida *et al.* (1994); Majumdar *et al.* (1996)]. These are associated with the probability $p(t)$, that the local or global order parameter has not changed sign in time $t$ following a quench to the critical point. In many systems of physical interest these exponents decay algebraically as

$$p(t) \propto t^{-\theta} \tag{1.30}$$

(see however example Sect. 6.1). It turned out that in systems where the scaling relation

$$\theta_g Z = \lambda_c - d + 1 - \eta_\perp/2 \tag{1.31}$$

is satisfied the dynamics of the global order parameter is a Markov process. In contrast in systems with non-Markovian global order parameter $\theta_g$ is in general a new, non-trivial critical exponent [Majumdar *et al.* (1996)], meaning that in this case **the persistence probability cannot be expressed by finite $n$-point correlators**. For example it was shown that while in the $d = 1$ Glauber Ising model the magnetization is Markovian and the scaling relation (1.31) is fulfilled, at the critical point of the $d = 1$ NEKIM Ising model this is not satisfied and the persistence behavior there is characterized by a different, non-trivial $\theta_g$ exponent [Menyhárd and Ódor (1997)] (see discussion in Sect. 4.6.2). As we can see the *universality classes of static models are split by the dynamical exponents.*

## 1.4 Crossover between classes

A model with some additional condition like an external field, surface ... may exhibit different universal scaling behavior by varying the parameters (in the field theoretical language (see Sect. 1.6) it may posses several competing fixed points in the phase space).[1] By tuning the control parameters close to these points we may see corrections to scaling. In terms of scaling forms, crossover phenomena are described by additional relevant scaling fields, characterized by a so-called crossover exponent [Riedel and Wegner (1969)].

For example let us consider a RD model exhibiting a phase transition at the control parameter value $p = p_0$, and add an additional reaction with a small probability $w$. If we define the relative distances in the phase diagram $\Delta = (p_0 - p)/p_0$ and $\Delta_c(w) = (p_0 - p_c(w))/p_0$ the phase boundary is well fitted by

$$\Delta_c \sim w^{1/\phi} \tag{1.32}$$

which means that the crossover exponent describes the shape of the critical line as it approaches the crossover point in the phase diagram. In this case a crossover scaling is described by the scaling function (see for example [Lawrie and Sarbach (1984)])

$$\rho(\Delta, w; t) = t^{-\alpha} F(\Delta^{\nu_{||}} t, w^{\mu_{||}} t), \tag{1.33}$$

where $\mu_{||} = \nu_{||}/\phi$ with $\phi$ estimated in the above. Further scaling forms and relations to other exponents are discussed later.

[1] It is matter of taste to call a model modified in such a way a different one.

Although crossover is well understood by field theory, several aspects are still open and are discussed in the literature. For example, the question whether the so-called effective exponents fulfill scaling laws over the entire crossover region has been revisited several times.

## 1.5 Critical exponents and relations of spreading processes

In the previous section I defined quantities describing dynamical properties of the bulk of a material. In a dual way to this one may also consider cluster properties arising by initiating a process from an ordered (correlated) state with a small cluster of activity. Here I define a basic set of critical exponents that occur in spreading processes and show the scaling relations among them. In such processes phase transition may exist to absorbing state(s) where the density of spreading entity (particle, agent, epidemic etc.) disappears. The order parameter is usually the density of active sites $\{s_i\}$ (particles, kinks, spins ... etc.) of a $d$-dimensional lattice of size $L$

$$\rho(t) = \frac{1}{L^d}\langle \sum_i s_i(t) \rangle \,, \tag{1.34}$$

which in the supercritical phase vanishes with the leading power behavior

$$\rho^\infty \propto |p - p_c|^\beta \,, \tag{1.35}$$

as the control parameter $p$ is varied. The "dual" quantity is the ultimate survival probability $P_\infty$ of an infinite cluster of active sites that scales in the active phase as

$$P_\infty \propto |p - p_c|^{\beta'} \tag{1.36}$$

with some critical exponent $\beta'$ [Grassberger and de la Torre (1979)]. In field theoretical description of reacting particles (see Sect. 1.6) $\beta$ is associated with the annihilation, $\beta'$ with the particle creation operator and in case of rapidity reversal symmetry (see Eq. (4.41)) they are equal.

The critical long-time behavior of these quantities are described by

$$\rho(t) \propto t^{-\alpha} f(\Delta\, t^{1/\nu_{||}}) \,, \qquad P(t) \propto t^{-\delta} g(\Delta\, t^{1/\nu_{||}}) \,, \tag{1.37}$$

where $\alpha$ and $\delta$ are the critical exponents for decay and survival, $\Delta = |p-p_c|$, $f$ and $g$ are *universal scaling functions* [Grassberger and de la Torre (1979); Muñoz *et al.* (1997); Janssen (2003)]. The obvious scaling relations among them are

$$\alpha = \beta/\nu_{||} \,, \qquad \delta = \beta'/\nu_{||} \,. \tag{1.38}$$

For *finite systems* (of size $N = L^d$) these quantities scale as

$$\rho(t) \propto t^{-\beta/\nu_{||}} f'(\Delta t^{1/\nu_{||}}, t^{d/Z}/N) \;, \tag{1.39}$$

$$P(t) \propto t^{-\beta'/\nu_{||}} g'(\Delta t^{1/\nu_{||}}, t^{d/Z}/N) \,, \tag{1.40}$$

leading to finite size scaling laws

$$t_R \propto L^Z, \quad \rho_s \propto L^{-\beta/\nu_\perp}, \quad \chi_s \propto L^{-\gamma/\nu_\perp} \;, \tag{1.41}$$

where $\rho_s$ is the steady state density and $\chi_s$ is its fluctuation averaged over the surviving samples.

For *"relatively short times"* or for initial conditions with a single active seed the the number of active sites $N(t)$ and its mean square of spreading distance $(x_i)$ from the origin

$$R^2(t) = \frac{1}{N(t)} \langle \sum_i x_i^2(t) \rangle \tag{1.42}$$

follow the "initial slip" scaling laws [Grassberger and de la Torre (1979)]

$$N(t) \propto t^{\eta} \quad , \tag{1.43}$$

$$R^2(t) \propto t^{z} \quad . \tag{1.44}$$

The spreading exponents in general are related by the **hyperscaling relation** derived in [Mendes *et al.* (1994)]

$$2\left(1 + \frac{\beta}{\beta'}\right)\delta + 2\eta = dz \tag{1.45}$$

but in case of symmetries special versions of it hold (see example (4.43)).

The derivation of (1.45) goes along the following lines. It starts with the general expression for the density of particles at space-point $r$ in the absorbing phase ($\Delta < 0$) and at large fixed value of $t$ (for $d = 1$):

$$\rho(r,t) = t^{\eta - z/2} F(r^2/t^z, \Delta t^{1/\nu_{||}}) \tag{1.46}$$

and with the relations (1.36),(1.37),(1.38) for the survival probability. Since the stationary distribution is unique

$$\rho(x,t) \to P_\infty \Delta^\beta \sim \Delta^{\beta + \beta'} \tag{1.47}$$

as $t \to \infty$. Hence $F(0,y) \sim y^{\beta+\beta'}$ which entails Eq. (1.45). If for example $\beta' = \beta$ one gets the special form of Eq.(1.45)

$$2\delta + \eta = z/2 \; . \tag{1.48}$$

In the absorbing phase $\rho(r,t)$ is expected to decrease exponentially as $\rho(r,t) \propto \exp(-r/\xi)$ usually. This implies for $F(u,v)$ (with $v < 0$) the form

$$F(u,v) \rightarrow \exp\left(-C\sqrt{u}|v|^{\nu_\perp}\right) \tag{1.49}$$

where $C > 0$ is constant. For $\xi$ to be time-independent the scaling law is required:

$$z = \frac{2\nu_\perp}{\nu_{||}} = \frac{2}{Z} , \tag{1.50}$$

i.e. the dynamical exponent of the model is related to the cluster spearing exponents of its components. In some anisotropic system (see example Sect. 4.6.9) this relation was found to be broken [Menyhárd and Ódor (2002)].

Another interesting quantity, which can be observed in nature relatively easily is the **fractal dimension** of spreading clusters. The average number of active sites per surviving clusters scales as $N_{sur}(t) = N(t)/P(t) \propto t^{\eta+\delta}$ and the fractal dimension $d_f$ is defined via

$$N_{sur}(t) \propto R(t)^{d_f} \tag{1.51}$$

resulting in the scaling law

$$d_f = Z(\eta + \delta) = d - \beta/\nu_\perp . \tag{1.52}$$

### 1.5.1 *Damage spreading exponents*

Phase transitions between chaotic and non-chaotic states may be described by damage spreading (DS). While DS was first introduced in biology [Kauffman (1969)] it has become an interesting topic in physics as well [Creutz (1986); Stanley *et al.* (1986); Derrida and Weisbuch (1987)]. The main question is if a damage introduced in a dynamical system survives or disappears. To investigate this the usual technique is to make replica(s) of the original system and let them evolve with the same dynamics and external noise. This method has been found to be very useful to measure accurately dynamical exponents of equilibrium systems [Grassberger (1995a)]. It has turned out however, that DS properties do depend on the applied dynamics. An example is the case of the two-dimensional Ising model with heat-bath algorithm versus Glauber dynamics [Mariz *et al.* (1990); Jan and de Arcangelis (1994); Grassberger (1995b)].

To avoid dependency on the dynamics a definition of "physical" family of DS dynamics was suggested by [Hinrichsen *et al.* (1997b)] according to the active phase may be divided to a sub-phase in which DS occurs for every member of the family, another sub-phase where the damage heals for every member of the family and a third possible sub-phase, where DS is possible for some members and the damage disappears for other members. The family of possible DS dynamics is defined to be consistent with the physics of the single replicas (symmetries, interaction ranges etc.).

Usually the order parameter of the damage is the Hamming distance between replicas

$$D(t) = \left\langle \frac{1}{L} \sum_{i=1}^{L} |s(i) - s'(i)| \right\rangle \tag{1.53}$$

where $s(i)$ and $s'(i)$ denote variables of the replicas. At continuous DS transitions $D$ exhibits power-law singularities as physical quantities at the critical point. For example one can follow the fate of a single difference between two (or more) replicas and measure the spreading exponents:

$$D(t) \propto t^{\eta_d} \tag{1.54}$$

Similarly the survival probability of damage variables behaves as:

$$P_D(t) \propto t^{-\delta_d} \tag{1.55}$$

and similarly to Eq. (1.42) the average mean square spreading distance of damage variables from the center scales as:

$$R_D^2(t) \propto t^{z_d} \; . \tag{1.56}$$

Grassberger conjectured, that DS transitions should belong to DP class (see Sect. 4.2) unless they coincide with other transition points and provided the probability for a locally damaged state to become healed is not zero [Grassberger (1995c)]. This hypothesis has been confirmed by simulations of many different systems.

## 1.6 Field theoretical approach to reaction-diffusion systems

In this review I define nonequilibrium systems formally by their (bosonic) field theoretical action where it is possible. Therefore in this subsection I give a brief introduction to the field theoretical formalism. This will be through the simplest example of reaction-diffusion systems, via the: $A + A \to \emptyset$ annihilating random walk (ARW) (see Sect. 4.5.1). Similar stochastic differential equation formalism can also be set up for growth processes in most cases. For a more complete introduction see [Cardy (1996, 1997); Täuber *et al.* (2005)].

A proper field theoretical treatment should start from the **Master equation** for the microscopic time evolution of probabilities $p(\alpha;t)$ of states $\alpha$

$$\frac{dp(\alpha;t)}{dt} = \sum_{\beta} R_{\beta\to\alpha} p(\beta;t) - \sum_{\beta} R_{\alpha\to\beta} p(\alpha;t) \ , \tag{1.57}$$

where $R_{\alpha\to\beta}$ denotes the transition matrix from state $\alpha$ to $\beta$. In field theory this can be expressed in Fock space formalism with annihilation ($a_i$) and creation ($c_i$) operators satisfying the commutation relation [Doi (1976)]

$$[a_i, c_j] = \delta_{ij} \ , \quad [a_i, a_j] = [c_i, c_j] = 0 \ . \tag{1.58}$$

The states are built up from the vacuum $|0>$ as the linear superposition

$$\Psi(t) = \sum_{\alpha} p(n_1, n_2, ...; t) c_1^{n_1} c_2^{n_2} ... |0> \ , \tag{1.59}$$

with occupation number coefficients $p(n_1, n_2, ...; t)$. The evolution of states can be described by a Schrödinger-like equation

$$\frac{d\Psi(t)}{dt} = -H\Psi(t) \tag{1.60}$$

with a generally non-hermitian (quasi-) Hamiltonian, which in case of the ARW process looks like

$$H = D\sum_{ij}(c_i - c_j)(a_i - a_j) - \lambda\sum_{j}(a_j^2 - c_j^2 a_j^2) \ , \tag{1.61}$$

here $D$ denotes the diffusion strength and $\lambda$ the annihilation rate. If we wanted to treat site restricted (sometimes called fermionic) models precisely — and in low dimensions particle exclusion turns out to be relevant in many cases (see Sect. 6.13) — we should use Pauli spin matrixes instead of bosonic operators. Such models are often integrable in one dimension [Alcaraz *et al.* (1994); Henkel *et al.* (1997); Schütz (2001)]. However fermionic field theories are much more difficult to solve. They involve non-commutative, Grassmann algebra and since near a transition to the absorbing state (which is very common in nonequilibrium models) the particle density is very low, they are expected to be equivalent to the bosonic theories.

By going to the continuum limit Eq.(1.61) turns into

$$H = \int d^d x \left[ D(\nabla\psi)(\nabla\phi) - \lambda(\phi^2 - \psi^2\phi^2) \right] \ , \tag{1.62}$$

and in the path integral formalism [Peliti (1985)] one can introduce a partition function over fields $\phi(x,t)$ and "response fields" $\psi(x,t)$ with the statistical weight $e^{-S(\phi,\psi)}$. The "response field $\psi(x,t)$ [Martin *et al.* (1973)] is defined by the two-point response function

$$R(t,s) = \left.\frac{\delta\langle\phi(t,\vec{r})\rangle}{\delta h(s,\vec{r})}\right|_{h=0} = \langle\phi(t,\vec{r})\psi(s,\vec{r})\rangle \tag{1.63}$$

where $h$ is the 'magnetic' field conjugate to $\phi$. For example the action $S(\phi,\psi)$ [Janssen (1976); de Dominics and Peliti (1978)], in case of ARW is

$$S = \int dt d^d x \left[\psi\partial_t\phi + D\nabla\psi\nabla\phi - \lambda(\phi^2 - \psi^2\phi^2)\right] \ . \tag{1.64}$$

The action is usually analyzed by renormalization group (RG) methods at criticality [Ma (1976); Amit (1984); Zinn-Justin (2002); Täuber *et al.* (2005); Täuber (to be published)]. The most common way of RG calculations is done via perturbative epsilon expansion (PRG) below the upper critical dimension $d_c = \epsilon + d$, which is the lower limit of the validity of the mean-field (MF) behavior of the system. However in nonequilibrium statistical physics no RG calculation is available at and above three loop order, preventing to use re-summation techniques and to compute accurately universal quantities in low dimensions.

In the field theoretical formalism the symmetries of a model can be expressed in terms of the $\phi(x,t)$ field and $\psi(x,t)$ response field variables and the corresponding hyper-scaling relations can be derived [Muñoz *et al.* (1997); Janssen (2003)].

Features not accessible to PRG may also play a crucial role and then the so-called non-perturbative renormalization group (NPRG) (or exact/functional renormalization) appears as a method of choice [Wegner and Houghton (1972); Bagnuls and Bervillier (2001); Berges *et al.* (2002a); Delamotte and Canet (2005); Delamotte (2007)]. Modern functional RG approaches are based on Wilson's idea [Wilson and Kogut (1974); Wegner and Houghton (1972)] to integrate out momentum modes within a path-integral representation of the theory. Without going deeply into the details of this method the main idea is to build a one-parameter family of models, indexed by a momentum scale $k$, interpolating smoothly between the short-distance physics at the (microscopic) scale $k = \Lambda$, where no fluctuation has been taken into account, and the long-distance physics at scale $k = 0$, where all fluctuations have been integrated out.

In the [Berges *et al.* (2002a)] version of the NPRG not the partition function $Z_k[J,\bar{J}]$ itself (as suggested originally by [Wegner and Houghton

(1972)]), but its Legendre transform, the state function $\Gamma_k$ (analogous to the Gibbs free energy at equilibrium)

$$\Gamma_k[\phi,\psi] + \log Z_k[J,\bar{J}] = \int J\phi + \int \bar{J}\psi - \int R_k\phi\psi \tag{1.65}$$

is followed under the infinitesimal change of the scale $s = \log(k/\Lambda)$. Here $J$ denotes the external current and $R_k$ describes the momentum cutoff applied. The exact functional differential equation governing the RG flow of $\Gamma$

$$\partial_s \Gamma_k = \frac{1}{2}\mathrm{Tr}\int_{q,\omega} \partial_s \hat{R}_k \left(\hat{\Gamma}_k^{(2)} + \hat{R}_k\right)^{-1}, \tag{1.66}$$

is set up by [Berges *et al.* (2002a); Canet *et al.* (2004)], where $\hat{R}_k$ is a symmetric, off-diagonal, $2\times 2$ matrix of element $R_k$ and $\hat{\Gamma}_k^{(2)}[\psi,\bar{\psi}]$ the $2\times 2$ matrix of second derivatives of $\Gamma_k$ with respect to $\phi$ and $\psi$. Usually the smooth cut-off function

$$R_k(q^2) = k^2(1 - q^2/k^2)\theta(1 - q^2/k^2) \tag{1.67}$$

is used, where $\theta(x)$ is the Heavyside step function. Since Eq. (1.66) cannot be solved exactly usually some truncation is applied. A standard truncation is the derivative expansion [Berges *et al.* (2002a)], in which $\Gamma_k$ is expanded as a power series of $\nabla$ and $\partial_t$. The local potential approximation (LPA) — which is the simplest such truncation — keeps only a potential term in $\Gamma_k$ while neglects any field renormalization:

$$\Gamma_k^{\mathrm{LPA}} = \int_{x,t} \{U_k(\phi,\psi) + \psi(\partial_t - D\nabla^2)\phi\}. \tag{1.68}$$

The NPRG equation for the effective potential in LPA, valid for all reaction-diffusion processes involving a single species, has been established in [Canet *et al.* (2004)]. Studying a particular model amounts to solve this equation in a sub-space defined by the symmetries of the problem, starting with the corresponding microscopic action $S$. The flow equation for the dimensionless potential $u = k^{d+2}U_k$, expressed in terms of the dimensionless fields $\phi \to k^{-d}\phi$ and $\psi$ looks as

$$\partial_s u = -(d+2)u + d\,\phi\, u^{(1,0)} - V_d\left[1 - \frac{u^{(2,0)}u^{(0,2)}}{(1+u^{(1,1)})^2}\right]^{-\frac{1}{2}}, \tag{1.69}$$

where $u^{(n,p)} = \dfrac{\partial^{n+p}u}{\partial^n\phi\partial^p\psi}$ and $V_d = \dfrac{2^{-d+1}\pi^{-d/2}}{d\,\Gamma(d/2)}$. By inserting combinations of fields invariant under the symmetries of the model one can try to follow the flows of the couplings and locate fixed points corresponding to different universality classes. These invariants and flows are related to those of the the topological phase spaces of the RD models discussed in Sect. 1.6.1.

By a Gaussian transformation of the action (1.64), using a suitable Gaussian distribution $P(\eta)$ one can integrate out the response field

$$e^{-\lambda\psi^2\phi^2} = \int e^{-\eta\psi} P(\eta) d\eta \ , \tag{1.70}$$

and we arrive to an alternative formalism: the **Langevin equation**, which in case of ARW is

$$\partial_t \phi(x,t) = D\nabla^2\phi(x,t) - 2\lambda\phi^2(x,t) + \eta(x,t) \tag{1.71}$$

with a Gaussian white noise, described by the correlations: $< \eta(x,t) >= 0$,

$$< \eta(x,t)\eta(x',t') >= -\Gamma\phi^2\delta^d(x-x')\delta(t-t') \ . \tag{1.72}$$

Here $\delta$ denotes the Dirac delta functional and $\Gamma$ is the noise amplitude, so the noise disappears for zero density but is imaginary causing complex solution. Equation (1.71) is just the diffusive mean-field equation supplemented with a stochastic noise. In case of general reactions — if the Langevin equation can be deduced — the noise term can be a more complex function. Such equations are usually derived from the master equation (1.57) through some kind of coarse-graining and the exact form of the noise term is often unclear, especially when the dynamics does not constrain it via the detailed balance symmetry (2.20).

### 1.6.1 *Classification scheme of one-component, bosonic RD models, with short ranged interactions and memory*

The classification of nonequilibrium systems according to their critical exponents is not so straightforward as in case of equilibrium models. Especially the factors affecting a given universality class turned out to be more complex. There has been several attempt to classify universality classes of nonequilibrium phase transitions in the past decades. In Sect. 4.1 I show how the mean-field classes of one-component, bosonic reaction-diffusion systems of type $mA \to (m+k)A$, $nA \to (n-l)A$ can be obtained. Complete field theoretical description below $d_c$ is still an open question in most cases.

Very recently a new strategy has been proposed [Elgart and Kamenev (2006)] for classifying bosonic one-component RD models exhibiting critical phase transition to absorbing states based on the topology of the phase portraits of the corresponding Hamiltonians.[2] These "Hamiltonians" are

[2]This treatment is restricted to homogeneous, isotropic, models with short ranged interactions and memory. Furthermore in low dimensional fermionic systems topological constraints can also be relevant (see for example Sect. 6.2).

derived from the master equation by introducing a quantum Hamiltonian for the generating function of the $n$-particle probabilities followed by a semi-classical limit of coordinates. Starting from a general action

$$S = \int dtd^dx \left[\psi\partial_t\phi + D\nabla\psi\nabla\phi + H_R(\phi,\psi)\right] , \tag{1.73}$$

where the first term describes the diffusion of particles and $H_R(\phi,\psi)$ is determined by the specific set of reactions one finds an analogy with the action (energy) of an equilibrium system in the following way. From $H_R(\phi,\psi)$ of Eq. (1.73) one can arrive to an effective non-Hermitian "Hamiltonian" $\hat{H}(p_i,q_i)$ operator by making a semi-classical (saddle point approximation) limit.

This operator depends on the canonical coordinates $p_i$ and $q_i$ derived from the generating function set up for $n$-particle occupancy probabilities $P_n(t)$ at a given site $i$

$$G(p_i,t) = \sum_{n=0}^{\infty} p_i^n P_n(t) . \tag{1.74}$$

Here the parameter $p_i$ plays the role of a "canonical momentum", such that for $p_i = 1$

$$\rho(t) = \sum_n nP_n(t) = \partial_{p_i} G(p,t)|_{p_i=1} , \tag{1.75}$$

and a conjugated "coordinate" operator can be introduced as

$$q_i = \partial/\partial p_i . \tag{1.76}$$

With these parameters the bosonic Master equation at site $i$ is

$$\partial_t G = \hat{H}_R(p_i,q_i)G . \tag{1.77}$$

and the Keldysh [Keldysh (1964)] "Hamiltonian" looks as

$$\hat{H}(p_i,q_i) = -D\sum_{<i,j>}(p_i - p_j)(q_i - q_j) + \sum_i \hat{H}_R(p_i,q_i) \tag{1.78}$$

exhibiting a kinetic part with the diffusion constant $D$ and a reaction part. The equations of motion, conserving the "energy" are

$$\partial_t q_i = \frac{\partial H_R}{\partial p_i} \tag{1.79}$$

$$\partial_t p_i = \frac{-\partial H_R}{\partial q_i} . \tag{1.80}$$

For example in case of ARW the reaction part looks as

$$\hat{H}_R = \lambda/2(1 - p^2)q^2 \ , \tag{1.81}$$

which by $\phi \to q$, $\psi \to p$ transcription and a continuum limit looks just as Eq. (1.62). In case more general reactions of type

$$kA \xrightarrow{\lambda} mA, \tag{1.82}$$

the non-Hermitian, normally ordered reaction Hamiltonian stays for

$$\hat{H}_R(p, \hat{q}) = \frac{\lambda}{k!}(p^m - p^k)\, \hat{q}^k \tag{1.83}$$

from which one can easily read-off the recipe of writing up such terms.

Due to the resemblance of $H_R(p, q)$ to the effective potential $V(q)$ in equilibrium statistical physics it is plausible that $H_R(p, q)$ encodes the information about possible nonequilibrium transitions as $V(q)$ does in equilibrium.[3] In particular the the potential minima (given by $\partial_q V = 0$) might be analogous to the $H_R(p, q) = 0$ zero energy curves in the phase space. According to Elgart and Kamenev these curves and the corresponding topology of the phase portrait classify possible phase transitions of reaction-diffusion models. In other words the **web of these curves** play the role of minima of $V(q)$ of equilibrium statistical mechanics.

Considering now the Hamiltonian system of (1.78) from the normalization condition $G(1, t) = 1$ follows that

$$\hat{H}_R(p, q)|_{p=1} = 0 \ , \tag{1.84}$$

furthermore any Hamiltonian describing a system with the empty absorbing state must satisfy

$$\hat{H}_R(p, q)|_{q=0} = 0 \ . \tag{1.85}$$

Substituting $p = 1$ in (1.79),(1.80) one obtains the mean-field equation for $\rho(t)$, neglecting all fluctuation effects. Therefore the zero "energy" lines of the phase space of this Hamiltonian are the $p = 1$ (corresponding to mean-field) and the $q = 0$ (corresponding to the absorbing phase). These two lines make up the separatrix, i.e. they divide the entire phase space into isolated sectors. All other trajectories described by (1.79),(1.80) cannot intersect them and are confined to one of the sectors (see Fig. 1.3).

By varying the control parameters of a model the corresponding phase-space trajectories also change and a phase transition is associated to the change of the topology of the that portrait (see example Fig. 1.4).

[3] However the diffusion has been found to be a relevant in many, mainly fermionic models. See for example [Jensen and Dickman (1993a); Ódor (2000, 2004b,c)] Sect. 4.7.

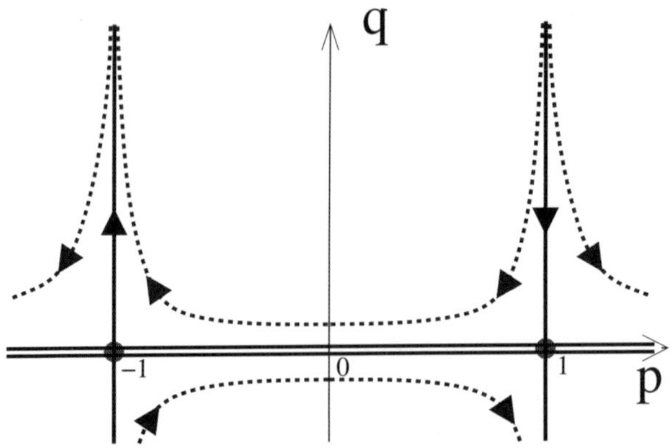


Fig. 1.3 Phase portrait of the ARW system ($2A \xrightarrow{\lambda} \emptyset$). The corresponding classical Hamiltonian is given by $H_R(p,q) = \frac{\lambda}{2}(1-p^2)q^2$. Solid black lines show zero-energy trajectories: generic lines $p=1$ and double degenerate $q=0$ and the "accidental" line $p=-1$. Dashed colored curves indicate trajectories with nonzero energy. The arrows show the evolution direction (from [Elgart and Kamenev (2006)].

However near to the transition point the saddle point approximation may lose its validity. Using RG one integrates out the small scale fluctuations, but the coupling constants and the functional form of $H_R(p,q)$ changes. Still around the transition the topology may be fully encoded in a relatively simple polynomial, which in turn provides a full characterization of the transition (at least for small $\epsilon = d_c - d$)[4] Hence considering distinct topologies, stable upon RG transformations one may classify the possible transitions.

This will be a general guideline to be followed in Chapter 4. of this book. Using (1.83) one can easily set up $\hat{H}_R(p,\hat{q})$ for a given family of RD models and find the generic phase space topology corresponding to it. The validity of this approach will be discussed by comparing the predictions of the phase portrait method to numerical results of other approaches.

### 1.6.2 *Ageing and local scale invariance (LSI)*

In equilibrium models conformal invariance (CI) has been proven to be a very powerful tool for classifying universality classes [Friedan *et al.* (1984); Cardy (1987)]. This occurs in certain two dimensional critical systems as

[4]We will show later that these polynomials formally look the same as invariants in the effective potential of NPRG [Canet *et al.* (2005)], however there the trajectories of the RG evolve in the momentum space.

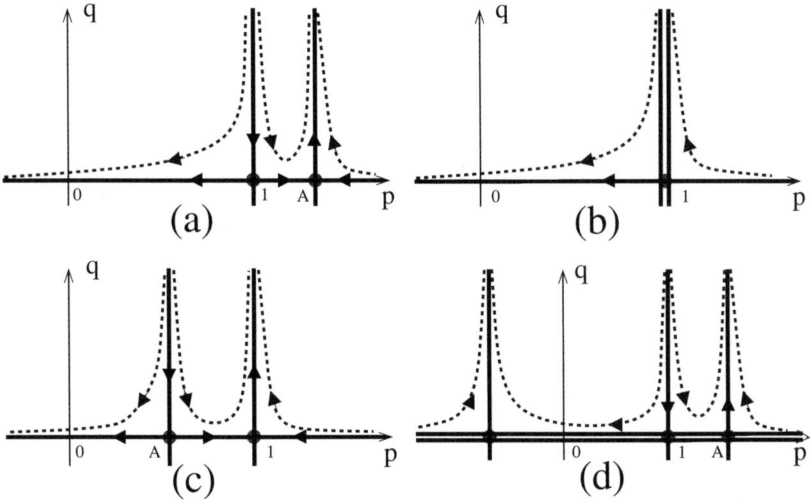


Fig. 1.4 Phase portraits of the system, $kA \xrightarrow{\lambda} (k-l)A, kA \xrightarrow{\mu} (k+m)A$ exhibiting the first-order phase transition. Thick solid lines represent the zero-energy trajectories. (a)–(c) Reaction set with $k = l = m = 1$: (a) absorbing phase, (b) transition point, and (c) unlimited proliferation phase. (d) Reaction set with $k = l = 2$ and $m = 1$ in the extinction phase. (from [Elgart and Kamenev (2006)]

the consequence of a larger group (the CI group) than the mere scale transformations [Belavin *et al.* (1984)]. In principle conformal transformation is a scale transformation: $\mathbf{r} \to b(\mathbf{r})\mathbf{r}$, with a space dependent rescaling factor, such that angles are kept fixed. In two dimensions CI allows to derive the possible set of critical exponents from the representation theory of the conformal (Virasoro) algebra. Conformal field theory provides us with a simple means of calculating critical exponents as well as the $n$-point correlation functions of the theory at the critical point [Henkel (1999)].

Recently the generalization of the generators of CI (albeit without invariance under time-translations) are proposed for anisotropic, dynamical models [Henkel (2002); Henkel and Pleimling (2005)]. The corresponding invariance is the so-called local scale-invariance (**LSI**). Since it is supposed to be the extension of the dynamical scale transformations for such systems it may serve as a convenient tool for classifying universality classes of nonequilibrium systems as well.

The quantities of main interest are the two-time autocorrelation function

$$C(t,s) = \langle \phi(t,\vec{r})\phi(s,\vec{r}) \rangle \tag{1.86}$$

and the auto-response function $R(t,s)$ (1.63), which describe ageing phenomena (for recent reviews see [Bouchaud and Bray (2004); Bray (1994)]). For $t, s \to \infty$ *and* $y = t/s > 1$ one expects the scaling forms

$$C(t,s) = s^{-b} f_C(t/s) \tag{1.87}$$

$$R(t,s) = s^{-1-a} f_R(t/s), \tag{1.88}$$

where $a$ and $b$ are ageing exponents and $f_C$ and $f_R$ are scaling functions such that $f_{C,R}(y) \sim y^{-\lambda_{C,R}/Z}$ for $y \gg 1$. Here $\lambda_C$ and $\lambda_R$ are the autocorrelation [Fisher and Huse (1988)] and auto-response [Picone and Henkel (2002)] exponents respectively and independent of equilibrium exponents and the dynamical exponent $Z$ (defined as usual $Z = \nu_{||}/\nu_{\perp}$).

As in case of CI one expects that **LSI fully determines the functional form of the scaling functions**. Henkel *et al.* derived $R(t,s)$ in general and the form of $C(t,s)$ for $Z = 2$ by identifying the quasi-primary operators of the theory [Henkel *et al.* (2001, 2006)]. The generalized form of $R(t,s)$ takes into account the difference between physical observable defined in lattice models and the associated quasi-primary scaling operators of the underlying field theory as well. This ansatz, derived for $m_0 = 0$ **disordered** initial order parameter density looks as

$$R(t,s) = s^{-1-a} \left(\frac{t}{s}\right)^{1+a'-\lambda_R/Z} \left(\frac{t}{s} - 1\right)^{-1-a'}, \tag{1.89}$$

where $a' \neq a$ is an independent ageing exponent in general. Some systems with detailed balance symmetry (2.20) has been analyzed recently and found to satisfy (1.89) [Henkel and Pleimling (2005); Berthier *et al.* (1999); Mazenko (2004); Godrèche and Luck (2000a); Lippiello and Zannetti (2000); Henkel and Schütz (2004); Mayer *et al.* (2006); Pleimling and Gambassi (2005)] with $a \neq a'$.

On the other hand renormalization-group results for some important universality classes concluded that $a = a'$ should be hold. In particular explicit two-loop field-theoretical computation of $R(t,s)$ for the $O(N)$ universality class and Model A dynamics at the critical point claim $a = a'$ [Calabrese and Gambassi (2002, 2005)] (see Sect. 2.3.2).

Very recently more nontrivial models with $Z \neq 2$ have been found exhibiting LSI. These involve the exactly solved spherical model with $Z = 4$ [Baumann and Henkel (2007)], the nonequilibrium kinetic Ising model in one dimension [Ódor (2006a)] and the phase-ordering kinetics of two dimensional disordered Ising models [Baumann *et al.* (2007)]. For a more detailed discussion of these results see a very recent review [Henkel (2007)]. On the other hand novel series of numerical studies and argumentation claim [Hinrichsen (2007b,c); Corberi *et al.* (2007)] that the observed accurate fitting with the LSI functional form (1.89) for models with $Z \neq 2$ cannot be considered exact in general, which would mean the existence of a **generalized Galilei-invariance for** these models.

## 1.7 The effect of disorder

A question of theoretical and experimental interest is whether and how the critical behavior is changed by introducing a small amount of uncorrelated impurity leading to models with quenched disorder. Realistic systems may contain a certain amount of quenched disorder. This disorder can take the form of vacancies or impurity atoms in a crystal lattice, or it can consist of extended defects such as dislocations or grain boundaries. The static critical behavior of equilibrium systems with bond or site disorder is well understood thank to the **Harris criterion** [Harris (1974a)]. It states that the addition of impurities to a system which undergoes a second-order phase transition does not change the critical behavior if

$$\alpha_H < 0 \; . \tag{1.90}$$

If $\alpha_H$ is positive, the transition is altered. Using the hyper-scaling relation (1.7) this means that the **disorder is irrelevant** for

$$\nu_\perp > 2/d \; . \tag{1.91}$$

This criterion can be derived heuristically by the following scaling arguments. Let's consider the effective local critical temperature in blocks of linear size $\xi$, which is given by an average of $T + \delta T(\mathbf{x})$ over the volume $V = \xi^d$. A sharp phase transition can only occur if the variation $\Delta T$ of these local critical temperatures from block to block is smaller than the global distance from the critical point $T$. For short-range correlated disorder, the central limit theorem yields $\Delta T \propto \xi^{-d/2} \propto T^{d\nu_\perp/2}$. Thus, a clean critical point is perturbatively stable, if the clean critical exponents fulfill the inequality $T^{d\nu_\perp/2} < T$ for $T \to 0$. This implies the exponent inequality (1.91). Critical systems have been divided into three categories according to the response of a pure system for quenched disorder

- Irrelevant disorder : The critical behavior is unchanged, the macroscopic variables are self-averaging (the relative width of their probability distributions vanishes as $L \to \infty$).

- Relevant (marginal) disorder: New critical behavior fixed point (line) with conventional power-law scaling. Lack of self-averaging (the width of their probability distributions approaches a size dependent constant as $L \to \infty$).
- Relevant disorder with infinite randomness critical point: Activated scaling behavior. The relative width of their probability distributions diverges with the system size.

The Harris criterion is a *necessary but not sufficient condition for the stability of an impure fixed point.* It only deals with the average behavior of the disorder at large length scales. However, effects due to qualitatively new physics at finite length scales (and finite disorder strength) are not described by the Harris criterion. Nonequilibrium models with quenched spatial disorder were considered first by Kinzel [Kinzel (1985)].

The Harris criterion has been rigorously established for a large class of disordered systems [Chayes *et al.* (1986)]. In nonequilibrium systems the situation is less clear and violations of the Harris criterion are reported (see Sects. 4.2.10, 4.5.6 and 4.6.8). In nonequilibrium phase transitions **rare regions in with large disorder fluctuations** may occur, which can exert much stronger effects, ranging from from strong power-law singularities in the free energy to a complete destruction of the phase transition (for recent reviews see [Iglói and Monthus (2005); Vojta (2006)]). In classical equilibrium systems, they lead to the so-called Griffiths singularity [Griffiths (1969)], of the free energy in the vicinity of the phase transition.

Rare regions occur for example in diluted systems, (see Sect. 4.5.6), which reduces $T_c$ from its clean bulk value $T_c^0$. In an infinite system one can find arbitrarily large spatial regions without impurities. For temperatures between $T_c^0$ and $T_c$, these regions show local magnetic order, even though the bulk system is globally in the paramagnetic phase. The probability $p_r$ for finding a rare region is exponentially small in its volume $V_\text{r}$ and in the impurity concentration $p$. Up to a pre-exponential factors it is given by

$$p_r \propto e^{-pV_\text{r}} \; . \qquad (1.92)$$

Rare regions are thus non-perturbative degrees of freedom that are not accounted for in conventional approaches to phase transitions and critical points that are based on perturbation theory and on PRG. The importance of large rare regions depends on how rapidly the contribution of a single region to observable quantities increases with its size. This is controlled by the relation between $d_\text{r}$ (the effective dimensionality of the region) and the lower critical dimension $d_c^-$ of the phase transition.[5] One can distinguish three cases:

**(a)** If $d_\mathrm{r} < d_c^-$ an isolated rare region cannot undergo a phase transition, its contribution can at most grow as a power of its linear size, which can't overcome the exponential drop in the rare region density. This causes a weak effect and the critical behavior is of conventional power-law type (see for example the classical, randomly diluted Ising model (Sect. 2.3.7), where $d_\mathrm{r} = 0$ and $d_c^- = 1$).

**(b)** If $d_\mathrm{r} = d_c^-$ the rare regions still cannot undergo phase transition independently, but their contribution increases exponentially with size and can overcome the exponential decrease of the rare region density. This results in Griffiths singularities with non-universal, continuously varying exponents as follows.

**Inside the Griffiths region** ($T_c < T < T_c^0$) the long-time behavior is dominated by these regions and any finite region decays exponentially

$$\rho(t) \propto \int dL_r \; L_r^d \; e^{-t/t_R} \; p_r \;, \tag{1.93}$$

where $t_R$ is the characteristic decay time of a region of size $L_r$. The average lifetime of such active regions grows as

$$t_R \propto e^{aV_r} \tag{1.94}$$

because a coordinated fluctuation of the entire region is required to flip it (the constant $a$ vanishes at the clean critical point and increases with decreasing $T$ (see, e.g., [Goldenfeld (1992)])). The saddle point analysis of (1.93) results in power-law decay

$$\rho(t) \propto t^{-p_r/a} \;, \tag{1.95}$$

hence one gets continuously changing decay exponents (see for example the NEKIM model in $1+1$ dimension (Sect. 4.6.8)).

**At the clean critical temperature** $T = T_c^0$ the $t_R$ is determined by finite-size scaling, which diverges as

$$t_R \sim L_\mathrm{r}^Z \;. \tag{1.96}$$

By inserting it to Eq. (1.93) the saddle-point method results in stretched exponential decay of the density.

$$\ln \rho(t) \sim -\tilde{p}^{Z/(d+Z)} \; t^{d/(d+Z)} \;, \tag{1.97}$$

where $\tilde{p} = -\ln(1-p)$. The scaling behavior at the **dirty critical point** $T = T_c$ is dominated by the rare regions, resulting in **'activated' scaling,**

[5] where the fluctuations are so strong that they completely destroy the ordered state.

characterized by the **'infinite randomness critical'** point. Here the dynamics is extremely slow. The power-law scaling (1.16) gets replaced by activated dynamical scaling

$$\ln(t_R) \propto \xi^{\psi}, \tag{1.98}$$

characterized by a new exponent $\psi$. This exponential relation between time and length scales implies that the dynamical exponent $Z$ is formally infinite.[6] The activated scaling leads to logarithmic time dependencies at the critical point. For example the density decays as

$$\rho(t) \propto [\ln(t)]^{-\bar{\alpha}} , \tag{1.99}$$

while the average number of particles of a cluster started from a single seed site increases like

$$N(t) \propto [\ln(t)]^{\bar{\eta}} \tag{1.100}$$

with $\bar{\alpha} = \beta/(\nu_\perp \psi)$ and $\bar{\eta} = d/\psi - 2\beta/(\nu_\perp \psi)$ (see for example the directed percolation in $1+1$ dimension (Sect. 4.2.10)).

**(c)** If $d_r > d_c^-$ the rare regions can undergo the phase transition, the dynamics of the locally ordered regions completely freezes, and they develop a truly static order parameter. As a result, the global phase transition is destroyed by smearing [Vojta (2003)] because different spatial parts of the system order at different values of the control parameter.

[6]In contrast, the static scaling behavior remains of power law type.

# Chapter 2

# Out of Equilibrium Classes

In this chapter I begin introducing basic nonequilibrium classes starting with the simplest dynamical extensions of equilibrium models. These dynamical systems **exhibit hermitian Hamiltonian** and starting from a nonequilibrium state they evolve into a Gibbs state. Such nonequilibrium models include, for example, phase ordering systems, spin glasses, glasses etc. In these cases one is usually interested in the 'nonequilibrium dynamics' at the equilibrium critical point $T_c > 0$ or at $T = 0$. The field theoretical free-energy functional to describe the ordered phase of a model with $N$-component field $\phi_i$ ($i = 1 \dots N$) is

$$F[\phi_i] = \int d^d x \left( \frac{1}{2} (\nabla \phi_i)^2 + V(\phi_i) \right) , \tag{2.1}$$

where $V(\phi_i)$ is the potential of the static model and the gradient term associates an energy cost to an interface between the phases.

## 2.1 Field theoretical description of dynamical classes at and below $T_c$

It is an important universal phenomenon that scaling behavior can be observed far away from criticality as well. If a magnet is rapidly brought (quenched) from some initially disordered state to some temperature $T < T_c$ (below its critical temperature $T_c > 0$) to the ordered state, compact domains, separated by topological defects such as interfaces or vortices slow down the dynamics [Bray (1994)]. The different domains corresponding to (at least two) macroscopically ordered states compete, the system cannot relax within finite time to any of them and the size of the domains for

$t \to \infty$ grow with a power law behavior

$$L(t) \propto \xi \propto t^{x/Z} , \tag{2.2}$$

where $Z$ is the dynamical exponent. The size of domains can be related to the density of the topological defects $\rho(t) \propto t^{-\alpha}$. For a coarsening system with interface defects

$$\alpha = 1/Z , \tag{2.3}$$

thus $x = 1$ in Eq. 2.2, while in case of vortices or others in $O(N)$ $(N \geq 2)$ models

$$\alpha = 2/Z , \tag{2.4}$$

(thus $x = 2$ in Eq. 2.2), with logarithmic corrections for $N = 2$.

If the **order parameter is not conserved** the equation for the time evolution of the field $\phi_i$ can be deduced from (2.1) as a functional derivative

$$\begin{aligned} \partial\phi_i/\partial t &= -\delta F/\delta\phi_i \\ &= \nabla^2 \phi_i - V'(\phi_i) , \end{aligned} \tag{2.5}$$

where $V'(\phi_i) \equiv dV/d\phi_i$, which is called Time-Dependent-Ginzburg-Landau (TDGL) equation describing the $T = 0$ behavior. With an additional Gaussian white noise term on the right-hand side

$$< \eta_i(x,t) > = 0, \tag{2.6}$$

$$< \eta_i(x,t)\eta_j(x',t') > = 2D\delta^d(x-x')\delta(t-t')\delta_{ij} . \tag{2.7}$$

It is called **'model-A'** critical dynamics in Langevin representation [Hohenberg and Halperin (1977)]. One can also set up an action as outlined in Sect. 1.6:

$$\mathcal{S}[\phi,\psi] = \int d^dx\, dt \left\{ \psi \left[ \partial_t\phi + D\,\frac{\delta\mathcal{H}[\phi]}{\delta\phi} \right] - D\psi^2 \right\} , \tag{2.8}$$

where $D$ denotes the diffusion constant and $\mathcal{H}$ is the usual Landau-Ginzburg-Wilson Hamiltonian (more details will be shown at specific models). Using RG formalism the dynamic response to external perturbations has been determined for several models (for a review see [Hohenberg and Halperin (1977)]). The linear response theory, which is applicable for infinitesimal disturbances gives different singularity behavior for the relaxation than what is observed in case of non-linear responses to finite changes in general [Rácz (1976)].

When model-A systems quenched from high temperature to $T < T_c$ the characteristic length (and the domain size) in the late time regime grows with an universal power-law

$$\xi \propto t^{x/2} . \tag{2.9}$$

For a quench to $T_c > 0$ the growth depends on the details of the model to be discussed later. The long-time autocorrelation and autoresponse exponents are not independent

$$\lambda_C = \lambda_R \tag{2.10}$$

which is not true in general. Careful analysis reveals [Janssen (1992)] that for $m_0 = 0$ the autocorrelation exponent is related to the initial slip exponent.

$$\lambda_C / Z = \frac{d}{Z} - \eta . \tag{2.11}$$

When the **order parameter is conserved**, as in phase separation, a different dynamics occurs:

$$\begin{aligned} \partial\phi/\partial t &= \nabla^2 \, \delta F/\delta\phi \\ &= -\nabla^2 \left[ \nabla^2 \phi - V'(\phi) \right] , \end{aligned} \tag{2.12}$$

which is called Cahn-Hilliard equation, describing $T = 0$ behavior. The difference to TDGL (2.5) is the $\nabla^2$ term, which ensures the conservation of $\int dx \, \phi(x,t)$. With an additional Langevin noise term on the right-hand side it is called **'model B'** critical dynamics. The corresponding dynamical functional can be derived

$$\mathcal{S}[\phi, \psi] = \int d^d x \, dt \left\{ \psi \left[ \partial_t \phi - \kappa \, \nabla^2 \frac{\delta \mathcal{H}[\phi]}{\delta \phi} \right] + \kappa \psi \nabla^2 \psi \right\} , \tag{2.13}$$

where $\kappa$ is related to the mobility. The presence of the gradient term makes the difference with Model A dynamics and according to PRG the critical exponents are exactly obtained on the sole basis of power-counting. In this case for a quench to $T < T_c$ the correlation length grows as

$$\xi \propto t^{x/3} , \tag{2.14}$$

while for critical quenches

$$Z = 4 - \eta_\perp \tag{2.15}$$

exactly [Hohenberg and Halperin (1977)]. Model-B is much less understood than model-A because due to the conservation law the critical slowing-down is more severe, hence it it very difficult to reach the asymptotic behavior.

It was argued that $\lambda_C = d$ [Majumdar *et al.* (1994)] and has been verified in several numerical studies.

In **model-C** systems a conserved secondary density is coupled to the non-conserved order parameter. Such models may exhibit **model-A** behavior or

$$\xi \propto t^{1/(2+\alpha_H/\nu_\perp)} \tag{2.16}$$

depending on the model parameters. with logarithmic corrections for $N = 2$.

## 2.2 Dynamical classes at $T_c > 0$

In this chapter basic models of statistical mechanics with additional dynamics will be discussed. Since percolation is a central topic in reaction-diffusion systems (discussed in detail in Sects. 3.1.4 and 5.2) for the sake of completeness I present related percolation results obtained for these models at criticality. Another novel feature of dynamic phase transitions is the emergence of a chaotic state. Therefore I shall discuss damage spreading transitions and behavior in these systems.

By combining different competing dynamics (for example by connecting two reservoirs with different temperatures to the system, or by external currents) we can arrive to nonequilibrium models having no hermitian Hamiltonian and no equilibrium Gibbs state. These models frequently exhibit novel kind of critical behavior (those will be discussed in Sect. 2.6.2) or insensitive to these extensions and remain in the same universality class.

Field theoretical investigations have revealed that **model-A systems are robust** against the introduction of various **competing dynamics**, which are *local* and *do not conserve the order parameter* [Grinstein *et al.* (1985)]. Furthermore it was shown, that this robustness of the critical behavior persists if the competing dynamics breaks the discrete symmetry of the system [Bassler and Schmittman (1994)] or it comes from reversible mode coupling to a non-critical conserved field [Täuber and Rácz (1997)].

On the other hand if **competing dynamics are coupled to model-B** systems by an external drive [Schmittman and Zia (1996)] or by local, anisotropic order-parameter-conserving process [Schmittman and Zia (1991); Schmittman (1993); Bassler and Rácz (1994, 1995)] long-range interactions are generated in the steady state with angular dependence. The universality class will be the same as that of the kinetic version of the equilibrium Ising model with **dipolar long-range interactions**.

As the number of neighboring interaction sites decreases by lowering the spatial dimensionality of a system with short-ranged interactions the relevance of fluctuations increases. In equilibrium models finite-range interactions cannot maintain long-range order in $d < 2$. This observation is known as the Landau-Peierls argument [Landau and Lifshitz (1981)]. According to the Mermin-Wagner theorem [Mermin and Wagner (1996)], for systems with continuous symmetry long-range order do not exist even in $d = 2$. Hence in equilibrium models phase transition universality classes exist for $d \geq 2$ only. One of the main open questions to be answered is whether there exist a class of nonequilibrium systems with restricted dynamical rules for which the Landau-Peierls or Mermin-Wagner theorem can be applied.

## 2.3 Ising classes

The equilibrium Ising model was introduced by [Ising (1925)] as the simplest model for an uni-axial magnet but it is used in different settings for example binary by fluids or alloys as well. It is defined in terms of spin variables $s_i = \pm 1$ attached to sites $i$ of some lattice with the Hamiltonian

$$H = -J \sum_{i,i'} s_i s_{i'} - B \sum_i s_i \tag{2.17}$$

where $J$ is the coupling constant and $B$ is the external field. In one and two dimensions it is solved exactly [Onsager (1944)], hence it plays a fundamental test-ground for understanding phase transitions.

The Hamiltonian of this model exhibits a global, so called $Z_2$ (up-down) symmetry of the state variables. While in one dimension a first order phase transition occurs at $T = 0$ only (see Sect. 4.10) in two dimensions there is a continuous phase transition where the system exhibits conformal symmetry [Henkel (1999)] as well. The critical dimension is $d_c = 4$. Table (2.1) summarizes some of the known critical exponents of the equilibrium Ising model. The quantum version of the Ising model, which in the simplest cases might take the form (in one dimension)

$$H = -J \sum_i (t\sigma_i^x + \sigma_i^z \sigma_{i+1}^z + h\sigma_i^z) , \tag{2.18}$$

— where $\sigma^{x,z}$ are Pauli matrices and $t$ and $h$ are couplings — for $T > 0$ has been shown to exhibit the same critical behavior as the classical one (in the same dimension). For $T = 0$ however quantum effects become important

Table 2.1 Static critical exponents of the Ising model

| exponent | $d=2$ | $d=3$ | $d=4$(MF) |
|---|---|---|---|
| $\alpha_H$ | 0(log) | 0.110(1) | 0 |
| $\beta$ | 1/8 | 0.3258(14)[a] | 1/2 |
| $\gamma$ | 7/4 | 1.235(5)[a] | 1 |
| $\eta_\perp$ | 1/4 | 0.0364(5) | 0 |
| $\nu_\perp$ | 1 | 0.6495(1)[b] | 1/2 |

[a] from Ref. [Guida and Zinn-Justin (1998)], [b] from Ref. [Bervillier *et al.* (2007)],

and the quantum Ising chain can be associated with the two dimensional classical Ising model such that the transverse field $t$ plays the role of the temperature. In general a mapping can be constructed between classical $d+1$ dimensional statistical systems and $d$ dimensional quantum systems without changing the universal properties, which has been widely utilized [Suzuki (1971); Fradkin and Susskind (1978)]. The effects of disorder and boundary conditions are not discussed here (for recent reviews see [Alonso and Munoz (2001); Iglói *et al.* (1993)]).

### 2.3.1 *Correlated percolation clusters at* $T_C$

If we generate clusters in such a way that we join nearest neighbor spins of the same sign we can observe percolation at $T_c$ in two dimensions. While the order parameter percolation exponents $\beta_p$ and $\gamma_p$ of this percolation (defined in Sect. 4.4.1) were found to be different from the exponents of the magnetization ($\beta$, $\gamma$) the correlation length exponent is the same: $\nu = \nu_p$.

For two dimensions models with $Z_2$ symmetry the universal percolation exponents are [Fortunato and Satz (2001); Fortunato (2002); Janke and Schakel (2006)]:

$$\beta_p = 0.0527(4), \qquad \gamma_p = 1.8951(5) \ . \tag{2.19}$$

These exponents are clearly different from those of the ordinary percolation classes (Table 4.6) or from Ising class magnetic exponents, but agree with tricritical point values, which can be reached by adding vacancies to the model [Janke and Schakel (2006)].

On the other hand by Fortuin-Kasteleyn cluster construction [Fortuin and Kasteleyn (1972)] the percolation exponents of the Ising model at $T_c$ coincide with those of the magnetization of the model.

### 2.3.2 *Dynamical Ising classes*

Kinetic Ising models such as the spin-flip Glauber Ising model [Glauber (1963)] and the spin-exchange Kawasaki Ising model [Kawasaki (1966)] were originally intended to study relaxational processes near equilibrium states. In order to assure the arrival to an equilibrium state the detailed balance condition for transition rates ($w_{i\to j}$) and probability distributions ($P(s)$) are required to satisfy

$$w_{i\to j}P(s(i)) = w_{j\to i}P(s(j)) \quad . \tag{2.20}$$

Knowing that $P_{eq}(s) \propto \exp(-H(s)/(k_BT))$ this entails the

$$\frac{w_{i\to j}}{w_{j\to i}} = \exp(-\Delta H(s)/(k_BT)) \tag{2.21}$$

condition which can be satisfied in many different ways.

#### 2.3.2.1 *Model-A*

Assuming **spin-flips** (which do not conserve the magnetization (**model-A**)) Glauber formulated the most general dynamical Ising model [Glauber (1963)]. In general the field theoretical action of this model looks as

$$\mathcal{S}[\phi,\psi] = \int d^dx\,dt\,\left\{\psi\left[\partial_t\phi + D\,\frac{\delta\mathcal{H}[\phi]}{\delta\phi}\right] - D\psi^2\right\} \tag{2.22}$$

where $D$ denotes a constant and uniform relaxation rate and $\mathcal{H}$ is the usual Landau-Ginzburg-Wilson Hamiltonian with Ising symmetry

$$\mathcal{H}[\phi] = \int d^dx\left(\frac{1}{2}[\nabla\phi(x)]^2 + V(\phi) - h(x)\phi(x)\right), \tag{2.23}$$

$$V(\phi) = \frac{r}{2}\phi^2 + \frac{u}{4!}\phi^4, \tag{2.24}$$

($k_BT$ has been set to unity). The time-reversal symmetry (corresponding to the detailed balance) can be conveniently expressed as an invariance of the action (2.22) under the specific field transformation

$$\begin{cases} \phi \to \phi \\ \psi \to \psi - \frac{1}{D}\partial_t\phi. \end{cases} \tag{2.25}$$

Indeed, one can straightforwardly check that after performing a time inversion $t \to -t$ in (2.22), which switches the sign of the kinetic term $\psi\,\partial_t\phi$, the field transformation (2.25) yields additional contributions from the latter term and from the noise term $D\psi^2$ that cancel out. Besides, the transformation of the Hamiltonian part under (2.25) produces an additional term $\propto \partial_t \phi \, \delta\mathcal{H}/\delta\phi$, which vanishes upon time integration in the stationary regime.

For a **one-dimensional, model-A** spin system in a magnetic field ($h$) the transition probabilities are

$$w_i{}^h = w_i(1 - \tanh h s_i) \approx w_i(1 - h s_i) \tag{2.26}$$

$$w_i = \frac{\Gamma}{2}(1 + \tilde{\delta} s_{i-1} s_{i+1}) \left(1 - \frac{\gamma}{2} s_i (s_{i-1} + s_{i+1})\right) \tag{2.27}$$

where $\gamma = \tanh 2J/kT$, $\Gamma$ and $\tilde{\delta}$ are further parameters. The $d = 1$ Ising model with Glauber kinetics is exactly solvable. In this case the critical temperature is at $T = 0$ and the transition is of first order. We recall that $p_T = e^{-\frac{4J}{kT}}$ plays the role of $\frac{T-T_c}{T_c}$ in one dimension and in the vicinity of $T = 0$ critical exponents can be defined as powers of $p_T$. For example the coherence length, $\xi$ satisfies $\xi \propto p_T{}^{-\nu_\perp}$ (see Sect. 4.10). In the presence of a magnetic field $B$, the magnetization is known exactly. At $T = 0$

$$m(T = 0, B) = sgn(B). \tag{2.28}$$

Moreover, for $\xi \gg 1$ and $B/kT \ll 1$ the exact solution reduces to

$$m \sim 2h\xi\,; \qquad h = B/k_B T. \tag{2.29}$$

In scaling form one writes

$$m \sim \xi^{-\frac{\beta_s}{\nu_\perp}} g(h\xi^{\frac{\Delta}{\nu_\perp}}) \tag{2.30}$$

where $\Delta$ is the static magnetic critical exponent. Comparison of Eqs. (2.29) and (2.30) results in $\beta_s = 0$ and $\Delta = \nu_\perp$. These values are well known for the one dimensional Ising model. It is clear that the transition is discontinuous at $B = 0$, also when changing $B$ from positive to negative values, see Eq.(2.28). The order of limits are meant as: $B \to 0$ and then $T \to 0$. The $\tilde{\delta} = 0$, $\Gamma = 1$ case is usually referred as the **Glauber-Ising model**. The dynamical exponents are [Glauber (1963); Majumdar *et al.* (1996)]:

$$Z_{1d\ Glauber} = 2, \quad \theta_{g,1d\ Glauber} = 1/4\,. \tag{2.31}$$

For **disordered initial state** ($m_0 = 0$), the exact two-time autocorrelation and autoresponse functions are known [Godrèche and Luck (2000b); Lippiello and Zannetti (2000)]

$$C(t,s) = \frac{2}{\pi} \arctan \sqrt{\frac{2}{t/s - 1}}, \quad R(t,s) = \frac{1}{\pi}\sqrt{\frac{1}{2s(t-s)}}\,. \tag{2.32}$$

Comparing $R(t,s)$ with the aging and **LSI** formulas (1.87,1.89) one can read off

$$\lambda_R = \lambda_C = 1, \quad a = b = 0, \quad a' = -1/2 \,. \tag{2.33}$$

[Picone and Henkel (2004)]. These results remain valid for long-ranged initial conditions $\langle \sigma_r \sigma_0 \rangle \sim r^{-\nu}$ with $\nu > 0$ [Henkel and Schütz (2004)].

**For spatial dimensions** $d > 1$ the critical temperature is $T_c > 0$ and the dynamical two-point functions scale with different exponents at and below $T_c$. The numerical values of $\lambda_C$ **at the phase transition point** can be seen in Table 2.2 (following [Henkel and Baumann (2007)]). When the model is quenched into the coexistence phase ($T < T_c$) simulations in **two dimensions** have shown that the value $\lambda_C = 1.25(1)$ [Fisher and Huse (1988)] is independent of the lattice realization and temperature. An analytical study [Mazenko (2004)] derived the following scaling relation in this case

$$\lambda_R = \lambda_C = 5/4, \quad a = a' = 1/2 \,, \tag{2.34}$$

hence $a = a'$ is satisfied again. In **three spatial dimensions** simulations resulted in

$$\lambda_R = 1.60(2) \,, \quad a = a' = 1/2 \tag{2.35}$$

in numerical agreement with LSI hypothesis of [Henkel and Pleimling (2003)].

### 2.3.2.2 *Model-B*

Applying **spin-exchange Kawasaki dynamics**, which conserves the magnetization (**model-B**) the transition rates in **one dimension** look as

$$w_i = \frac{1}{2\tau}\left[1 - \frac{\gamma_2}{2}(s_{i-1}s_i + s_{i+1}si + 2)\right] \,, \tag{2.36}$$

where $\gamma_2 = \tanh(2J/k_B T)$. In this case the dynamical exponent is different. According to linear response theory [Zwerger (1981)] in one dimension, at the critical point ($T_c = 0$) it is

$$Z_{1d\ Kaw} = 5 \,. \tag{2.37}$$

Note however, that in case of fast quenches to $T = 0$ coarsening with scaling exponent $1/3$ is reported [Cornell *et al.* (1991)]. Hence another dynamic Ising universality class appears with the same static but different dynamical exponents.

Table 2.2 Critical dynamical exponents in the Ising model. Columns denoted by (A) and (B) refer, model-A and model B dynamics.

| | $d=1$ | | $d=2$ | | $d=3$ | $d=4$ | |
|---|---|---|---|---|---|---|---|
| | A | B | A | B | A | A | B |
| $Z$ | 2 | 5 | 2.165(10) | 3.75(20) | 2.032(4) | 2 | 4 |
| $\eta$ | 1/2 | | 0.191(3) | | 0.104(3) | 0 | |
| $\lambda_C = \lambda_R$ | 1 | | 1.595(2) | 1.95(11) | 2.767(6) | 4 | |
| $a$ | 0 | | 0.115 | 0.107(1) | 0.506 | 1 | 1/2 |
| $a'$ | -1/2 | | -0.302(2) | | -0.528(5) | | |
| $b$ | 0 | | 0.115 | 0.107(1) | 0.506 | 1 | 1/2 |
| $\theta_g$ | 1/4 | | 0.27(1) | | 0.41(2) | 1/2 | |
| $\theta_l$ | 3/8 | | 0.22(1) | | 0.26(1) | | |

Two and three dimensional data are from [Zwerger (1981); Zheng (2000, 2001); Grassberger (1995a); Jaster *et al.* (1999); Stauffer (1996); Zheng (1998); Henkel *et al.* (2006); Pleimling and Gambassi (2005)] or obtained by the scaling relation (1.28).

Interestingly while the two dimensional equilibrium Ising model is solved the exact values of the dynamical exponents are not known. Table 2.2 summarizes the known dynamical exponents of the Ising model in $d = 1, 2, 3, 4$. The $d = 4$ results are mean-field values. In Sect. 4.6.2 I shall discuss another fully nonequilibrium critical point of the $d = 1$ Ising model with competing dynamics (the NEKIM model), in which the dynamical exponents break the scaling relation (1.31) and therefore the magnetization is a non-Markovian process. For $d > 3$ there is no non-Markovian effect (hence $\theta$ is not independent) but for $d = 2, 3$ the situation is still not completely clear [Zheng (1998)].

In one dimension the domain walls (kinks) between up and down regions can be considered as particles. The spin-flip dynamics can be mapped onto particle movement

$$\uparrow\downarrow\downarrow \rightleftharpoons \uparrow\uparrow\downarrow \quad \sim \quad \bullet\circ \rightleftharpoons \circ\bullet \tag{2.38}$$

or to the creation or annihilation of neighboring particles

$$\uparrow\uparrow\uparrow \rightleftharpoons \uparrow\downarrow\uparrow \quad \sim \quad \circ\circ \rightleftharpoons \bullet\bullet \; . \tag{2.39}$$

Therefore the $T = 0$ Glauber dynamics is equivalent to the diffusion limited annihilation (ARW, see Sect. 4.10). When we map the spin-exchange dynamics in the same way, more complicated particle dynamics emerges, for example

$$\uparrow\uparrow\downarrow\downarrow \rightleftharpoons \uparrow\downarrow\uparrow\downarrow \quad \sim \quad \circ\bullet\circ \rightleftharpoons \bullet\bullet\bullet \; . \tag{2.40}$$

One particle may give birth of two others or three particle may coalesce to one. Therefore these models are equivalent to branching and annihilating random walks to be discussed in Sect. 4.6.1.

### 2.3.3 *Competing dynamics added to spin-flip*

Competing dynamics in general break the detailed balance symmetry (2.20) and cause the kinetic Ising model to relax to a nonequilibrium steady state (if it exists). Generally these models become unsolvable, for an overview see [Rácz (1996)]. It was argued by [Grinstein *et al.* (1985)] that stochastic **spin-flip** models with two states per site and updating rules of a short-range nature with $Z_2$ symmetry should belong to the (kinetic) Ising model universality class. Their argument is based on the stability of the dynamic Ising fixed point in $d = 4 - \epsilon$ dimensions with respect to perturbations that retain both the spin inversion and lattice symmetries. This hypothesis has received extensive confirmation from Monte Carlo simulations [Castro *et al.* (1998)], [Oliveira *et al.* (1993)], [Tamayo *et al.* (1995); H. W. J. Blote and Zia (1990); Santos and Teixeira (1995)] as well as from analytic calculations [Marques (1989, 1990); T. Tomé and Santos (1991)]. The models investigated include Ising models with a competition of two (or three [Tamayo *et al.* (1995)]) Glauber-like rates at different temperatures [Marques (1989); T. Tomé and Santos (1991); Gonzaléz-Miranda *et al.* (1987); Blote *et al.* (1990)], or a combination of spin-flip and spin-exchange dynamics [Garrido *et al.* (1989)], majority vote models [Castro *et al.* (1998); Santos and Teixeira (1995)] and other types of transition rules with the restrictions mentioned above [Oliveira *et al.* (1993)].

Note that in all of the above cases the ordered state is fluctuating and non-absorbing. By relaxation processes this allows fluctuations in the bulk of a domain. It is also possible (and in one dimension it is the only choice) to generate nonequilibrium two-state spin models (with short ranged interactions) in which the ordered states are frozen (absorbing), hence by the relaxation to the steady state fluctuations occur at the boundaries only. In this case non-Ising universality appears, which is called the voter model (VM) universality class (see Sect. 4.5).

In two dimensions **general $Z_2$ symmetric update rules** were investigated by [Oliveira *et al.* (1993); Drouffe and Godréche (1999); Dornic *et al.* (2001); Achahbar *et al.* (1996)]. These rules are summarized below following [Drouffe and Godréche (1999)]. Let us consider a two dimensional lattice of spins $s_i = \pm 1$, evolving with the following dynamical rule. At each evolution step, the spin to be updated flips with the heat bath rule: the probability that the spin $s_i$ takes the value $+1$ is $P(s_i = 1) = p(h_i)$, where the local field $h_i$ is the sum over neighboring sites $\sum_j s_j$ and

$$p(h) = \frac{1}{2}\left(1 + \tanh[\beta(h)h]\right). \qquad (2.41)$$

The functions $p(h)$ and $\beta(h)$ are defined over integral values of $h$. For a square lattice, $h$ takes the values 4, 2, 0, $-2$, $-4$. We require that $p(-h) = 1 - p(h)$, in order to keep the up down symmetry, hence $\beta(-h) = \beta(h)$ and this fixes $p(0) = 1/2$. The dynamics therefore depends on two parameters

$$p_1 = p(2), \quad p_2 = p(4), \qquad (2.42)$$

or equivalently on two effective temperatures

$$T_1 = \frac{1}{\beta(2)}, \quad T_2 = \frac{1}{\beta(4)}. \qquad (2.43)$$

Defining the coordinate system

$$t_1 = \tanh\frac{2}{T_1}, \quad t_2 = \tanh\frac{2}{T_2} \qquad (2.44)$$

with $0 \leq t_1, t_2 \leq 1$ this yields

$$p_1 = \frac{1}{2}\left(1 + t_1\right), \quad p_2 = \frac{1}{2}\left(1 + \frac{2t_2}{1 + t_2^2}\right), \qquad (2.45)$$

with $1/2 \leq p_1, p_2 \leq 1$. One can call $T_1$ and $T_2$ as two temperatures, associated with *interfacial noise*, and to *bulk noise* respectively. Each point in the parameter plane $(p_1, p_2)$, or alternatively in the temperature plane $(t_1, t_2)$, corresponds to a particular model. The class of models thus defined comprises as special cases the Ising model, the voter and anti-voter models [Ligget (1985)], as well as the majority vote [Ligget (1985); Castro *et al.* (1998)] model (see Fig. 2.1). The $p_2 = 1$ line corresponds to models with no bulk noise ($T_2 = 0$), hence the dynamics is only driven by interfacial noise, defined above. The $p_1 = 1$ line corresponds to models with no interfacial noise ($T_1 = 0$), hence the dynamics is only driven by bulk noise. In both cases effects due to the curvature of the interfaces are always present. For these last models, the local spin aligns in the direction of the majority of its neighbors with probability one, if the local field is $h = 2$, i.e. if there is no consensus amongst the neighbors. If there is consensus amongst them, i.e. if $h = 4$, the local spin aligns with its neighbors with a probability $p_2 < 1$.

Simulations [Oliveira *et al.* (1993)] revealed that the transition line between the low and high temperature regions is Ising type except for the endpoint ($p_2 = 1, p_1 = 0.75$), which is first-order and corresponds to the voter model class. The **local persistence exponent** was also found to be

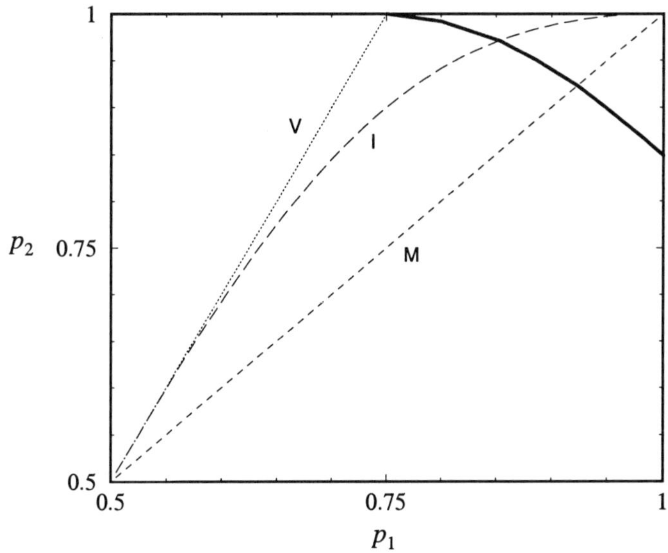


Fig. 2.1 Phase diagram of two dimensions, $Z_2$ symmetric nonequilibrium spin models from [Drouffe and Godrèche (1999)]. Broken lines correspond to the noisy voter model (V), the Ising model (I) and the majority vote model (M). The low temperature phase is located in the upper right corner, above the transition line (full line).

constant along the line $\theta_l \sim 0.27$ [Drouffe and Godrèche (1999)] in agreement with that of the Ising model-A (see Table 2.2), except for the VM point, where it is [Drouffe and Godrèche (2001)]

$$\theta_l \sim 0.37 \,. \tag{2.46}$$

The dynamics of this class of models may be described formally in terms of reaction diffusion processes for a set of coalescing, annihilating, and branching random walkers [Drouffe and Godrèche (1999)]. There are simulation results for other models exhibiting absorbing ordered state indicating VM critical behavior [Lipowski and Droz (2002a); Hinrichsen (1997)].

It is important to note that nontrivial, nonequilibrium phase transition may occur even in one dimension if spin-exchange is added to spin-flip dynamics, the details will be discussed in Sect. 4.6.2.

### 2.3.4 *Competing dynamics added to spin-exchange*

As mentioned in Sect. 2.1 **model B** systems are more sensitive to competing dynamics. *Local and anisotropic* order parameter conserving processes generate critical behavior that coincides with that of the kinetic dipolar-interaction Ising model. The universal behavior of such driven lattice gas models are discussed in Sect. 3.1.

### 2.3.5 *Long-range interactions and correlations*

Universal behavior is due to the fact at criticality long-range correlations are generated that make details of short ranged interactions irrelevant. However one can also investigate the scaling behavior in systems with long-range interactions or with dynamically generated long-range correlations. If the Glauber Ising model (with non-conserving dynamics) is changed to a nonequilibrium one in such a way that one couples **nonlocal dynamics** [Droz *et al.* (1990)] to it, long-range isotropic interactions are generated and mean-field critical behavior emerges. For example if the nonlocal dynamics is a random Lévy flight with a spin exchange probability distribution

$$P(r) \propto \frac{1}{r^{d+\sigma}} \tag{2.47}$$

then effective long range interactions of the form $V_{eff} \propto r^{-d-\sigma}$ are generated and the critical exponents change continuously as the function $\sigma$ and $d$ [Bergersen and Rácz (1991)]. Similar conclusions for other nonequilibrium classes will be discussed later (Sect. 4.2.7).

The effect of power-law correlated initial conditions $\langle\phi(0)\phi(r)\rangle \sim r^{-(d-\sigma)}$ in case of a **quench to the ordered phase in systems with non-conserved order parameter** was investigated by [Bray *et al.* (1991)]. An important example is the $(2+1)$-dimensional Glauber-Ising model quenched to zero temperature. It was observed that long-range correlations are relevant if $\sigma$ exceeds a critical value $\sigma_c$. Furthermore, it was shown that the relevant regime is characterized by a continuously changing exponent in the autocorrelation function $A(t) = [\phi(r,t)\phi(r,0)] \sim t^{-(d-\sigma)/4}$, whereas the usual short-range scaling exponents could be recovered below the threshold. These features are in agreement with the simulations of the two-dimensional Ising model quenched from $T = T_c$ to $T = 0$.

### 2.3.6 *Damage spreading behavior*

The high temperature phase of the Ising model is chaotic. When $t$ is lowered a non-chaotic phase may emerge at $T_d$ with different types of transitions depending on the dynamics. The dynamics-dependent DS critical behavior in different Ising models is in agreement with a conjecture by [Grassberger (1995c)]. Dynamical simulations with the heat-bath algorithm in two and three dimensions [Grassberger (1995a); Gropengiesser (1994); Wang and Suzuki (1996)] resulted in $T_d = T_c$ with a DS dynamical exponent coinciding with that of the $Z$-s of the replicas. In this case the DS transition picks up the Ising class universality of its replicas. With Glauber dynamics in two dimensions $T_d < T_c$ and DP class (see Sect. 4.2 DS exponents were found [Grassberger (1995b)]. With Kawasaki dynamics in two dimensions on the other hand the damage always spreads [Vojta (1998)]. With Swendsen-Wang dynamics in two dimensions $T_d > T_c$ and DP class DS behavior was observed [Hinrichsen *et al.* (1998)].

In one-dimensional, nonequilibrium Ising models it is possible to design different dynamics showing either PC (Sect. 4.6) or DP (Sect. 4.2) class DS transition as the function of some control parameter. This depends on the DS transition coincides or not with the critical point. In the PC class DS case damage variables follow BARW2 dynamics (see Sects. 4.6.1 and 4.6.2) and $Z_2$ symmetric absorbing states occur [Hinrichsen and Domany (1997); Ódor and Menyhárd (1998)].

### 2.3.7 *Disordered Ising classes*

The static critical behavior of the Ising model with bond or site disorder (the effect of a random field is more complicate, for latest experimental results see, e.g., [Belanger (2000)]) is well understood thanks to the Harris criterion (1.90). The specific-heat exponent of the **three-dimensional Ising model** is positive, hence according to (1.90) the introduction of a small amount quenched **disorder is a relevant** perturbation.

In order to appreciate the effects of disorder, let's consider the **random-bond Ising model** defined by the classical Hamiltonian

$$H_{\text{dis}} = -\sum_{(i,j)} J_{i,j} s_i s_j \ , \tag{2.48}$$

where the $J_{i,j}$ are random variables uniformly distributed over the interval $[1-\varepsilon/2, 1+\varepsilon/2]$ with $0 < \varepsilon \leq 2$ [Huse and Henley (1985); Paul *et al.* (2004)]. The randomly placed impurities that alter the local exchange couplings, but do not generate random fields or destroy the long-range order, roughen domain walls in Ising systems for dimensionality $5/3 < d < 5$. They also pin (localize) the walls in energetically favorable positions. This drastically **slows down the kinetics of ordering.** The pinned domain wall is a new critical phenomenon governed by a zero-temperature fixed point. Coarsening domains are trapped by energy barriers $E_B(R) \simeq E_0 R^{\psi}$, with exponent $\psi = \chi/(2-\zeta)$, where $\chi$ and $\zeta$ are the pinning and roughening exponents (see Chapter 7).

For $d = 2$, these exponents are known to be $\chi = 1/3$ and $\zeta = 2/3$, yielding $\psi = 1/4$ and the model exhibits a second-order phase transition with a disorder dependent critical temperature. Provided that on average $< J_{i,j} >= 1$, the critical temperature in the presence of bond disorder does not change much. For $d = 3$, perturbative calculation gives $\psi \simeq 0.55$ [Huse and Henley (1985)]. Following a quench from $m_0 = 0$ to the ferromagnetic phase $T < T_C(\varepsilon)$ a system with model-A exhibits phase ordering with a dynamical exponent obtained via scaling arguments [Paul *et al.* (2004)] and field theory [Schehr and Rieger (2005)]

$$Z = 2 + \frac{\epsilon}{T} \,. \tag{2.49}$$

The random-bond Ising model ages in much the same way as a simple magnet. The prediction of Table (1.1) for the exponents $a, b$ in class S systems hold for this non-glassy disordered system as well, but the value of the dynamical exponent $Z = Z(T, \varepsilon)$ (and also of $\lambda_{C,R}$) becomes dependent on temperature and on the distribution of the couplings. The form of the response function agrees with the prediction of LSI, for a large range of values of $Z$ [Baumann *et al.* (2007)] (see however [Hinrichsen (2007c)]).

The addition of non-magnetic impurities (**dilution**) leads to a new universality class (as confirmed by RG analyses, Monte Carlo simulations, and experiments, see [Pelissetto and Vicari (2000a); Folk *et al.* (2003)]) to which the weakly dilute Random Ising Model (**RIM**) belongs. The RIM is a randomly site diluted lattice spin ($s_i$) model with the nearest-neighbor interaction Hamiltonian

$$H_{RIM} = -J \sum_{\langle ij \rangle} \kappa_i \kappa_j \, s_i s_j \,, \tag{2.50}$$

where $\kappa_i$ are uncorrelated, quenched random variables such that

$$\kappa_i = \begin{cases} 1 & \text{with probability } \; p \\ 0 & \text{with probability } \; 1-p \,. \end{cases} \tag{2.51}$$

In this model the phase transition disappears once the non-magnetic sites (where $\kappa_i = 0$) percolate (see Sect. 4.4.1). For $p < p_p$, magnetic order survives up to a nonzero temperature $T_c(\rho)$; but at $p_p$, it is destroyed immediately by thermal fluctuations (see Fig. 2.2). The rare regions are of

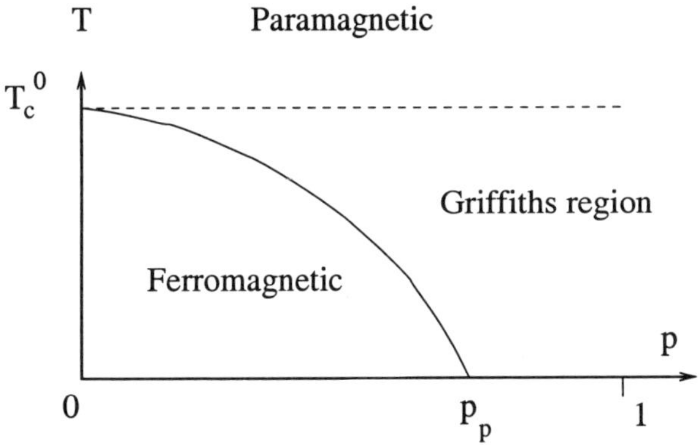


Fig. 2.2 Schematic phase diagram of RIM the as function of temperature $T$ and impurity concentration. $p_p$ is the isotropic percolation threshold of the lattice, and $T_c^0$ is the clean critical temperature.

finite size in all directions and their effective dimensionality is $d_r = 0$. Since the lower critical dimension of the Ising model is $d_c^- = 1$, this system falls in class (a) according to the classification outlined in Sect. 1.7. Their Griffiths singularities are exponentially weak and can be estimated asymptotically via the contribution to the the spin-spin autocorrelation function (1.21).

$$A_r(t,0) \propto \int_0^\infty dL_r\ w(L_r)\ \exp[-t/t_R(L_r, r_0)]\ . \tag{2.52}$$

Here $w(L_r)$ is the probability for finding a rare region of size $L_r$ as and $t_R(L_r, r_0)$ is the relaxation time of such a rare region at the reduced temperature $r_0$.

In the **paramagnetic phase** $T > T_c^0$ the relaxation time of rare regions remains finite even for the largest islands. Therefore the spin-spin autocorrelation function decays in an exponential manner

$$A(t,0) \propto e^{-t/\xi^Z}\ . \tag{2.53}$$

At the **clean critical temperature** $T = T_c^0$ ($r_0 = 0$) the relaxation time of the large rare regions is given by finite-size scaling, hence the the autocorrelation exhibits stretched exponential time dependence asymptotically

$$A(t,0) \propto e^{-\tilde{p}^{Z/(d+Z)}\, t^{d/(d+Z)}}\ , \tag{2.54}$$

where $\tilde{p} = -\ln(1-p)$.

Table 2.3 Critical dynamical exponents of the model-A RIM in three dimensions from Ref. [Pelissetto and Vicari (2000b)]

| $\nu_\perp$ | $\eta_\perp$ | $\gamma$ | $Z$ | $\lambda_C$ | $\eta$ |
|---|---|---|---|---|---|
| 0.668(10) | 0.0327(19) | 1.313(14) | 2.2(2) | 2.73(30) | 0.10(3) |

**Inside the Griffiths region,** $T_c(p) < T < T_c^0$, the relaxation time is limited by the time taken to coherently reverse the entire rare region, which increases exponentially with $L_r$, therefore the saddle point approximation of the the integral (2.52) results in power law decay, with continuously changing exponent

$$A(t,0) \propto t^{-d/(d-1)} . \tag{2.55}$$

At the **dirty critical point** the large scale properties of RIM are described by the field theory of a O(1) model with a random mass term

$$H^{\psi}[\varphi] = \int d^d x \left[\frac{1}{2}(\nabla\varphi)^2 + \frac{1}{2}[r_0 + \psi(x)]\varphi^2 + \frac{g_0}{4!}\varphi^4\right] \tag{2.56}$$

where $\varphi \equiv \varphi(x)$ and $\psi(x)$ is a Gaussian random variable such that

$$< \psi(x)\psi(x') >= \Delta\delta^d(x - x') \tag{2.57}$$

and $r_0$, the bare mass, is adjusted so that the renormalized one is zero. Above the percolation threshold $p_p$, the critical (both static and dynamic) properties of the RIM model has been found to be independent of the actual impurity concentration [Pelissetto and Vicari (2000a); Folk *et al.* (2003)]. The latest critical exponent estimates are summarized in Table 2.3.

The relaxational dynamics of the model-A RIM in dimension $d = 4 - \epsilon$ is described by the Langevin equation:

$$\tilde{\eta}\frac{\partial}{\partial t}\varphi(x,t) = -\frac{\delta H^{\psi}[\varphi]}{\delta\varphi(x,t)} + \zeta(x,t) , \tag{2.58}$$

where $\zeta(x,t)$ is a thermal noise with correlators

$$\langle\zeta(x,t)\rangle = 0 , \tag{2.59}$$

$$\langle\zeta(x,t)\zeta(x',t')\rangle = 2\tilde{\eta}T\delta(x - x')\delta(t - t') \tag{2.60}$$

and $\tilde{\eta}$ is the 'friction' coefficient. The dynamical critical exponent $Z$ is known up to two loops in a $\sqrt{\epsilon}$ [Janssen *et al.* (1995)] and up to three loops in fixed-dimension expansion [Prudnikov *et al.* (1998)] in three dimensions, but the agreement with the numerical simulation data is rather poor. The critical dynamical exponents determined by field theory and simulations are summarized in Table 2.3.

Table 2.4 Static exponents of the $q$ states Potts model in two dimension.

| exponent | $q=0$ | $q=1$ | $q=2$ | $q=3$ | $q=4$ |
|---|---|---|---|---|---|
| $\alpha_H$ | $-\infty$ | $-2/3$ | 0(log) | 1/3 | 2/3 |
| $\beta$ | 1/6 | 5/36 | 1/8 | 1/9 | 1/12 |
| $\gamma$ | $\infty$ | 43/18 | 7/4 | 13/9 | 7/6 |
| $\nu_\perp$ | $\infty$ | 4/3 | 1 | 5/6 | 2/3 |

## 2.4 Potts classes

The generalization of the two-state equilibrium Ising model was introduced by [Potts (1952)], for an overview see [Wu (1982)]. In the $q$-state Potts model the state variables can take $q$ different values $s_i \in (0, 1, 2, ...q)$ and the Hamiltonian is a sum of Kronecker delta function of states over nearest neighbors

$$H = -J \sum_{<i,i'>} \delta(s_i - s_{i'}) \,. \tag{2.61}$$

This Hamiltonian exhibits a global symmetry described by the permutation group of $q$ elements ($S_q$). The Ising model is recovered in the $q = 2$ case (discussed in Sect. 2.3). The $q$-state Potts model exhibits a disordered high-temperature phase and an ordered low-temperature phase. The transition is first-order, mean-filed-like for $q$-s above the $q_c(d)$ curve (shown in Fig. 2.3) (and for $q > 2$ in high dimensions).

The $q = 1$ **limit** can be shown [Fortuin and Kasteleyn (1972)] to be equivalent to the isotropic percolation (see Sect. 4.4.1) which is known to exhibit a continuous phase transition with $d_c = 6$. The problem of finding the effective resistance between two node points of a network of linear resistors was solved by Kirchoff in 1847. [Fortuin and Kasteleyn (1972)] showed that Kirchoff's solution can be expressed as a $q = 0$ **limit** of the Potts partition function. Further mappings were discovered between the spin glass [Edwards and Anderson (1975)] and the $q = 1/2$ Potts model and between the two dimensional $q = 3, 4$ cases and vertex models (see [Baxter (1982)]). From this figure we can see that for $q > 2$ Potts models continuous transitions occur in two dimensions only ($q = 3, 4$). Fortunately these models are exactly solvable (see [Baxter (1982)]) and exhibit conformal symmetry as well as topological, Yang-Baxter invariance. The static exponents in two dimensions are known exactly (Table 2.4)

The Potts model has many applications, examples include the isotropic to nematic transition of liquid crystals [de Gennes (1969)], martensitic tran-

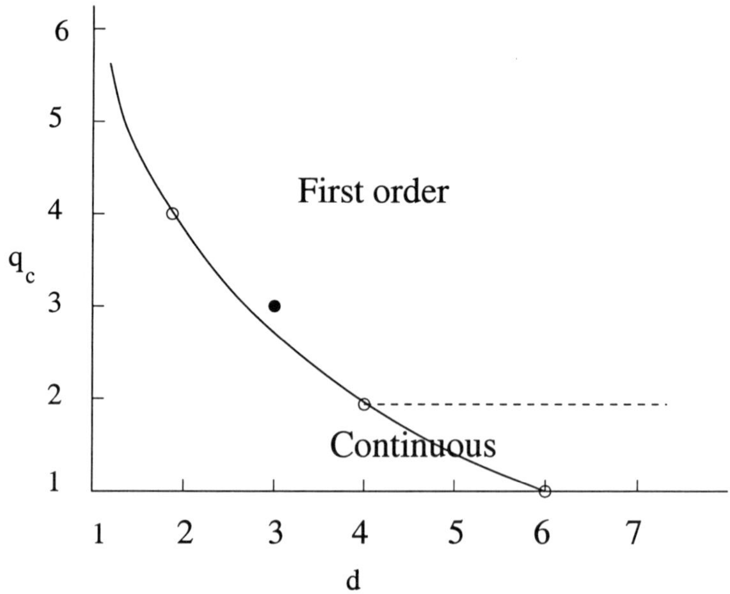


Fig. 2.3 Schematic plot from [Wu (1982)] for $q_c(d)$ (solid line). Open symbols correspond to continuous phase transition, filled symbol to a known first order transition. Below the dashed line the transition is continuous too.

sition of $\beta$ tungsten [Weger and Goldberg (1973)] and quantum field theory models of particle physics [McKane (1977)].

### 2.4.1 *Correlated percolation at $T_c$*

In two dimensions geometrical clusters (defined in Sect. 4.4.1) percolate right at $T_c$ of the model. In higher dimensions geometrical clusters percolate below $T_c$ and their fractal structure is unrelated to any thermodynamic singularity. In Potts models with $Z_3$ symmetry the following percolation exponents are reported [Fortunato and Satz (2001)]:

$$\beta_p = 0.075(14), \qquad \gamma_p = 1.53(21) \ . \tag{2.62}$$

In case of Fortuin-Kasteleyn cluster construction [Fortuin and Kasteleyn (1972)] the percolation exponents of the $q$ state Potts model at $T_c$ coincide with those of the magnetization of the model.

More generally in two dimensions [Janke and Schakel (2006)] found that the Fortuin-Kasteleyn clusters describe the standard critical behavior, and the geometrical clusters describe the **tricritical behavior** which arises, when we include vacant sites in the pure Potts model.

### 2.4.2 *The vector Potts (clock) model*

A variant of the $q$-state Potts model is the vector Potts model (or clock model), which exhibits a $Z_q$ cyclic symmetry (a subgroup of $S_q$) of the variables $\Theta = (2\pi/q)r,\ r = 0, ...q-1$. Its Hamiltonian is

$$H = -J \sum_{i,i'} \cos(\Theta_i - \Theta_{i'}) \quad . \tag{2.63}$$

Note that **for q = 2, 3 the clock model is equivalent to the ordinary Potts** model. In two dimensions for $q \leq 4$ it exhibits second order phase transition but **for q > 4 it has several distinct critical points**. These separate a high temperature, disordered phase from an intermediate phase with conventional, second-order transition and the intermediate phase from a low-temperature vortex phase with an **XY model type transition** (see Sect. 2.5).

Application of this model include effective theories of high energy physics for example [Margaritis *et al.* (1988)], where it was shown using NPRG method that in the $Z_3$ symmetrical model for $d = 3$ no fixed point solution exists a implying first order transition.

### 2.4.3 *Dynamical Potts classes*

The **model-A** dynamical exponents of the Potts classes are determined in two dimensions for $q = 3, 4$ by short-time Monte Carlo simulations [de Silva *et al.* (2002)] (Table 2.5). The exponents were found to be the same for heat-bath and Metropolis algorithms. As one can see the dynamical exponents of

Table 2.5 Model-A dynamical exponents of the $q$ states critical Potts model in two dimensions.

| exponent | $q = 3$ | $q = 4$ |
|---|---|---|
| $Z$ | 2.198(2) | 2.290(3) |
| $\eta$ | 0.075(3) | |
| $\lambda_C$ | 1.830(2) | |
| $\theta_g$ | 0.350(8) | |

the $q = 2$ (Ising), $q = 3, 4$ are rather close, which raised the hypothesis of a "weak universality" [Suzuki (1974)] with a common value $Z = 2.17(4)$ [Tang and Landau (1987)], which is still debated [de Silva *et al.* (2002)], generating more and more precise exponent estimates [Fan and Zhong (2007)].

If we quench the system into the coexistence phase ($T < T_c$) from a disordered state the ageing exponents satisfy $a = a'$ and **LSI** and for non-conserved dynamics the following ageing exponents have been determined [Lorenz and Janke (2007)] (Table 2.6). For the **zero temperature local persistence** exponent in **one dimension** exact formulas have been determined. For sequential dynamics [Derrida *et al.* (1995)]

Table 2.6 Model-A dynamical exponents of the $q$ states $T = 0$ Potts model in two dimensions (for quenches from disordered state).

| | $q = 3$ | $q = 8$ |
|---|---|---|
| $Z$ | 2 | 2 |
| $\lambda_R$ | 1.19(3) | 1.25(1) |
| $a = a'$ | 0.49 | 0.51 |

$$\theta_{l,s} = -\frac{1}{8} + \frac{2}{\pi^2}\left[\cos^{-1}(\frac{2-q}{\sqrt{2}q})\right]^2 \tag{2.64}$$

while for parallel dynamics $\theta_{l,p} = 2\theta_{l,s}$ [Menon and Ray (2001)]. In deterministic coarsening this exponent is different again (see [Bray *et al.* (1994); Gopinathan (2001)]. As we can see *the dynamical universality class characterized by the growth of the length scale $\xi \propto t^{1/2}$ divided into sub-classes, as reflected by the persistence exponent.*

The phase ordering properties of the two-dimensional, **model-A** clock model has been investigated for $q = 3, 6$ by [Corberi *et al.* (2006)]. For a quench from a disordered state to $T = 0$ numerical evidence was found for the universality with

$$Z = 2, \quad \lambda_C = 5/4, \quad a = 1/2 \tag{2.65}$$

dynamical exponents. This suggests that the **exponents are independent of $q$ finite.** For $q \to \infty$ crossover to the **XY model** class behavior (see Sect. 2.5) is expected.

As in the Ising model in nonequilibrium systems symmetry in the Potts model turned out to be a relevant factor for determining the universal behavior of transitions to **fluctuating ordered states** [Brunstein and Tomé (1998); Crisanti and Grassberger (1994); Szabó and Czárán (2001)]. On the other hand in the case of nonequilibrium transitions to **absorbing**

**states** a simulation study [Lipowski and Droz (2002a)] suggests first order transition for all $q > 2$ state Potts models in $d > 1$ dimensions. The $q = 2$, $d = 2$ case corresponds to the Voter model class (Sect. 4.5) and in one dimension either PC class transition ($q = 2$) or N-BARW2 class transition ($q = 3$) (Sect. 6.12) occurs.

The **DS transition** of a $q = 3$, $d = 2$ Potts model with heat-bath dynamics was found to belong to the DP class (Sect. 4.2) because $T_d > T_c$ [da Silva *et al.* (1997)].

### 2.4.4 *Long-range interactions*

The effect of long-range interaction has also been investigated in the case of the one dimensional $q = 3$ Potts model [Glumac and Uzelac (1998)] with the Hamiltonian

$$H = -\sum_{i<j} \frac{J}{|i-j|^{1+\sigma}} \delta(s_i, s_j) \, . \tag{2.66}$$

For $\sigma < \sigma_c \sim 0.65$ a crossover from second order (nonuniversal, $\sigma$ dependent) to a first order (mean-field) transition was located by simulations. As in the Ising model here we can expect to see this crossover if we generate long-range interactions by the addition of a Lévy type Kawasaki dynamics.

## 2.5 XY model classes

The classical XY model is defined by the Hamiltonian

$$H = -J \sum_{i,i'} \cos(\Theta_i - \Theta_{i'}) \tag{2.67}$$

with continuous $\Theta_i \in [0, 2\pi]$ state variables. This model has a global $U(1)$ symmetry. Alternatively the XY model can be defined as a special $N = 2$ case of $O(N)$ symmetric models such that the spin vectors $S_i$ are two dimensional with absolute value $\mathbf{S}_i^2 = 1$

$$H = -J \sum_{<i,i'>} \mathbf{S}_i \mathbf{S}_{i'} \quad . \tag{2.68}$$

In this continuous model in **two dimensions** no local order parameter can take zero value according to the Mermin-Wagner theorem [Mermin and Wagner (1996)]. The appearance of free vortexes (which are non-local) cause an unusual transition mechanism that implies that most of the thermodynamic quantities do not show power-law singularities. The singular behavior of the correlation length ($\xi$) and the susceptibility ($\chi$) is described by the forms for $T > T_c$

$$\xi \propto \exp\left(C(T-T_c)^{-1/2}\right), \quad \chi \propto \xi^{2-\eta_\perp}, \tag{2.69}$$

where $C$ is a non-universal positive constant. Conventional critical exponents cannot be used, but one can define scaling dimensions. At $T_c$ the two-point correlation function has the following long-distance behavior

$$G(r) \propto r^{-1/4}(\ln r)^{1/8} \tag{2.70}$$

implying $\eta_\perp = 1/4$, and in the entire low-temperature phase

$$G(r) \propto r^{-\eta_\perp(T)} \tag{2.71}$$

such that the exponent $\eta_\perp$ is a continuous function of the temperature, i.e. the model has a line of critical points starting from $T_c$ and extending to $T = 0$. This is the so called Kosterlitz-Thouless (**KT**) critical behavior [Kosterlitz and Thouless (1973)] and corresponds to the conformal field theory with $c = 1$ [Itzykson and Drouffe (1989)]. This kind of transition can experimentally observed in many effectively two-dimensional systems with $O(2)$ symmetry, such as thin films of superfluid helium and describes roughening transitions of SOS models at crystal interfaces.

In **three dimensions** the critical exponents of the O(N) ($N = 0,1,2,3,4$) symmetric field theory have been determined by perturbative expansions up to seventh loop order [Guida and Zinn-Justin (1998)]. The Table 2.7 summarizes these results for the XY case (for a more detailed overview see [Pelissetto and Vicari (2000a)]).

Table 2.7 Static exponents of the XY model in three dimensions.

| $\alpha_H$ | $\beta$ | $\gamma$ | $\nu_\perp$ | $\eta_\perp$ |
|---|---|---|---|---|
| -0.011(4) | 0.347(1) | 1.317(2) | 0.670(1) | 0.035(2) |

At the KT phase transition the **critical dynamics** is slow due to the existence of vortices. The power-law growth of the nonequilibrium spatial correlation length and the autocorrelation function undergo strong correction to scaling making difficult to identify aging phenomena and extracting critical exponents. Spin-wave or similar assumptions have been setup [Picone and Henkel (2004); Abriet and Karevskia (2004)] for the scaling functions, but those forms could fit the numerical data at low temperatures only.

**In two dimensions, model-A** dynamical relaxation exponents and the autocorrelation properties have been investigated by simulations [Ying *et al.* (2001); Zheng *et al.* (2003); Lei and Zheng (2007)]. For a disordered initial condition logarithmic correction to scaling is present [Bray *et al.* (2000)],

$$\xi(t) \sim (t/\ln t)^{1/2} \,, \tag{2.72}$$

but a power-law correction for an ordered vortex free start was found. Table 2.8 summarizes the known dynamical exponents

Table 2.8 Model-A dynamical exponents of the critical XY model

| $d$ | $Z$ | $\lambda_C$ | $a = a'$ |
|---|---|---|---|
| 2 | 2.04(1) | 1.49(2) | |
| 3 | 2 | 2.68(10) | 0.52 |

For a quench to $T < T_c$ in $d = 3$ simulations gave the ageing exponents (Table 2.8) [Abriet and Karevski (2004)].

$$a = a' = 0.5, \quad \lambda_R = \lambda_C = 1.7 \,. \tag{2.73}$$

These values satisfy the ansatz for LSI $R(t, s)$ (1.29) numerically, but field theory [Calabrese and Gambassi (2002)] seems to contradict LSI.

### 2.5.1 *Long-range correlations*

Similarly to the Ising model [Bassler and Rácz (1995)] studied the validity of the Mermin-Wagner theorem by transforming the two dimensional XY model to a nonequilibrium one using **two-temperature, model-A** dynamics. They found that the Mermin-Wagner theorem does not apply for this case since effective long-range interactions are generated by the local nonequilibrium dynamics. The universality class of the phase transition of the model coincides with that of the two temperature driven Ising model (see Table 3.1).

## 2.6 $O(N)$ symmetric model classes

As already mentioned in the previous section the $O(N)$ symmetric models are defined on spin vectors $\mathbf{S}_i$ of unit length $\mathbf{S}_i^2 = 1$ with the Hamiltonian

$$H = -J \sum_{<i,i'>} \mathbf{S}_i \mathbf{S}_{i'} . \tag{2.74}$$

The best known of these models is the classical Heisenberg model, which corresponds to $N = 3$. This is the simplest model of isotropic ferromagnets. The $N = 4$ case corresponds to the Higgs sector of the Standard Model at finite temperature. The $N = 0$ case is related to polymers and the $N = 1$ and $N = 2$ cases are the Ising and XY models respectively (therefore they are not discussed here). The critical dimension is $d_c = 4$ and by the Mermin-Wagner theorem we cannot find a finite temperature phase transition in the short range equilibrium models for $N > 2$ below $d = 3$.

The static critical exponents have been determined by $\epsilon = 4 - d$ expansions up to five loop order [Gorishny *et al.* (1984)], by exact RG methods (see [Margaritis *et al.* (1988); Berges *et al.* (2002b); Bervillier *et al.* (2007)] and the references therein), by simulations [Hasenbuch (2001)] and by series expansions in three dimensions (see [Guida and Zinn-Justin (1998)] and the references there). In Table 2.9 I show the latest estimates from [Guida and Zinn-Justin (1998)] in three dimensions for $N = 0, 3, 4$. The $N \to \infty$

Table 2.9 Static exponents of the $O(N)$ model in three dimensions

| $N$ | $\alpha_H$ | $\beta$ | $\gamma$ | $\nu_\perp$ | $\eta_\perp$ |
|---|---|---|---|---|---|
| 0 | 0.235(3) | 0.3024(1) | 1.597(2) | 0.588(1) | 0.028(2) |
| 3 | -0.12(1) | 0.366(2) | 1.395(5) | 0.707(3) | 0.035(2) |
| 4 | -0.22(2) | 0.383(4) | 1.45(1) | 0.741(6) | 0.035(4) |

limit is the exactly solvable spherical model [Berlin and Kac (1952); Stanley (1968)]. It's critical behavior is not mean-field like in $2 < d < 4$. For a detailed discussion of the static critical behavior of the $O(N)$ models see [Pelissetto and Vicari (2000a)].

The dynamical exponents for **model-A** are known exactly for the ($N \to \infty$) spherical model [Janssen *et al.* (1989); Majumdar *et al.* (1996)] case. The critical dynamics of this model for $d > 2$ has been determined for long-range (L) [Picone and Henkel (2002)]

$$C_{ini} \propto |r|^{-d-\sigma} , \tag{2.75}$$

Table 2.10 Critical dynamical exponents of the model-A spherical model in $d > 2$

| $Z$ | $d$ | $a = a'$ | $\lambda_R$ | range |
|---|---|---|---|---|
| 2 | $< 4$ | $d/2 - 1$ | $d - \sigma/2 - 1$ | L |
| 2 | $> 4$ | $d/2 - 1$ | $(d - \sigma)/2 + 1$ | L |
| 2 | $< 4$ | $d/2 - 1$ | $3d/2 - 2$ | S |
| 2 | $> 4$ | $d/2 - 1$ | $d$ | S |

with ($d+\sigma < 2$) and short range (S) initial correlations [Godrèche and Luck (2000a)] (see Table 2.10). For the quench to $T = 0$ both the long-range initial correlation (2.75) [Picone and Henkel (2002)] and the long-range interaction case

$$J_R \propto |r|^{-d-\sigma'} , \tag{2.76}$$

[Cannas *et al.* (2001)] has been investigated. The results are summarized in Table 2.11. These exponents imply that LSI is satisfied in the spherical

Table 2.11 $T \to 0$ dynamical exponents of the model-A spherical model.

| $Z$ | $d$ | $a = a'$ | $\lambda_R$ | type |
|---|---|---|---|---|
| 2 | $> 2$ | $d/2 - 1$ | $(d - \sigma)/2$ | (2.75) |
| $\sigma'$ | $> 2$ | $d/\sigma' - 1$ | $d/2$ | $0 < \sigma' < 2$ |
| $\sigma'$ | $\leq 2$ | $d/\sigma' - 1$ | $d/2$ | $0 < \sigma' < d$ |

model. For other cases $\epsilon = 4 - d$ expansions up to two loop order exist [Majumdar *et al.* (1996); Oerding *et al.* (1997)]. For the critical initial slip a closed formula has been derived [Janssen *et al.* (1989)]

$$\eta = \frac{n+2}{n+8}\left[1 + \frac{6}{n+8}\left(\frac{n+3}{n+8} + \ln(\frac{3}{2})\right)\epsilon\right]\frac{\epsilon}{4} . \tag{2.77}$$

For a discussion about the combination of different dynamics see the general introduction Sect. 2.1 and [Täuber *et al.* (1999)].

### 2.6.1 *Correlated percolation at $T_c$*

In three dimensions for $O(2)$, $O(3)$ and $O(4)$ symmetric models the Fortuin-Kasteleyn cluster construction [Fortuin and Kasteleyn (1972)] results in percolation points and percolation exponents which coincide with the corresponding $T_c$-s and magnetization exponent values [Blanchard *et al.* (2000)].

Table 2.12 Model A critical dynamical exponents of the $O(N)$ model in three dimensions.

| $N$ | $Z^{b}$ | $\lambda_R = \lambda_C$ | $a = a'$ | $\eta^{a}$ | $\theta_g$ |
|---|---|---|---|---|---|
| $\infty$ | 2 | 5/2 | 1/2 | 1/4 | 1/4 |
| 3 | 2.032(4) | 2.789(6) | | 0.1725 | 0.38 |

[a] from (2.77), [b] from Ref. [Zheng (1998)],

### 2.6.2 *Disordered O(N) classes*

In **three-dimensional** $O(N)$**-vector models** the exponent of the specific heat is $\alpha_H < 0$ for $N \geq 2$ [Pelissetto and Vicari (2000a)] and the critical behaviour is unchanged in the presence of weak quenched disorder (apart from large crossover effects). The $N = 1$ (Ising) case is already disscussed in Sect. 2.3.7.

# Chapter 3

# Genuine Basic Nonequilibrium Classes with Fluctuating Ordered States

In this chapter I discuss nonequilibrium models, which are obtained from equilibrium ones either by the addition of an other heath bath (competing dynamics) or by the introduction of external currents. These system has no hermitian Hamiltonian in general (the detailed balance condition (2.20) is broken) and at their critical phase transitions novel type of scaling behavior emerges. The exploration of such "genuine nonequilibrium universality classes" has been a major topic of statistical physics and I shall present the most important results in the rest of the book.

## 3.1 Driven lattice gas (DLG) classes

The driven lattice gas (DLG), a model of super-ionic currents was introduced by [Katz *et al.* (1984)] and has become theoreticians first prototype for anisotropic, nonequilibrium phase transitions. In the DLG particles may move preferentially along one of the lattice directions, say $\hat{x}$. One may imagine this is induced by an external drive, $E\hat{x}$, e.g., an applied electric field assuming the particles are positive ions. Consequently, for periodic boundary conditions, a particle current and an anisotropic interface set up along $\hat{x}$ at low temperature, $T < T_E$. That is, a liquid-like phase, which is striped then coexists with its gas. More specifically, assuming a square lattice half filled with particles, simulations show that the function $T_E$ monotonically increases with $E$ from the Onsager value $T_0 = T_{\text{Onsager}} = 2.269 J k_B^{-1}$ to $T_\infty \simeq 1.4 T_{\text{Onsager}}$. This limit corresponds to a *nonequilibrium* critical point. It was numerically shown to belong to a universality class other than the Onsager one (see Table 3.1) [Albano and Saracco (2002)]. Note that for anisotropic systems such as the DLG model, the following modified hyperscaling relation is expected to hold [Binder and Wang. (1989)]:

$$\nu_{||} + (d-1)\nu_{\perp} - 2\beta = \gamma, \tag{3.1}$$

### 3.1.1 *Driven lattice gas model in two-dimensional (DDS)*

As already pointed out in Chapter 2, if additional dynamics are coupled to spin-exchange Ising models novel kind of universality classes may emerge. In two dimensions both simulations and field theory [Praestraad *et al.* (1994); Praestgaard *et al.* (2000)] predicts the critical exponents shown in Table (3.1). The upper critical dimension is $d_c = 3$. It has been proven that the Langevin equation (3.2) suggested by [Garrido *et al.* (2000)]:

$$\partial_t \phi(\mathbf{r}) = -\nabla_{\perp}^4 \phi + \tau_{\perp} \nabla_{\perp}^2 \phi + \frac{g}{6} \Delta_{\perp} \phi^3 + \tau_{||} \nabla_{||}^2 \phi + \zeta(\mathbf{r}, t) \tag{3.2}$$

(and therefore the critical behavior) of the anisotropic diffusive system (**ADS**) coincides with that of the randomly driven lattice gas.

Table 3.1 Critical exponents of the $d = 2$ randomly driven lattice gas. This first row corresponds to the anisotropic (ADS, the second to the external current driven (DDS) model.

| | $\beta$ | $\gamma$ | $\eta_{\perp}$ | $\nu_{\perp}$ | $Z$ |
|---|---|---|---|---|---|
| ADS | 0.33(2) | 1.16(6) | 0.13(4) | 0.62(3) | 2.016(40) |
| DDS | 1/2 | 1 | | 1/2 | 4/3 |

Other systems in the **ADS** universality class are the two-temperature model by [Garrido *et al.* (1990)], the ALGA model [Binder (1981)] and the infinitely fast driven lattice gas model [Achahbar *et al.* (2001)]. In the randomly driven lattice gas model particle current does not occur but an **anisotropy** can be found. Therefore it was argued [Achahbar *et al.* (2001)] that the particle current is not a relevant feature for this class.

This argument offers an explanation for why some set of simulations of driven lattice systems [Vallés and Marro (1987)] lead to different critical behavior than that of the canonical coarse-grained representative of this class (**DDS**), in which an explicit particle **current** $j\hat{\mathbf{x}}$ is added to the continuous, model-B Ising model Hamiltonian:

$$\frac{\partial \phi(\mathbf{r}, t)}{\partial t} = -\nabla \left[ \eta \frac{\delta H}{\delta \phi} + j\hat{\mathbf{x}} \right] + \nabla \zeta \tag{3.3}$$

[Janssen and Schmittman (1986); Leung and Cardy (1986); Schmittman and Zia (1996)] (here $\eta$ is a parameter, $\zeta$ is a Gaussian noise). The Langevin equation describes the stochastic evolution of the local particle density $\rho(\mathbf{x}, t)$ and in terms of $\phi = 2\rho - 1$, it reads

$$\partial_t \phi(\mathbf{r}) = -\nabla_{\perp}^4 \phi + \tau_{\perp} \nabla_{\perp}^2 \phi + \frac{g}{6} \Delta_{\perp} \phi^3 + \tau_{||} \nabla_{||}^2 \phi - \epsilon \nabla_{||} \phi^2 + \zeta(\mathbf{r}, t), \quad (3.4)$$

Comparing Eqs. (3.2) and (3.4) one can see that they only differ in the current term $\epsilon \nabla_{||} \phi^2$ that vanishes in (Eq. (3.4)). This difference has the consequences that for the latter model one obtains mean-field exponents for $d \geq 5$ and the renormalization group analyses resulted in different critical exponents (for example $\beta = 1/2$) exactly (see Table (3.1)).

### 3.1.2 *Driven lattice gas model in one-dimensional (ASEP, ZRP)*

Formerly, combinations of Glauber [Glauber (1963)] and Kawasaki [Kawasaki (1972)] kinetics were introduced with the aim of investigating temperature-driven nonequilibrium phase transitions in one dimension [Droz *et al.* (1989); Rácz and Zia (1994)]. Since in one dimension any $T > 0$ bulk dynamics destroys the ordered state the only possibility for a phase transition is the addition of a ($T > 0$) spin-exchange dynamics, which acts at the boundaries of spin domains. In this case the ordered phase becomes a frozen, absorbing one [Menyhárd (1994)] (see Sect. 4.6.2) and a continuous phase transition belonging to the parity conserving (PC) class emerges. Its discussion is postponed to Sect. 4.6.2, where classes of models with transition to absorbing states are provided.

Current driven lattice gases are systems of classical particles with hard-core interactions moving in a preferred direction. They can model ions diffusing in a narrow channel under the influence of an electric field [Schütz (2001)], cars proceeding on a long road [Chowdhury *et al.* (2000)], or ribosomes moving along a m-RNA [McDonald *et al.* (1968); Chowdhury *et al.* (2007)]. These systems are usually defined on an open lattice coupled with two reservoirs at both ends or on a lattice with periodic boundary conditions. For long times they settle into nonequilibrium steady states characterized by the bulk density and the corresponding particle current. Besides the boundary conditions these states are strongly affected by localized inhomogeneities as well.

The matrix product formulation is one of the most powerful techniques to study such model. In this method one assigns an operator to every state of a lattice and every configuration is a product of such operators. For systems with nearest neighbor interactions it turned out that the algebra of these operators is quadratic. During the last decade several quadratic algebras have been introduced to study the critical behavior of one dimensional DDS. Both continuous and discontinuous transitions have been found between different (flowing, jammed, etc.) phases. Many exact solutions were derived, still little attention has been paid to the scaling exponents and universality classes in these models. Furthermore, one observes long-range (power-law) correlations not only at criticality, but generally in the steady state. These properties are absent from their one-dimensional, classical counterparts.

One of the simplest DDS is the one-species, asymmetric simple exclusion process (**ASEP**) (see Fig. 3.1). This is a site restricted, RD model with

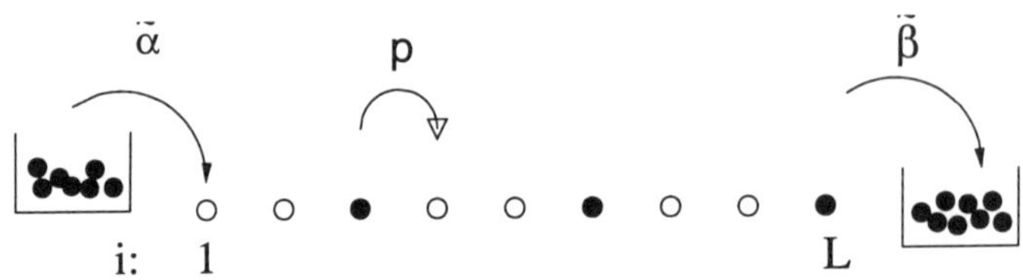


Fig. 3.1 Dynamics of the ASEP with particle injection (rate $\tilde{\alpha}$) at the left and removal (rate $\tilde{\beta}$) at the right boundary.

hard-core repulsion: each site $i$ ($1 \leq i \leq L$) is either occupied by a particle, or it is empty. The model follows the dynamics

- *Hopping*: A pair of sites $(i, i+1)$ is selected randomly or by stochastic cellular automata (SCA) sub-lattice updates. If we find a particle at site $i$ and a hole at site $i+1$, we move the particle one site to the right with probability $p$, otherwise nothing happens.
- *Injection* and *removal*: At the boundaries particles/holes can be injected and/or removed. At the left boundary a particle can be injected with probability $\tilde{\alpha}$ if site 1 is empty. At the right end of the chain a particle will be removed with probability $\tilde{\beta}$.

The model can also be generalized to allow hopping in both directions by introducing a probability $q$ for hopping onto an unoccupied site to the left. Furthermore, one can also inject particles at the right end and remove particles at the left end. These modifications do not change the basic features of the model and will not be considered in the following.

It is well known that the ASEP model can be mapped onto a surface growth model known as the single-step model [Meakin *et al.* (1986)] and in the appropriate hydrodynamical limit its density profile obeys the Burgers equation [Burgers (1974)]

$$\frac{\partial v}{\partial t} = \frac{\partial}{\partial x}\left[\nu\frac{\partial v}{\partial x} + \lambda v^2 + \eta(x,t)a\ ,\right] \tag{3.5}$$

which represents the evolution of a (vortex free) velocity field $v(x,t)$, **conserving momentum** only. This form is closely related to the KPZ equation (7.24) (see Sect. 7.4).

The ASEP exhibits *boundary-induced phase transitions* [Krug (1991)]: the boundaries, which can inject and remove particles from the system, govern – in a subtle interplay with the local dynamical rules – the *macroscopic* behavior of the model and can produce different phases and phase transitions. These phases can be characterized by the value of the current. A detailed analysis of ASEP with different update schemes [Rajewsky *et al.* (1998)] has shown that the phase diagram has the same basic structure (see Fig. 3.2).

One finds three different phases and the system is governed by two independent length scales $\xi_{\tilde{\alpha}}$ and $\xi_{\tilde{\beta}}$, which represent the influence of the boundaries [Derrida *et al.* (1992)]. For $\tilde{\alpha} > \tilde{\beta}$ and $\tilde{\beta}/p < 1/2$ the system is in the so-called *high density phase.* This is separated by a first order transition line at $\tilde{\alpha}/p = \tilde{\beta}/p < 1/2$ from the *low-density phase*, were the relaxation time is finite. On the coexistence line the relaxation time $t_R$ scales as

$$t_R \propto L^2\ . \tag{3.6}$$

For $\tilde{\alpha}/p > 1/2$ and $\tilde{\beta}/p > 1/2$ a *maximum current phase* emerges with a continuous phase transition. Here $t_R$ scales with the dynamic exponent $Z = 3/2$ belonging to the KPZ class and the current becomes independent of $\tilde{\alpha}$ and $\tilde{\beta}$.

It was shown that imposing additional conserved degrees of freedom (both mass and momentum) [Arndt *et al.* (1998, 1999)] one obtains a **two species ASEP**, with particles of opposite charge hopping in opposite directions and with a variable passing probability. From the hydrodynamics perspective this can be viewed as two coupled Burgers equations, with the number of positive and negative momentum quanta individually conserved. This process remains in the KPZ universality class but only in the sense of a factorization, as (KPZ $\otimes$ KPZ). The two Burgers equations decouple at large length scales due to the perfect screening.

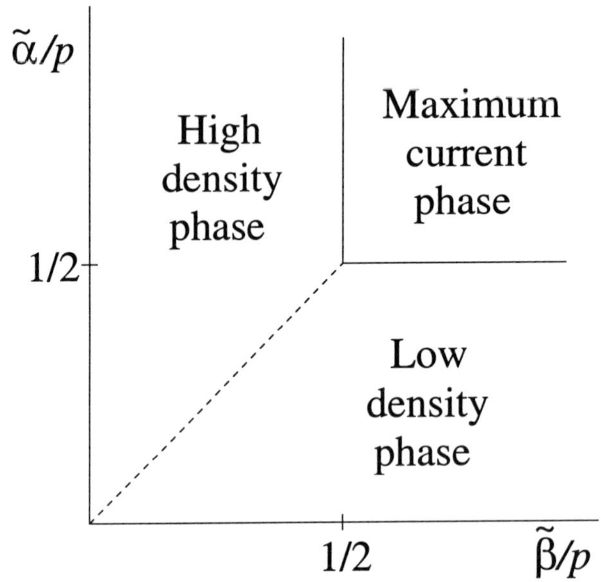


Fig. 3.2 Generic phase diagram of the one dimensional ASEP model. The dashed line denotes a first order, the solid line a second order phase transition.

The bosonic version of ASEP is the zero range process (**ZRP**). It is defined on $d$-dimensional lattices (for a recent review see [Evans and Hanney (2005)]): each site $l$ can hold an integer number of particles, which can hop to the neighboring sites with a rate that depends on the number particle number of the departure $n_l$. In one dimension ZRP is usually defined on a ring of size $L$, where $N$ particles can hop asymmetrically from site $l$ to $l+1$ with a rate $u(n_l)$. The ZRP can be mapped onto the ASEP (see Fig. 3.3), such that a particle of ZRP corresponds to a vacancy of the ASEP. Thus depending on the form $u(n_l)$ there may be long-range interactions in the corresponding ASEP. The ZRP exhibits a condensation transition, which appears in a number of unexpected context such as wealth condensation in macroeconomics, jamming in traffic [Chowdhury *et al.* (2000)], coalescence in granular systems or gelation in networks (for further references see [Evans and Hanney (2005)]).

Besides the exactly solved cases the dynamics of the condensation condensation transition of **homogeneous, single component ZRP** has been instigated by mean-field and heuristic scaling arguments. The heuristic, random walk approach is based on the observation that the condensation

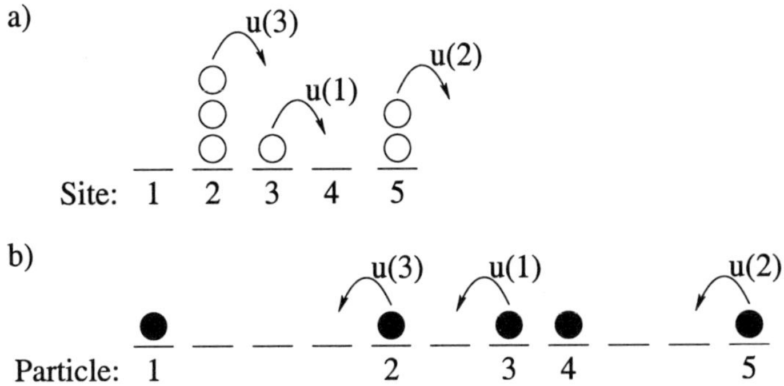


Fig. 3.3 Mapping between the zero-range process and the asymmetric exclusion process, from [Evans and Hanney (2005)]

dynamics can be divided into two regimes:

- Nucleation, during which particles accumulate at condensate sites;
- Coarsening, when the condensate size $\langle m(t)\rangle$ increases via particle exchange.

The coarsening proceeds until only a single condensate site remains and the domain growth is described by the scaling law

$$\langle m(t)\rangle \sim t^{\eta} , \tag{3.7}$$

characterized by the exponent $\eta$.[1] The coarsening depends on two time scales: the escape time, which scales as $\mathcal{O}(L)$ and the transit time between domains. According to the random walk argument this depends on the symmetry of the hopping dynamics.

- Asymmetric dynamics:
  Since the particle reaches the next condensate site to the right with certainty, the typical transit time scales as $\mathcal{O}(L)$. The growth of $\langle m(t)\rangle$ is determined only by the time it takes a condensate site to lose all of its particles of particles. This time scales as $\mathcal{O}(L \times L)$, hence the condensate grows with the exponent $\eta = 1/2$ in agreement with the mean-field analysis.
- Symmetric dynamics:
  The probability that a particle reaches the next condensate site is proportional to the inverse separation of condensate sites. The typical time it takes a particle to escape from a condensate site

[1] in the literature this domain growth exponent may be denoted by exponent $\delta$.

is therefore $\mathcal{O}(L^2)$. Therefore the domain growth time scales as $\mathcal{O}[L \times L^2]$, hence the domain size increase with the exponent $\eta = 1/3$ for symmetric dynamics.

The results are consistent with the simulations for both asymmetric and symmetric cases [Godrèche (2003); Grosskinsky *et al.* (2003)]. The upper critical dimension for the coarsening of the ZRP is $d_c = 2$ and above two dimensions, we expect the coarsening to be determined by the mean-field results for the scaling function and exponent, $\eta = 1/2$.

### 3.1.3 *Driven lattice gas with disorder*

Driven lattice gas models have been shown to be sensitive to spatial inhomogeneities or quenched disorder of any kind. Due to a mapping sketched on see Fig. 7.1 this has a strong consequence on the surface growth models too. Disorder can be introduced in several way. It has been argued long time ago that strong enough random site dilution can substantially affect the transport properties of particles with hard-core interactions, and can make the system respond non-monotonically to the driving field [Ramaswamy and Barma (1987); Barma and Ramaswamy (1993)].

This sensitivity has been investigated in case of boundary induced phase transitions [Krug (1991)], where besides an external field particles are fed from one side and leave on the opposite. Two types of phase transitions involving non-analytic changes in the density profiles and the particle number fluctuation spectra are encountered upon varying the feeding rate and the particle interactions, and associated diverging length scales are identified. The fluctuations of the diffusion constant can be neglected for $d \geq 3$, but for $d \leq 2$ it is a relevant perturbation leading to super-diffusive spreading ($Z < 2$) of density fluctuations and scale dependent diffusion constant. At $d = d_c = 2$ the effect of fluctuations is marginal and logarithmic corrections to mean-field profile scaling is expected. In **one dimension** a complex phase diagram emerges with first and second order phase transitions.

Besides the **site-wise (ST) disorder** one can also introduce **particle-wise (PT) disorder**, in which particle $i$ may hop to empty neighboring sites with rates $p_i$ to the right and $q_i$ to the left (where $p_i$ and $q_i$ are independent and identically distributed random variables). In case of the **totally asymmetric** case ($q_i = 0$) (TASEP) exact studies [Krug and Ferrari (1996); Evans (1996); Goldstein and Speer (1998)], mean-field calculations and simulations [Tripathy and Barma (1997)] discovered dynamical, Bose-

Einstein condensation like phase transitions between homogeneous and an inhomogeneous states (for ST and PT both).

The so-called 'drop-push' model and the ASEP on quenched, spatially disordered, **one dimensional** lattices were investigated by [Tripathy and Barma (1997)]. It was conjectured and supported by simulations that the behavior of a system with current $J$ and density $\rho$ is largely determined by $c = \partial J/\partial \rho$ (where $c$ is the mean speed of kinematic waves which transport density fluctuations through the system). In particular for nonzero $c$ the quenched disorder does not affect the asymptotic behavior of the decay of fluctuations. On the other hand for $c = 0$ disorder-induced phase separation with coexisting macroscopic regions of different density was discovered. Since the model is equivalent to that of a moving interface in the presence of a certain type of columnar disorder (the interface height $h(x,t)$ is related to the density fluctuations by $\Delta\rho = \partial_x h$) the change of surface critical exponents (see Chapter 7) from that of the one dimensional KPZ class (Sect. 7.4) have been determined.

In the strongly disordered, **one dimensional** partially asymmetric model ($q_i > p_i$ for a finite fraction of $i$) a stationary state, is found, in which the average drift is zero and the system exhibits diffusive motion. This state can be described by an infinite randomness fixed point [Juhász *et al.* (2005)] (characterized by activated scaling see Sect. 1.7). While this behavior emerges in case of left-right symmetry, for asymmetrical cases a Griffiths phase, with dynamical exponents

$$Z_{ST} = Z_{PT}/2 \tag{3.8}$$

is derived, both for ST and PT type of impurities.

### 3.1.4 *Critical behavior of self-propelled particles*

Another XY-like nonequilibrium model that exhibits ordered state for $d \leq 2$ dimensions is motivated by the description of the 'flocking" behavior among living things, such as birds, slime molds and bacteria. In the simplest version of the self-propelled particle model [Vicsek *et al.* (1995)] each particle's velocity is set to a fixed magnitude, $v_0$. Interaction with the neighboring particles changes only the direction of motion: the particles tend to align their orientation to the local average velocity. In one dimension the lattice dynamics is defined as

$$\begin{aligned} x_i(t+1) &= x_i(t) + v_0 u_i(t), \\ u_i(t+1) &= G\Big(\langle u(t)\rangle_i\Big) + \xi_i \ , \end{aligned} \tag{3.9}$$

where the particles are characterized by their coordinate $x_i$ and dimensionless velocity $u_i$. The function $G$ incorporates both the propulsion and friction forces, which set the average velocity to a prescribed value $v_0$: $G(u) > u$ for $u < 1$ and $G(u) < u$ for $u > 1$. The distribution function $P(x = \xi)$ of the noise $\xi_i$ is uniform in the interval $[-\eta/2, \eta/2]$. Keeping $v_0$ constant, the adjustable control parameters of the model are the average density of the particles, $\rho$, and the noise amplitude $\eta$. The order parameter is the average velocity $\phi \equiv \langle u \rangle$ which vanishes as

$$\phi(\eta, \rho) \sim \begin{cases} \left(\frac{\eta_c(\rho)-\eta}{\eta_c(\rho)}\right)^\beta & \text{for } \eta < \eta_c(\rho) \\ 0 & \text{for } \eta > \eta_c(\rho) \end{cases}, \tag{3.10}$$

at a critical $\eta_c(\rho)$ value. This model is similar to the $XY$ model of classical magnetic spins because the velocity of the particles, like the local spin of the $XY$ model, has fixed length and continuous rotational $U(1)$ symmetry. In the $v_0 = 0$ and low noise limit the model reduces *exactly* to the Monte-Carlo dynamics of the $XY$ model.

A field theory that included the nonequilibrium effects in a self-consistent way the was proposed by [Tu and Toner (1995)]. Their model is different from the XY model for $d < 4$. The essential difference between the self-propelled particle model and the equilibrium XY model is that at different times, the "neighbors" of one particular "bird" will be different depending on the velocity field itself. Therefore, two originally distant "birds" can interact with each other at some later time. Tu and Toner found an upper critical dimension $d_c = 4$, below which linearized hydrodynamics breaks down, but owing to **Galilean invariance** they could obtain **exact scaling exponents in two dimensions**. For the dynamical exponent they got $Z = 6/5$. Numerical simulations [Vicsek *et al.* (1995); Czirók *et al.* (1997)] indeed found a long range ordered state with a continuous transition characterized by $\beta = 0.42(3)$ in two dimensions. In **one dimension** field theory and simulations [Czirók *et al.* (1999)] provided evidence for a continuous phase transition with $\beta = 0.60(5)$, which is different from the mean-field value 1/2 [Stanley (1971)].

Recently simulation studies [Grégorie and Chaté (2004)] reported that this noise induced phase transition is first order at least for high velocities ($v \geq 0.3$) [Nagy *et al.* (2006)]. Another analytical and numerical study [Aldana *et al.* (2007)] pointed out that the nature of the phase transition depends on how the noise is introduced in the physical situations considered. The situation is not clear and resembles to the debate on some diffusive RD model discussed in Chapter 5.

# Chapter 4

# Genuine Basic Nonequilibrium Classes with Absorbing State

In this chapter I introduce "genuine nonequilibrium" universality classes that do not occur in dynamical generalizations of equilibrium systems and emerge at phase transitions without relevant fluctuations. Naturally in these models there is no hermitian Hamiltonian and they are defined by transition rates not satisfying the detailed balance condition (2.20). They can be described by a master equation and the deduced stochastic action or Langevin equation if it exists. The best known cases are reaction-diffusion systems with order-disorder transitions in which the ordered state may exhibit only small fluctuations, hence they trap a system falling in it (absorbing state). Examples occur in models of population [Albano (1994)], epidemics [Ligget (1985); Mollison (1977)], catalysis [Ziff *et al.* (1986)] or enzyme biology [Berry (2003)].

Phase transitions in such models may occur in low dimensions in contrast with equilibrium models [Marro and Dickman (1999)]. As was already shown in Sect. 2.3.2 reaction-diffusion particle systems may be mapped onto spin-flip systems, stochastic cellular automata [Chopard and Droz (1998)] or interface growth models (see Sect. 6.13.2). The mapping however can lead to nonlocal interactions that may not have real physical relevance. The universality classes of the simple models presented in this chapter constitute the fundamental building blocks of more complex systems.

For a long time, phase transitions with completely frozen absorbing states were investigated. A few universality classes of this kind were known [Grassberger (1996)],[Hinrichsen (2000a)], of which the most prominent one was that of the directed percolation (DP) process [Kinzel (1983)]. An early hypothesis [Janssen (1981); Grassberger (1982a); Grinstein *et al.* (1989)] was confirmed by all models up to now. This conjecture known as **DP hy-**

**pothesis** claims that **in one component systems exhibiting continuous phase transitions to a single absorbing state (without extra symmetry and inhomogeneity or disorder) short ranged interactions can generate DP class transition only**. Despite the robustness of this class experimental observation is still sparse [Grassberger (1996); Hinrichsen (2000b)], probably owing to the high sensitivity to disorder that cannot be avoided in real materials and the unconventional finite size scaling [Pruessner (2007)], (see however a very recent experimental study [Takeuchi *et al.* (2007)]).

A major problem with the analysis of these models is that they are usually far from the critical dimension and critical fluctuations destroy a mean-field like behavior. A further complication is that bosonic field theoretical methods cannot describe particle exclusion that may obviously happen in $d = 1$. The success of the application of bosonic field theory in many cases is the consequence of the asymptotically low density of particles near the critical point. However in multi-component systems, where the exchange between different types is non-trivial, bosonic field theoretical descriptions may fail. In case of the binary production (BP) models (Sect. 6.7) the bosonic formalism predicts diverging density in the active phase contrary to the lattice model version of hard-core particles [Ódor (2000); Carlon *et al.* (2001); Hinrichsen (2001a); Ódor (2001a); Hinrichsen (2001b); Park *et al.* (2001)]. Fermionic field theories on the other hand have the disadvantage that they are non-local, hence results exist for very simple reaction-diffusion systems only [Park *et al.* (2000); Wijland (2001); Brunel *et al.* (2000)]. Other techniques like independent interval approximation [Krapivsky and Ben-Naim (1997)], the empty interval method [ben Avraham *et al.* (1990)], series expansion [Essam *et al.* (1996)] or density matrix renormalization (DMRG) have been applied successfully in some cases.

The universal scaling-law behavior in these models is described by the critical exponents in the neighborhood of a steady state, hence the generalization of dynamical exponents introduced for "out of equilibrium classes" (Sect. 1.7) (like $Z$, $\theta$, $\lambda$ ...etc.) is used. Besides these exponents there are genuinely nonequilibrium dynamical exponents as well to characterize spreading behavior, defined in Sect. 1.5. For each class I discuss the damage spreading behavior, the effects of different boundary conditions, crossovers, anisotropy, disorder and long-correlations generated by anomalous diffusion or by special initial states.

According to the phase-space topology scheme introduced in Sect. 1.6.1 there are 5 basic, non-trivial ($d_c > 1$), independent families of universality classes of homogeneous, one-component, bosonic RD systems exhibiting short range interactions and memory:

- **DP** (Sect. 4.2)
- **PCPD** (Sects. 4.7 and 6.7)
- **3CPD** (Sect. 4.7)
- **PC** (Sect. 4.6)
- **1R** (Sect. 4.8)

I shall discuss them one by one in the sections shown in the parentheses, but first let's see what the simplest mean-field approach tells us about these classes.

## 4.1 Mean-field classes of general $nA \to (n+k)A$, $mA \to (m-l)A$ processes

For a long time reaction-diffusion models with offsprings of a single parent have been investigated. Note that there was and early forgotten numerical study by [Grassberger (1982b)] claiming a non-DP type of continuous phase transition in a model where particle production can occur by the reaction of two parents. Later it turned out than in such binary production lattice models, where the solitary particles follow random walk (hence they behave like a coupled system) different universal behavior emerges indeed (see Sect. 6.7). In this section I discuss the mean-field classes of models

$$nA \stackrel{\sigma}{\to} (n+k)A, \qquad mA \stackrel{\lambda}{\to} (m-l)A, \tag{4.1}$$

with $n, m, k, l$ positive integers and $m - l \geq 0$. By tuning the reaction parameters $\sigma$ and $\lambda$ this model exhibits a phase transition between an active and absorbing state. In the active phase the particles move and react with each other, while in the absorbing state the density is zero or the state is frozen (see PCP (Sect. 6.6). The lack of $\emptyset \to nA$ reaction makes the absorbing phase a trapped, non-fluctuating one. The reactions can generate diffusion of particles, or such such motion may be added explicitly. However diffusion does not play a role in the simple, site-mean-field theory. Note that this theory gives different result in case of bosonic and site-restricted (fermionic) models.

### 4.1.1 *Bosonic models*

In bosonic models $m > n$ is needed to stop infinite proliferation of particles in the active phase. In the framework of mean-field approximation all spatial fluctuations are neglected and the particle density $\rho$ is governed by the rate equation

$$\dot{\rho} = A\rho^n - B\rho^m \ , \tag{4.2}$$

where $A$ and $B$ are functions of the reaction rates, the dimensions and topology of the lattices. This equation shows a phase transition to absorbing state for $A = A_c = 0$. In the active phase, near the critical point $\epsilon = A - A_c$ the steady state solution $\rho(t \to \infty)$ of the simplified equation:

$$\dot{\rho} = \epsilon\rho^n - \rho^m \ , \tag{4.3}$$

exhibits the leading order singularity

$$\rho(\infty) \propto \epsilon^{1/(m-n)} \tag{4.4}$$

hence the mean-field order parameter is

$$\beta_{MF} = 1/(m-n) \ . \tag{4.5}$$

Putting $\rho = \rho(\infty)\rho' = \epsilon^{1/(m-n)}\rho'$ one can rewrite Eq. (4.3) as

$$\epsilon^{-(m-1)/(m-n)}\dot{\rho}' = (\rho')^n - (\rho')^m \ . \tag{4.6}$$

Since $\rho'$ is of the order of unity at all times the relaxation time of the process shows the singularity

$$\xi_t \propto \epsilon^{-(m-1)/(m-n)} \ , \tag{4.7}$$

and the mean-field value of the temporal correlation length exponent is

$$\nu_{||,MF} = (m-1)/(m-n) \ . \tag{4.8}$$

As a consequence, by definition the density decay exponent is

$$\alpha_{MF} = \beta/\nu_{||} = 1/(m-1) \ . \tag{4.9}$$

### 4.1.2 *Site restricted (fermionic) models*

Next I derive the basic mean-field exponents for site restricted models of type Eq. (4.1) ([Ódor (2003a)]). Now the lattice site mean-filed approximation can be parameterized by the creation probability ($\sigma$) and the annihilation or coagulation probability ($\lambda = 1 - \sigma$)

$$\frac{\partial \rho}{\partial t} = K(k)\sigma\rho^n(1-\rho)^k - L(l)(1-\sigma)\rho^m, \tag{4.10}$$

where $\rho$ denotes the site occupancy probability, $K(k)$ and $L(l)$ are combinatorial constants depending on the dimension (coordination number) of the approximation and the concrete details of the reactions. Each empty site has a probability (1-$\rho$) in the mean-field approximation, hence the need for $k$ empty sites at a creation brings in a $(1-\rho)^k$ probability factor (in bosonic mean-field theories one should explicitly add $m' > n$ annihilation terms if $m \leq n$ by definition of the model to regulate the theory). By expanding $(1-\rho)^k$ and keeping the lowest order contribution one can see that for site restricted lattice systems a $\rho^{n+1}$-th order term appears automatically with negative coefficient that regulates Eq. (4.10). In the inactive phases one expects a dynamical behavior described by the

$$mA \to \emptyset \tag{4.11}$$

process, for which

$$\rho \propto t^{1/(m-1)} \tag{4.12}$$

is known [Lee (1994); Cardy and Täuber (1996)]. The steady state solution of (4.10) can be found analytically in many cases and may result in different, continuous or discontinuous phase transitions. Here I split the discussion of the solutions into three parts

- $n = m$
- $n > m$
- $n < m$

### 4.1.3 *The $n = m$ symmetric case*

The steady state solution in this case can be obtained by solving

$$K(k)\sigma(1-\rho)^k = L(l)(1-\sigma), \tag{4.13}$$

where the trivial ($\rho = 0$) solution (corresponding to the absorbing state) has been factored out. For the active phase one gets

$$\rho = 1 - \left[\frac{L}{K}\frac{1-\sigma}{\sigma}\right]^{1/k}, \tag{4.14}$$

which vanishes

$$\sigma_c = \frac{L}{K+l} > 0 \tag{4.15}$$

with the leading order singularity

$$\rho \propto |\sigma - \sigma_c|^{\beta^{MF}}, \tag{4.16}$$

and order parameter exponent exponent

$$\beta_{MF} = 1 \ . \tag{4.17}$$

At the critical point the time dependent behavior is described by

$$\frac{\partial \rho}{\partial t} = -2K^2 \rho^{n+1} + O(\rho^{n+2}) \ , \tag{4.18}$$

which gives a leading order power-law solution

$$\rho \propto t^{-1/n} \tag{4.19}$$

hence the corresponding mean-field exponents are

$$\alpha_{MF} = \beta/\nu_{||} = 1/n \ . \tag{4.20}$$

These classes generalize the mean-field class of DP (Sect. 4.2) and binary production models (Sect. 6.7). For referencing it in Table 8.2 of the Summary I call it: PARWs (symmetric production and annihilating random walk class).

### 4.1.4 *The $n > m$ asymmetric case*

In this case besides the $\rho = 0$ absorbing state solution we can get an active state if

$$K\sigma\rho^{n-m}(1-\rho)^k = L(1-\sigma) \tag{4.21}$$

is satisfied. Both sides are linear functions of $\sigma$ such that for $\sigma \to 0$ only the $\rho = 0$ is a solution. The left hand side is a convex function of $\rho$ (from above) with zeros at $\rho = 0$ and $\rho = 1$. Therefore by increasing $\sigma$ from zero the left hand side meets the right hand side at $\sigma_c$, $\rho_c > 0$ (see Fig. 4.1(a)). If this solution is stable, a discontinuous transition of the steady state density (order parameter) takes place by varying the reaction rates.

So in the $n > m$ case the mean-field solution provides a **first order transition** transition (see [Ódor (2003a)]) and it does not imply anything with respect to possible classes for models below the critical dimension ($d < d_c$). Note however, that by higher order cluster mean-field approximations, when the diffusion plays a role the transition may turn into a continuous one and simulations in one dimension confirm this (see for example [Dickman and Tomé (1991); Ódor *et al.* (1993); Menyhárd and Ódor (1995)]).

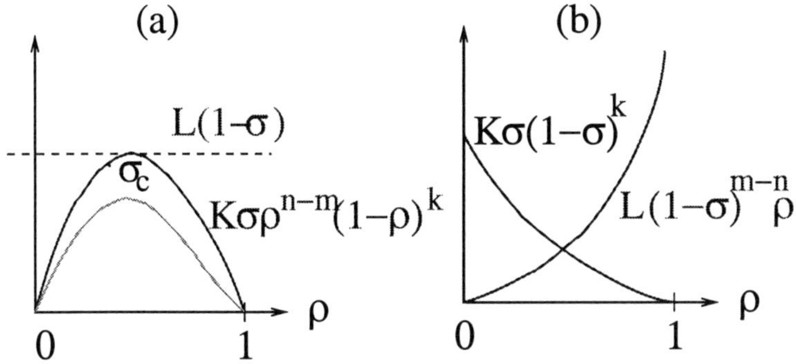


Fig. 4.1 Steady state mean-field solution for (a) $n > m$ and (b) $n < m$ cases

### 4.1.5 *The asymmetric $n < m$ case*

By factoring out the trivial $\rho = 0$ solution we are faced with the general condition for a steady state

$$K\sigma(1-\rho)^k = L(1-\sigma)\rho^{m-n}. \tag{4.22}$$

One can easily check that in this case the critical point is at

$$\sigma_c = 0 \ , \tag{4.23}$$

(see Fig. 4.1(b)) and the density decays as

$$\alpha = 1/(m-1) \ , \tag{4.24}$$

as for branching ($n = 1$) and $m = l$ particle annihilating models (BmARW) showed by Cardy and Täuber [Cardy and Täuber (1996)]. However the steady state solution for $n > 1$ gives different $\beta$ exponents than those of BmARW classes, namely

$$\beta = 1/(m-n) \ . \tag{4.25}$$

This implies novel kind of critical behavior in low dimensions, that should be a subject of further investigations (see Sect. 4.7 and [Ódor (2003b)]). For referencing it in Table 8.2 of the summary I call it: PARWa (asymmetric production and annihilating random walk class).

### 4.1.6 *Upper critical behavior and below*

The engineering, upper critical dimension can be obtained by dimensional analysis (power counting) of the action. However due to the fluctuations the bare (initial) action and the renormalized one may differ significantly. For example in case of the

$$A \to 2A, \quad 2A \to \emptyset \tag{4.26}$$

BARWo model (Sect. 4.2.4) the $A \to \emptyset$ reaction is also generated (via $A \to 2A \to \emptyset$) affecting the critical behavior and $d_c$.

Similarly, lower order terms can be generated by RG transformation in case of the general reactions (4.1). The relevant action is written as [Ohtsuki and Keyes (1986)]

$$S = \int dt d^d x [\psi(\partial_t - D\Delta)\phi + Ds\phi^n + Du(\psi\phi^m - \psi^2\phi^n)] \ , \tag{4.27}$$

where $s$ and $u$ are the renormalized couplings. Introducing a momentum scale $\kappa$ and time scale $t$ the scaling dimensions of fields are

$$[\psi] = \frac{m-n}{m-n+1} d \ , \quad [\phi] = \frac{1}{m-n+1} d \tag{4.28}$$

and couplings can be expressed, keeping the action dimensionless

$$[D] = \kappa^{-2} t^{-1} \ , \quad [s] = 2 - \frac{n-1}{m-n+1} d \ , \quad [u] = 2 - \frac{m-1}{m-n+1} d \ . \tag{4.29}$$

The upper critical dimension $d_c$ of the process can be obtained from the marginality condition $d_u(d_c) = 0$

$$d_c = 2(m-n+1)/(m-1) \ . \tag{4.30}$$

Note, however that the exploration of the renormalized action is a highly nontrivial task. In some cases the important physics is in the strongly coupled regime, which may be accessible by NPRG only. Such situation occurs for example in case of the BARWe models (see Sect. 4.6.1), where a second critical dimension $d_c' < d_c$ emerges or in the KPZ model (see Sect. 7.4), where the value of $d_c$ has been debated for a long time.

As one could see the site restricted and bosonic variants of the same RD models may exhibit different behavior. In the bosonic ones the density can diverge in the active phase if $n \geq m$ and the transition becomes first order. Since only bosonic field theory has been used up to now (the fermionic one would require tedious Grassmann variables) one usually avoids this problem by adding a regularizing annihilation reaction, which prevents the infinite proliferation of particles.

However in low dimensional, site restricted models the particles can block each other and novel kind of phase transitions and universality classes emerge due to this topological constraint in multi-particle reaction RD models (see [Ódor and Menyhárd (2002)], Sects. 6.13, 6.13.1).

The field theory of the $n = m = l = 2$, bosonic model was investigated in [Howard and Täuber (1997)] and concluded non-DP type of criticality with $d_c = 2$ (see Sect. 6.7), but the true fixed point behavior has not been described yet [Janssen *et al.* (2004b)]. For higher $n$ and $m$ values even less field theoretical results exist. Numerical simulations in one and two dimensions for various $n = 3$ and $n = 4$ models resulted in somewhat contradictory results [Park *et al.* (2002); Kockelkoren and Chaté (2003a); Ódor (2003a)]. There is a disagreement in the value of the upper critical dimension, but in any case $d_c$ seems to be very low ($d_c = 1-2$) in agreement with Eq. (4.30), hence the number of non-mean-field classes in such models is limited. On contrary there is a series of mean-field classes **depending on $n$ and $m$** [Ohtsuki and Keyes (1986); Park *et al.* (2002); Ódor (2003a)]

## 4.2 Directed percolation (DP) classes

The directed percolation (DP) introduced by [Broadbent and Hammersley (1957)] is an **anisotropic percolation** with a preferred direction $t$. This means that this problem should be $d \geq 2$ dimensional. If there is an object (bond, site etc.) at $(x_i, y_j, ..., t_k)$ it must have a nearest neighboring object at $t_{k-1}$ unless $t_k = 0$ (see Fig. 4.2). If we consider the preferred direction as the time we recognize a spreading process of an agent $A$ that cannot have spontaneous source: $\emptyset \not\to A$. This results in the possibility of a completely frozen, so called *absorbing state* from which the system cannot escape if it has fallen into it. As a consequence these kinds of models may have phase transitions in $d = 1$ spatial dimension already. By increasing the branching probability $p$ of the agent we can have a phase transition between the absorbing state and an active steady state with finite density of $A$-s. **In general if the lowest level reactions in RD model are unary and the transition is continuous at some finite branching rate it is very likely that it belongs to a robust universality (DP) class.** For a long time all models of such absorbing phase transitions were found to belong to the DP class and the DP conjecture was advanced (see Sect. 3.1.4). This hypothesis has been confirmed by most models up to now, but in some cases (like the famous PCPD model (see Sect. 6.7.1) or models

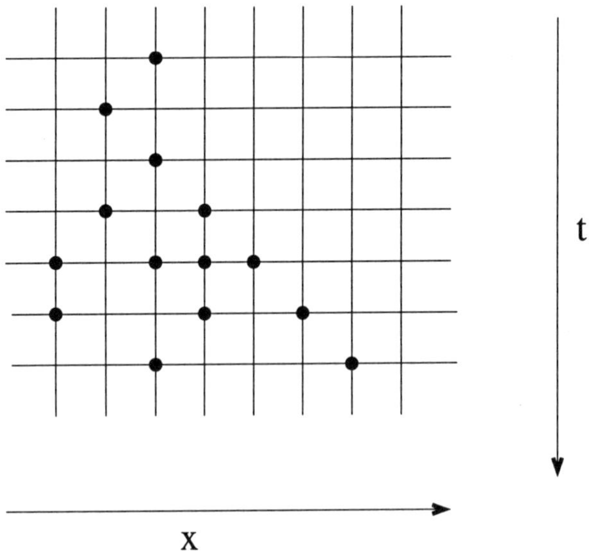


Fig. 4.2 Directed site percolation in $d = 1 + 1$ dimensions

discussed in Sect. 4.8), **where vacuum state is not accessible** i.e. no $A \to \emptyset$ is generated by fluctuation, the applicability of the DP hypothesis is not so clear. It is true that in models, where no direct channel to vacuum exists (by field theoretical language) a single particle can survive even in the absorbing state, so strictly speaking there is an absorbing state, which is not completely empty, but exhibits some 'fluctuations' due to the diffusing particle. Therefore these models literally do not break the DP hypothesis, which require a single (frozen) absorbing state.

Moreover DP class exponents were discovered in some systems with multiple absorbing states. For example in models with infinitely many frozen absorbing states [Jensen and Dickman (1993a); Jensen (1993a); Jensen and Dickman (1993b); Mendes *et al.* (1994)] the **static exponents** were found to coincide with those of DP.[1] Furthermore in models where the symmetry of the absorbing states is broken by an external field the emergence of DP behavior was reported [Park and Park (1995); Menyhárd and Ódor (1996); Ódor and Menyhárd (1998)] . So although the necessary conditions for the DP behavior seem to be confirmed the determination of sufficient condi-

[1]However the spreading exponents in case of non-natural initial conditions are non-universal (see Sect. 6.6).

tions is an open problem. There are many introductory works available now to DP [Kinzel (1983); Marro and Dickman (1999); Grassberger (1996); Hinrichsen (2000a)] therefore I shall not go very deeply into the discussions of details of various representations.

In the reaction-diffusion language the DP is built up from the following processes

$$A \xrightarrow{\gamma} \emptyset \quad A\emptyset \overset{D}{\leftrightarrow} \emptyset A \quad A \xrightarrow{\sigma} 2A \quad 2A \xrightarrow{\lambda} A\,. \tag{4.31}$$

The mean-field equation for the coarse-grained particle density $\rho(t)$ is

$$\frac{d\rho}{dt} = (\sigma - \gamma)\rho - (\lambda + \sigma)\rho^2 \quad . \tag{4.32}$$

This has the stationary stable solution

$$\rho(\infty) = \begin{cases} \frac{\sigma-\gamma}{\lambda+\sigma} & \text{for} \ : \ \sigma > \gamma \\ 0 & \text{for} \ : \ \sigma \leq \gamma \end{cases} \tag{4.33}$$

exhibiting a continuous transition at $\sigma = \gamma$. A small variation of $\sigma$ or $\gamma$ near the critical point implies a linear change of $\rho$. Therefore the order parameter exponent in the mean-field approximations

$$\beta_{MF} = 1\ . \tag{4.34}$$

Near the critical point the $O(\rho)$ term is the dominant one, hence the density approaches the stationary value exponentially. For $\sigma = \gamma$ the remaining $O(\rho^2)$ term causes power-law decay $\rho \propto t^{-1}$, indicating

$$\alpha_{MF} = 1\ . \tag{4.35}$$

Note that we could have get these exponents from the general forms (4.5, 4.9) with the substitutions $m = 2$ and $n - 1$ as well.

One can obtain the correlation length exponent by generalizing the mean-field rate equation (4.32) for a local particle density equation $\rho(x,t)$ and by adding a term $D\nabla^2\rho$, which describes particle diffusion. At the transition point $\sigma = \gamma$ and what is left behind is just an annihilating random walk process (ARW) (see Sect. 4.5.1). Let's make a scaling analysis under the rescaling $x' \to bx$, $t' \to b^Z t$ transformation. Since the density scales as $\rho' \to b^{-\beta/\nu_\perp}\rho$ with some rearrangement we get

$$\dot{\rho} = (\lambda + \sigma)\rho^2 b^{Z-\beta/\nu_\perp} + D b^{Z-2}\nabla^2\rho\,. \tag{4.36}$$

Since this equation should be invariant under such renormalization

$$Z_{MF} = 2\ , \quad \nu_{\perp,MF} = 1/2 \tag{4.37}$$

is a necessary condition in high dimensions. To take into account the fluctuations we must write up the full action. Following the description of [Elgart and Kamenev (2006)] (see Sect. 1.6.1) the reaction Hamiltonian corresponding to (4.31) is

$$\begin{aligned} H_R &= \gamma(1-p)\,q + \sigma(p^2-p)\,q + \frac{\lambda}{2}(p-p^2)\,q^2 \\ &= \left(\gamma - \sigma p + \frac{\lambda}{2}p\,q\right)(1-p)\,q\,. \end{aligned} \tag{4.38}$$

The phase portrait of this system is depicted in Fig. 4.3. The zero energy

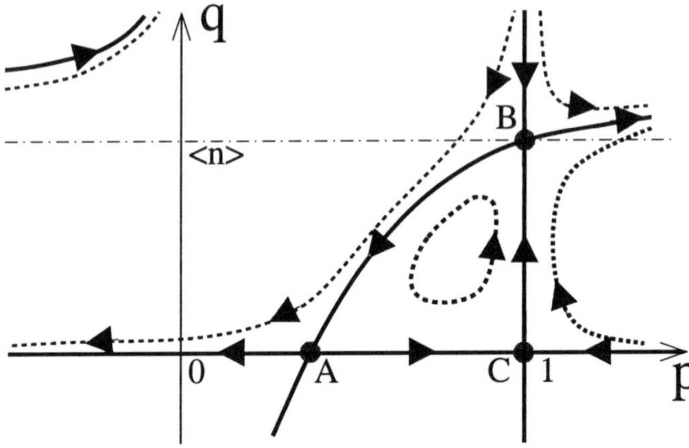


Fig. 4.3 Phase portrait of the DP system in the active phase. Thick solid lines represent zero-energy trajectories, which divide the phase space into a number of disconnected regions. The point $B = \{1, 2(\sigma-\gamma)/\lambda\}$ corresponds to the active mean-field fixed point. From [Elgart and Kamenev (2006)].

lines are the generic $p = 1$ (mean-field) and $q = 0$ (absorbing phase) trajectories, along with the $q = 2(\sigma p - \gamma)/\lambda p$ curve. According to the mean-field analysis [classical equations (4.32) and the equation of motion (1.80) with $p = 1$], the active phase mean-field phase corresponds to point $B$ in Fig. 4.3. The system can be brought to extinction by tuning the control parameter $m = \sigma - \gamma$ to zero. The transition is represented by the phase portrait with the three zero-energy trajectories intersecting at the point $(1, 0)$. To focus on the vicinity of this point the shift of variable

$$p - 1 \to p \tag{4.39}$$

is done usually. Using the formal transcription $p \to \psi(x,t)$, $q \to \phi(x,t)$ the

Table 4.1 Estimates for the critical exponents of directed percolation. Data are from: [Jensen (1999a)] (1d), [Voigt and Ziff (1997)](2d), [Jensen (1992)](3d), [Bronzan and Dash (1974); Janssen (1981)](4 − ϵ).

| critical exponent | $d = 1$ | $d = 2$ | $d = 3$ | $d = 4 - \epsilon$ |
|---|---|---|---|---|
| $\beta = \beta'$ | 0.276486(8) | 0.584(4) | 0.81(1) | $1 - \epsilon/6 - 0.01128\,\epsilon^2$ |
| $\nu_\perp$ | 1.096854(4) | 0.734(4) | 0.581(5) | $1/2 + \epsilon/16 + 0.02110\,\epsilon^2$ |
| $\nu_{\|\|}$ | 1.733847(6) | 1.295(6) | 1.105(5) | $1 + \epsilon/12 + 0.02238\,\epsilon^2$ |
| $Z = 2/z$ | 1.580745(10) | 1.76(3) | 1.90(1) | $2 - \epsilon/12 - 0.02921\,\epsilon^2$ |
| $\delta = \alpha$ | 0.159464(6) | 0.451 | 0.73 | $1 - \epsilon/4 - 0.01283\,\epsilon^2$ |
| $\eta$ | 0.313686(8) | 0.230 | 0.12 | $\epsilon/12 + 0.03751\,\epsilon^2$ |
| $\eta_\perp$ | 1.504144(19) | 1.159(15) | 1.783(16) | $2 + O(\epsilon)$ |
| $\lambda_R/Z = \lambda_C/Z$ | 1.9(1) | 2.75(10) | | $4 + O(\epsilon)$ |
| $\theta_g$ | 1.5(2) | | | $2 - 5\epsilon/24$ |
| $\gamma_P$ | 2.277730(5) | 1.60 | 1.25 | $1 + \epsilon/6 + 0.06683\,\epsilon^2$ |

full field theoretical action with the couplings $m$, $u = \sigma$, $v = \lambda/2$ looks like

$$S = \int d^d x dt [H_D + H_R] = \int d^d x dt \left[\psi(\partial_t - D\nabla^2)\phi + (m + u\psi - v\phi)\psi\phi\right] , \tag{4.40}$$

where $\phi$ is the density, $\psi$ is the response field (appearing in response functions). The action is invariant under the following rapidity-reversal symmetry

$$\phi(x,t) \to -\psi(x,-t) \quad , \quad \psi(x,t) \to -\phi(x,-t) \quad . \tag{4.41}$$

This symmetry yields [Grassberger and de la Torre (1979); Muñoz *et al.* (1997)] [2] the scaling relations

$$\beta = \beta' \tag{4.42}$$

$$4\delta + 2\eta = dz \ . \tag{4.43}$$

This field theory was found to be equivalent [Cardy and Sugar (1980)] to the Reggeon field theory [Abarbanel *et al.* (1975); Brower *et al.* (1978)], which is a model of scattering elementary particles at high energies and low-momentum transfers. The the upper critical dimension of directed percolation is can be deduced from the general form (4.30)

$$d_c = 4 \ . \tag{4.44}$$

[2] Note that sometimes (4.41) can only be seen after a coarse graining procedure procedure, asymptotically in models exhibiting DP class behavior.

This stochastic process can through standard techniques [Janssen (1976)] be transformed into a Langevin equation formalism. Below the critical dimension the RG analysis of the Langevin equation

$$\frac{\partial \rho(x,t)}{\partial t} = D\nabla^2 \rho(x,t) + (\sigma - \gamma)\rho(x,t) - \\ - (\lambda + \sigma)\rho^2(x,t) + \sqrt{\rho(x,t)}\eta(x,t) \quad (4.45)$$

is necessary [Janssen (1981)]. Here $\eta(x,t)$ is the Gaussian noise field, defined by the correlations

$$< \eta(x,t) > = 0 \quad (4.46)$$

$$< \eta(x,t)\eta(x',t') > = \Gamma\delta^d(x-x')\delta(t-t') \,. \quad (4.47)$$

The noise term is proportional to $\sqrt{\rho(x,t)}$ ensuring that in the absorbing state ($\rho(x,t) = 0$) it vanishes. The square-root behavior stems from the definition of $\rho(x,t)$ as a coarse-grained density of active sites averaged over some mesoscopic box size.

Note that DP universality occurs in many other processes like in odd offspring, branching and annihilating random walks (BARWo) (see Sec. 4.2.4) and in models described by field theory with higher order terms like $\rho^3(x,t)$ or $\nabla^4\rho(x,t)$, which are irrelevant under the RG transformation.[3] In the vicinity of the phase transition point they all exhibit the topology shown on Fig. 4.3.

Perturbative $\epsilon = 4 - d$ renormalization group analysis [Bronzan and Dash (1974); Janssen (1981)] up to two-loop order resulted in estimates for the critical exponents shown in the Table 4.2. The best results obtained by approximative techniques for DP like improved mean-field [Ben-Naim and Krapivsky (1994); Dickman (2007)], coherent anomaly method [Ódor (1995)], Monte Carlo simulations [Grassberger and de la Torre (1979); Grassberger (1989a, 1992a); Dickman and Jensen (1991)], series expansions [De'Bell and Essam (1983); Essam *et al.* (1988); Jensen and Dickman (1993c); Jensen (1996a); Jensen and Dickman (1993d); Jensen and Guttmann (1995, 1996); Jensen (1999a)], DMRG [Hieida (1998); Carlon *et al.* (1999)], and numerical integration of Eq. 4.45 [Dickman (1994)].

The **local persistence** probability defined as the probability ($p_l(t)$) that a particular site never becomes active up to time $t$. Numerical simulations [Hinrichsen and Koduvely (1998)] for this in the 1+1 d Domany-Kinzel SCA (see Sect. 4.2.3) found power-law behavior, with the exponent

$$\theta_l = 1.50(1) \,. \quad (4.48)$$

[3] As long as the sign of the quadratic coupling term is positive, otherwise first order transition and tricritical phenomena emerges (see Sect. 5.1).

The **global persistence** probability defined here as the probability ($p_g(t)$) that the deviation of the global density from its mean value does not change its sign up to time $t$. The simulations of [Hinrichsen and Koduvely (1998)] in 1+1 dimensions claim $\theta_g \simeq \theta_l$. This agrees with field theoretical $\epsilon = 4-d$ expansions [Oerding and van Wijland (1998)] who predict $\theta_g = 2$ for $d \geq 4$ and for $d < 4$:

$$\theta_g = 2 - \frac{5\epsilon}{24} + O(\epsilon^2) \ . \tag{4.49}$$

If we substitute this value into (1.31), as well as other exponents from Table 4.2 it is not clear whether the condition for Markovian global order parameter is fulfilled, because the scaling relation can be satisfied within the rather large numerical error margin of exponents. So it is an open question if $\theta_g$ is an independent exponent or not. Certainly for $d = 4$ (1.31) is satisfied with the mean-filed exponents suggesting a dependence of $\theta_g$ to other ones in lower dimensions.

Since DP is the most studied universality class in nonequilibrium systems, with phase transition to absorbing state, its **crossover to other classes** has extensively been investigated. First the crossover from **isotropic to directed percolation** was investigated by perturbative RG [Frey *et al.* (1993)] approach up to one-loop order. They found that while for $d > 6$ the percolation is isotropic, for $d > 5$ the directed Gaussian fixed point is stable. For $d < 5$ the asymptotic behavior is governed by the DP fixed point. On the other hand in $d = 2$ exact calculations and simulations [Fröjdh and den Nijs (1997)] found that the isotropic percolation is stable with respect to DP if we control the crossover by a spontaneous particle birth parameter. It is still an open question what happens for $2 < d < 5$. **Crossovers to mean-field behavior** generated by long-range interactions (see Sect. 4.2.8) and to **compact directed percolation** (see Sect. 4.5) are be discussed later. **Crossover to the parity conserving class** is postponed until Sect. 4.6.2. A very recent study [Park and Park (2007)] has investigated the **crossover of DP to the coupled PCP and PCPD classes** (the results are summarized in Sects. 6.6 and 4.102 respectively). Finally a very recent study has pointed out the possibility of a **crossover to the KPZ** class by surface growth, due to a conjugated external field (particle source) (discussed in Sect. 7.4.4).

It was conjectured [Deloubriére and van Wijland (2002)] that in one dimension "fermionic" (single site occupancy) and bosonic (multiple site occupancy) models may exhibit different critical behavior. An attempt using for fermionic field theory for DP in 1 + 1 dimensions made by [Brunel *et al.*

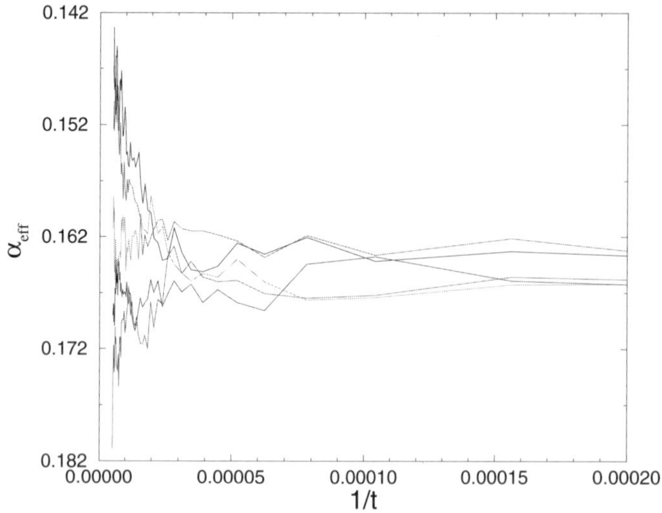


Fig. 4.4 Local slopes of the density decay in a bosonic BARW1 model (Sect. 4.2.4). Different curves correspond to $\lambda$ = 0.12883, 0.12882, 0.12881 0.1288, 0.12879 (from bottom to top) [Ódor and Menyhárd (2002)].

(2000); Wijland (2001)]. This ran into severe convergence problems and did not result in precise quantitative estimates for the critical exponents. Although the bosonic field theory is expected to describe the fermionic case owing to the asymptotically low density at criticality it has never been proven rigorously. Since analytic results are known only for bosonic field theory, which gives rather inaccurate critical exponent estimates [Ódor and Menyhárd (2002)] performed simulations of a single-offspring BARW process (see (4.66)) with unrestricted site occupancy to investigate the density ($\rho(t)$) decay from a random initial state. Figure 4.4 shows the local slopes of density decay defined as

$$\alpha_{eff} = -\frac{d \log \rho(t)}{d \log t} \tag{4.50}$$

around the critical point for several annihilation rates ($\lambda$). The critical point is estimated at $\lambda_c = 0.12882(1)$ (corresponding to straight line) with the extrapolated decay exponent $\alpha = 0.165(5)$, which agrees well with fermionic model simulation and series expansion results 0.1595(1) [De'Bell and Essam (1983)].

Note that in site restricted model the processes

$$A \overset{\gamma}{\to} \emptyset \quad A\emptyset \overset{D}{\leftrightarrow} \emptyset A \quad A \overset{\sigma}{\to} 2A \tag{4.51}$$

generate a DP class phase transition, while in the bosonic version the $2A \xrightarrow{\lambda} A$ process is also necessary to ensure an active steady state (without it the density diverges for $\sigma > \gamma$). Models exhibiting DP transitions have been reviewed in great detail in [Marro and Dickman (1999)], [Hinrichsen (2000a)] and [Lübeck and Willmann (2005)]. In the next sections I discuss only three important examples.

### 4.2.1 *The contact process*

The contact process is one of the earliest and simplest lattice model for DP with **asynchronous** update introduced by Harris [Harris (1974b); Grassberger and de la Torre (1979)] to model epidemic spreading without immunization. Its dynamics is defined by nearest-neighbor processes that occur spontaneously due to specific rates (rather than probabilities). In numerical simulations models of this type are usually realized by random sequential updates. In one dimension this means that a pair of sites $\{s_i, s_{i+1}\}$ is chosen at random and an update is attempted according to specific transition rates $w(s_{i,t+dt}\,,\, s_{i+1,t+dt}\,|\, s_{i,t}\,,\, s_{i+1,t})$. Each attempt to update a pair of sites increases the time $t$ by $dt = 1/N$, where $N$ is the total number of sites. One time step (sweep) therefore consists of $N$ such attempts. The contact process is defined by the rates

$$w(A,I\,|\,A,A) = w(I,A\,|\,A,A) = \mu\,, \tag{4.52}$$

$$w(I,I\,|\,A,I) = w(I,I\,|\,I,A) = \lambda\,, \tag{4.53}$$

$$w(A,A\,|\,A,I) = w(A,A\,|\,I,A) = 1\,, \tag{4.54}$$

where $\lambda > 0$ and $\mu > 0$ are two parameters (all other rates are zero). Equation (4.52) describes the creation of inactive ($I$) spots within active ($A$) islands. Equations (4.53) and (4.54) describe the shrinkage and growth of active islands. In order to fix the time scale, we chose the rate in Eq. (4.54) to be equal to one. The active phase is restricted to the region $\lambda < 1$, where active islands are likely to grow. In one dimension series expansions and numerical simulations have determined the critical point and critical exponents precisely [Dickman and Jensen (1991); De'Bell and Essam (1983); Essam *et al.* (1988); Jensen and Dickman (1993c); Jensen (1996a); Jensen and Dickman (1993d); Jensen and Guttmann (1995, 1996); Dickman and de Silva (1998)]. In two dimensions the order parameter moments and the cumulant ratios were determined by [Dickman and de Silva (1998)].

### 4.2.2 *Two-point correlations, ageing properties*

Recently the ageing property of the contact process has been investigated by numerical simulations [Enns *et al.* (2004); Ramasco *et al.* (2004); Hinrichsen (2006)] and field theory [Baumann and Gambassi (2007)] up to first order in the field-theoretical epsilon-expansion ($\epsilon = 4 - d$). A system with a homogeneous initial particle density (corresponding to the active phase) was quenched at $t = 0$ to the critical point and the aging functions $R(t, s)$ and $C(t, s)$ were determined numerically. The first simulations [Enns *et al.* (2004); Ramasco *et al.* (2004)] showed quite nice agreement with the scaling forms of the autocorrelation and autoresponse functions. In particular it was found that $\lambda_C/Z = 1.85(10)$ and $\lambda_R/Z = 1.85(10)$ in $d = 1$ [Enns *et al.* (2004); Ramasco *et al.* (2004)], whereas $\lambda_C/Z = 2.75(10)$ and $\lambda_R/Z = 2.8(3)$ in $d = 2$ [Ramasco *et al.* (2004)]. These results suggest that $\lambda_R = \lambda_C$ independently of the dimensionality. On the same numerical footing it was noticed that $1 + a = b$, in stark contrast to the case of slow dynamics of systems with detailed balance such as Ising ferromagnets, for which $a = b$ (see Eq. 2.34).

A more detailed analysis [Hinrichsen (2006)] found deviations from the full functional form (1.89) for $t/s < 0.1$ and Hinrichsen argued that LSI is not a generic property of ageing phenomena but is restricted to diffusive models ($Z = 2$) or above the upper critical dimension (see Fig. 4.5). Contrary to this [Henkel *et al.* (2006)] suggested that there is crossover in case of nonequilibrium critical dynamics because both the ageing regime ($t - s \sim O(s)$) and the quasi-stationary regime ($t - s \ll s$) display scaling with the same length scale $L(t) \propto t^{1/Z}$. Furthermore in case of the contact process all simulations begin with a non-vanishing order parameter in the initial state, contrary to the conditions for which the form (1.89) is set up. Thus probably a generalized form of the autoresponse function for $m_0 \neq 0$ should be compared with the simulations.

The field theory [Baumann and Gambassi (2007)] confirmed the relation $b = 1 + a$, due to the rapidity-reversal symmetry (4.41), and that $\lambda_R = \lambda_C$, holds beyond the numerical evidences provided in [Enns *et al.* (2004); Ramasco *et al.* (2004)]. Further consequence of the symmetry is that $\lambda_{R,C}$ are not independent, but related to the known static and dynamic exponents via

$$\lambda_{R,C} = Z(1 + \alpha) + d \; , \tag{4.55}$$

which is confirmed by the available numerical data (see Table 4.1 on page 90). Baumann and Gambassi found that the universal scaling function

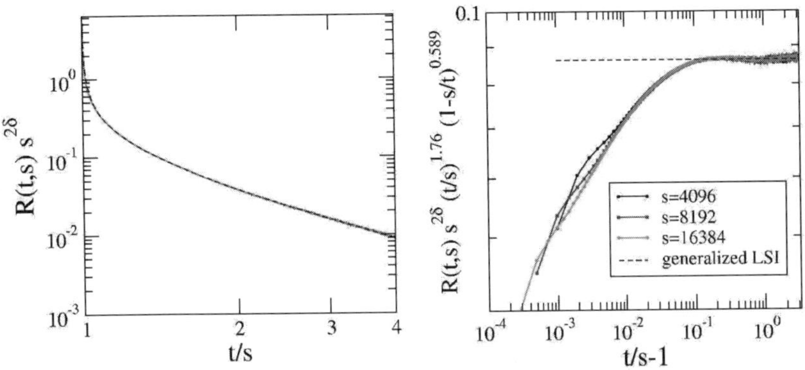


Fig. 4.5 Left: Fit of the generalized scaling function (1.89) on top of the collapsed numerical data. As can be seen, there is an excellent coincidence over a wide range of $t/s$. Right: The same data sets plotted against $t/s - 1$ scaled in such a way that the prediction according to Eq. (1.89) would lead to a horizontal line. The sharp decline for $t/s \to 1$ illustrates why this scaling form is incompatible with the quasi-stationary limit in DP. This decline is stable as $s$ is increased so that scaling corrections due to a microscopic reference time as its origin can be ruled out. From [Hinrichsen (2006)].

(1.89) *does not* agree with the prediction of the theory of local scale invariance, when the non-Gaussian fluctuations are taken into account.

In conclusion it is likely that the universal scaling function (1.89) (set up for $m_0 = 0$ initial order parameter density) does not fit the numerical data of ageing for small $t/s$, thus the validity of LSI for DP is an open question. The two-point, ageing properties can be described by scaling functions

$$R(t,s) = s^{-1-a} f_R(t/s), \qquad C(t,s) = s^{-b} f_C(t/s), \tag{4.56}$$

where

$$f_R(y \gg 1) \sim y^{-\lambda_R/Z}, \qquad f_C(y \gg 1) \sim y^{-\lambda_C/Z} . \tag{4.57}$$

Since $b = 1 + a = 2\beta/\nu_{||} = 2\alpha$ and $\lambda_{R,C} = Z(1+\alpha) + d$ there is **no need to introduce independent exponents to describe the dynamical scaling behavior of the ageing functions of DP**.

### 4.2.3 *DP-class stochastic cellular automata*

Cellular automata (CA) as simplest systems exhibiting synchronous dynamics have extensively been studied [Wolfram (1983)]. When the update rules are made probabilistic, phase transitions as the function some control parameter may emerge. There are many stochastic cellular automata

(SCA) that exhibit DP transition [Boccara and Roger (1993)] perhaps the first and simplest being the (1+1)-dimensional Domany-Kinzel (DK) model [Domany and Kinzel (1984)]. In this model the state at a given time $t$ is specified by binary variables $\{s_i\}$, which can have the values $A$ (active) and $I$ (inactive). At odd times the odd-indexed, whereas at even times the rest of the sites are updated according to specific conditional probabilities. This defines a cellular automaton with **parallel** updates (discrete time evolution) acting on two independent triangular sub-lattices (Fig. 4.6). The condi-

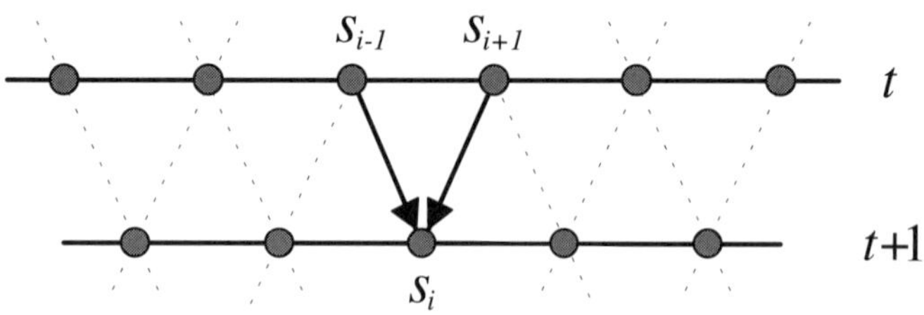


Fig. 4.6 Update in the Domany-Kinzel model.

tional probabilities in the Domany-Kinzel model $P(s_{i,t+1} \mid s_{i-1,t} \, , \, s_{i+1,t})$ are given by

$$P(I \mid I, I) = 1 \, , \tag{4.58}$$

$$P(A \mid A, A) = p_2 \, , \tag{4.59}$$

$$P(A \mid I, A) = P(A|A, I) = p_1 \, , \tag{4.60}$$

and $P(I|s_{i-1}, s_{i+1}) + P(A|s_{i-1}, s_{i+1}) = 1$, where $0 \leq p_1 \leq 1$ and $0 \leq p_2 \leq 1$ are two parameters. Equation (4.58) ensures that the configuration $\ldots, I, I, I, \ldots$ is the absorbing state. The process in Eq. (4.59) describes the creation of inactive spots within active islands with probability $1 - p_2$. The random walk of boundaries between active and inactive domains is realized by the processes in Eq. (4.60). A DP transitions can be observed only if $p_1 > \frac{1}{2}$, when active islands are biased to grow [Wolfram (1983)].

In the **site mean-field approximation** this reaction scheme leads to the differential equation for the density of active (occupied) sites:

$$\partial_t \, \rho(p_1, p_2) \;=\; 2\,(2p_1 - 1)\,\rho \;-\; 2\,(2p_1 - p_2)\,\rho^2 \, . \tag{4.61}$$

In the steady state there are two solutions for the order parameter

$$\rho(p_1, p_2) \;= 0, \qquad\qquad \rho(p_1, p_2) \;=\; \frac{2p_1 - 1}{2p_1 - p_2} \, . \tag{4.62}$$

The first one is stable for $p_1 < 1/2$ and corresponds to the absorbing phase. The active phase is described by the second solution which is stable for $p_1 > 1/2$. Expanding in the active phase as $p_1 \to 1/2$ in terms of

$$\delta p_1 = (p_1 - p_c)/p_c \tag{4.63}$$

one obtains asymptotically

$$\rho(p_1, p_2) = \frac{1}{1 - p_2}\,\delta p_1 + \mathcal{O}(\delta p_1^2) \,, \tag{4.64}$$

hence the leading order singularity is characterized by

$$\beta_{MF} = 1 \,. \tag{4.65}$$

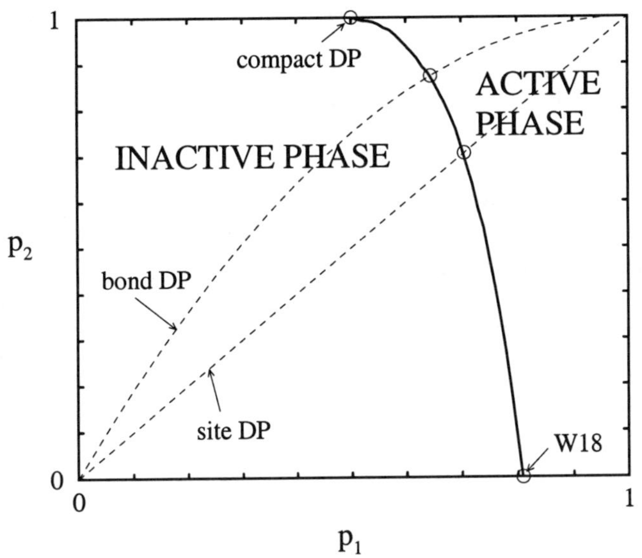


Fig. 4.7 Phase diagram of the 1d Domany-Kinzel SCA [Hinrichsen (1997)].

The phase diagram of the one-dimensional DK model is shown in Fig. 4.7. It comprises an active and an inactive phase, separated by a phase transition line (solid line) belonging to the DP class. The dashed lines corresponds to *directed bond percolation* (for $p_2 = p(2 - p_1)$) and *directed site percolation* (for $p_1 = p_2$) models. At the special symmetry end point ($p_1 = \frac{1}{2}$, $p_2 = 1$) compact domain growth, the so-called 'compact directed percolation' (CDP) occurs and the transition becomes first order (see Sect. 4.5). The transition on the $p_2 = 0$ axis corresponds to the transition of the stochastic version of Wolfram's rule-18 cellular automaton [Wolfram (1983)]. This range-1 SCA generates A at time $t$ only when the right or left neighbor is A at $t-1$:

```
t-1:       AII       IIA
  t:        A         A
```

with probability $p_1$ [Boccara and Roger (1993)]. The critical point was determined by precise simulations ($p_1^* = 0.80948(1)$) [Vesztergombi *et al.* (1997)]. In the $t \to \infty$ limit the steady state is built up from II and IA blocks [Eloranta and Nummelin (1992)]. This finding permits us to map this model onto an even simpler one, the rule-6/16 SCA with new variables: IA $\to$ A and II $\to$ I:

```
t-1:        I I    I A    A I    A A
  t:         I      A      A      I
```

By solving the GMF approximations and applying Padé approximations [Szabó and Ódor (1994)] or CAM method [Ódor (1995)] very precise order parameter exponent estimates were found: $\beta = 0.2796(2)$. The **DS phase structure** of the one-dimensional DK model was explored by [Hinrichsen *et al.* (1997b)] and DP class transitions were found.

Another SCA example of DP like transition I mention is the family of **range-4 SCA** with an acceptance rule

$$s(t+1,j) = \begin{cases} X \text{ if } \; y \leq \sum_{j-4}^{j+4} s(t,j) \leq 6 \\ 0 \;\; \text{otherwise} \end{cases},$$

where $X \in \{0,1\}$ is a two valued random variable such that $Prob(X=1) = p$ [Ódor and Szolnoki (1996)]. The $y = 3$ case was introduced and investigated by [Bidaux *et al.* (1989)] in $d = 1,2,3$. The very first simulations in one dimension [Bidaux *et al.* (1989)] suggested a counter-example to the DP conjecture. More precise spreading simulations of this model [Jensen (1991)], GMF + CAM calculations and simulations of the $y < 6$ family in one and two dimensions have proven that this does not happen for any case [Ódor and Szolnoki (1996)]. The transitions are either belong to DP class or first order.

### 4.2.4 *Branching and annihilating random walks with odd number of offspring*

Branching and annihilating random walks (BARW) introduced by [Takayasu and Tretyakov (1992)] can be regarded as generalizations of the DP process. They are defined by the following reaction-diffusion processes:

$$A \xrightarrow{\sigma} (k+1)A \qquad mA \xrightarrow{\lambda} \emptyset \qquad A\emptyset \overset{D}{\leftrightarrow} \emptyset A \,. \tag{4.66}$$

The $2A \to A$ and $2A \to \emptyset$ reactions dominating in the inactive phase are shown to be equivalent asymptotically in the long time limit [Peliti (1986)] (see Sect. 4.5.1). Therefore the $m = 2$ and $k = 1$ (BARW1) model differs from the DP process (4.31) in that spontaneous annihilation of particles is not allowed. The action of the BARW process was established by [Cardy and Täuber (1996, 1998)]

$$S = \int d^dxdt \; [\psi(\partial_t - D\nabla^2)\phi - \lambda(1-\psi^m)\phi^m$$
$$+\sigma(1-\psi^k)\psi\phi] \,. \tag{4.67}$$

The bosonic RG analysis of BARW systems [Cardy and Täuber (1996)] proved that for $m = 2$ all the lower branching reactions with $k-2, k-4, \ldots$ are generated via fluctuations involving combinations of branching and annihilation processes. As a consequence for **odd-$k$** the $A \to \emptyset$ reaction appears (via $A \to 2A \to \emptyset$). Therefore after the first coarse graining step in the BARW1 (and in general in the odd $m$ BARW (BARWo) cases) the action becomes the the same as that of the DP process. This prediction was confirmed by simulations (see for example [Jensen (1993b)]).

For even $k$ (BARWe), when the parity of the number of particles is conserved the spontaneous decay $A \to \emptyset$ is not generated. Hence there is an absorbing state with a lone wandering particle. This systems exhibits a non-DP class critical transition, which will be discussed in Sect. 4.6.1.

### 4.2.5 *DP with spatial boundary conditions*

In this section I begin discussing the effect of boundary conditions for universality classes of nonequilibrium systems exhibiting absorbing states. For a recent review of critical behavior at surfaces of *equilibrium* models see [Iglói *et al.* (1993)]). [Cardy (1983a)] suggested that surface critical phenomena may be described by introducing an additional *surface exponent* for the order parameter field, which is generally independent of the other bulk exponents. In nonequilibrium statistical physics one can introduce spatial, temporal (see Sect. 4.2.8) or mixed (see Sect. 4.2.6) boundary conditions.

In DP an absorbing wall may be introduced by cutting all bonds (the inactive boundary condition (**IBC**)) crossing a given $(d-1)$-dimensional hyper-plane in space. In case of reflecting boundary condition (**RBC**), the wall acts like a mirror, so that the sites within the wall are always a mirror image of those next to the wall. A third type of boundary condition is the active boundary condition (**ABC**), in which sites within the wall are forced to be active. The density at the wall is found to scale as

$$\rho_s^{stat} \sim (p - p_c)^{\beta_1} \tag{4.68}$$

with a surface critical exponent $\beta_1 > \beta$. Owing to the rapidity reversal symmetry of DP (4.41), only one extra exponent is needed to describe surface effects, hence the cluster survival exponent is the same

$$\beta_1' = \beta_1 \,. \tag{4.69}$$

The mean lifetime of finite clusters at the wall is defined as

$$\langle t \rangle \sim |\Delta|^{-\tau_1} \,, \tag{4.70}$$

where $\Delta_s = (p - p_c)$, and is related to $\beta_1$ by the scaling relation

$$\tau_1 = \nu_{||} - \beta_1 \,. \tag{4.71}$$

The average size of finite clusters grown from seeds on the wall is

$$\langle s \rangle \sim |\Delta|^{-\gamma_1} \,. \tag{4.72}$$

Series expansions [Essam *et al.* (1996)] and numerical simulations [Lauritsen *et al.* (1997)] in 1+1 dimensions indicate that the presence of the wall alters several exponents. However, the scaling properties of the correlation lengths (as given by $\nu_{||}$ and $\nu_\perp$) are *not* altered.

The field theory for DP in a semi-infinite geometry was first analyzed by [Janssen *et al.* (1988)]. They showed that the appropriate action for DP with a wall at $x_\perp = 0$ is given by $S = S_{\text{bulk}} + S_{\text{surface}}$, where

$$S_{\text{surface}} = \int d^{d-1}x \int dt \; \Delta_s \, \psi_s \, \phi_s , \tag{4.73}$$

with the definitions $\phi_s = \phi(\mathbf{x}_{||}, x_\perp = 0, t)$ and $\psi_s = \psi(\mathbf{x}_{||}, x_\perp = 0, t)$. The surface term $S_{\text{surface}}$ corresponds to the most relevant interaction consistent with the symmetries of the problem and the one that also respects the absorbing-state condition. The appropriate surface exponents were computed to first order in $\epsilon = 4 - d$ using renormalization group techniques:

$$\beta_1 = \frac{3}{2} - \frac{7\epsilon}{48} + O(\epsilon^2). \tag{4.74}$$

These calculations also showed that the corresponding hyperscaling relation is

$$\nu_{\|} + d\nu_{\perp} = \beta_1 + \beta + \gamma_1 \tag{4.75}$$

relating $\beta_1$ to

$$\gamma_1 = \frac{1}{2} + \frac{7\epsilon}{48} + O(\epsilon^2). \tag{4.76}$$

The schematic phase diagram for boundary DP is shown in Fig. 4.8, where $\Delta$ and $\Delta_s$ represent, the deviations of the bulk and surface , respectively, from criticality, respectively. For $\Delta > 0$ and for $\Delta_s$ sufficiently negative the

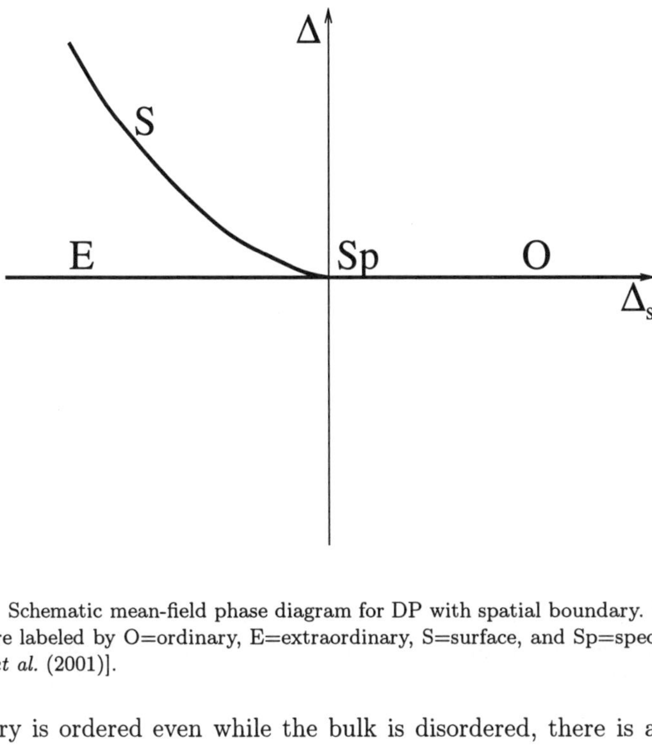


Fig. 4.8 Schematic mean-field phase diagram for DP with spatial boundary. The transitions are labeled by O=ordinary, E=extraordinary, S=surface, and Sp=special. From [Fröjdh *et al.* (2001)].

boundary is ordered even while the bulk is disordered, there is a *surface transition.* For $\Delta_s < 0$ and $\Delta \to 0$, the bulk is ordered in the presence of an already ordered boundary, and there is an *extraordinary transition* of the boundary. Finally at $\Delta = \Delta_s = 0$, where all the critical lines meet, and where both the bulk and the isolated surface are critical, we find a multi-critical point, i.e. the *special transition.* For $\Delta_s > 0$ and $\Delta \to 0$ there is an *ordinary transition*, since the bulk orders in a situation in which the boundary, if isolated, would be disordered. At the ordinary transition, one finds just **one extra independent exponent** related to the boundary. This can be the surface density exponent $\beta_{1,\mathrm{dens}}$.

In **one dimension** the two cases IBC and RBC belong to the same universality class [Lauritsen *et al.* (1998)] as was identified as the ordinary transition. There are numerical data for the exponents of the extraordinary and special transitions (however see [Janssen *et al.* (1988)] for an RG analysis). The best exponent estimates currently available were summarized in [Fröjdh *et al.* (2001)]. Some of them are shown in Table 4.2. In $d = 1$ the best results are from series expansions [Essam *et al.* (1996); Jensen (1999b)]; in all other cases are from Monte-Carlo data. [Lauritsen *et al.* (1997); Fröjdh *et al.* (1998); Lauritsen *et al.* (1998); Howard *et al.* (2000)] The exponent $\tau_1$ was conjectured to equal unity, [Essam *et al.* (1996)] although this has been challenged by the estimate $\tau_1 = 1.00014(2)$ [Jensen (1999b)].

Table 4.2 Critical exponents for DP in $d = 1$ and $d = 2$ for the ordinary transition at the boundary.

| | $d = 1$ | $d = 2$ | Mean Field |
|---|---|---|---|
| $\beta_1$ | 0.733 71(2) | 1.07(5) | 3/2 |
| $\delta_1 = \alpha_1$ | 0.423 17(2) | 0.82(4) | 3/2 |
| $\tau_1$ | 1.000 14(2) | 0.26(2) | 0 |
| $\gamma_1$ | 1.820 51(1) | 1.05(2) | 1/2 |

It has been known for some time that the presence of an **edge** introduces new exponents, independent of those associated with the bulk or with a surface (see [Cardy (1983a)]). For an investigation showing numerical estimates in two-dimensional and mean-field values see [Fröjdh *et al.* (1998)]. Table 4.3 summarizes results for the ordinary edge exponents. A closely related application is the study of spreading processes in narrow channels [Albano (1997)].

Table 4.3 Numerical estimates for the ordinary $\beta_2^{\mathrm{O}}$ exponents for edge DP together with the mean field values. Note that $\beta_2^{\mathrm{O}}(\pi) = \beta_1^{\mathrm{O}}$.

| Angle ($\alpha$) | $\pi/2$ | $3\pi/4$ | $\pi$ | $5\pi/4$ |
|---|---|---|---|---|
| $\beta_2^{\mathrm{O}}$ ($d = 2$) | 1.6(1) | 1.23(7) | 1.07(5) | 0.98(5) |
| $\beta_2^{\mathrm{O}}$ (MF) | 2 | 5/3 | 3/2 | 7/5 |

### 4.2.6 DP with mixed (parabolic) boundary conditions

Boundary conditions, which act both in space and time direction have also been investigated in dynamical systems. These turn out to be related to hard-core repulsion effects of one-dimensional systems (Sect. 6.2). [Kaiser and Turban (1994, 1995)] investigated the $1+1$-dimensional DP process confined in a parabola-shaped geometry. Assuming an absorbing boundary of the form $x = \pm C t^{\sigma}$ they proposed a general scaling theory. It is based on the observation that the coefficient of the parabola ($C$) scales as $C \to \Lambda^{Z\sigma - 1} C$ under rescaling

$$x \to \Lambda x\,, \quad t \to \Lambda^Z t\,, \quad \Delta \to \Lambda^{-1/\nu_\perp} \Delta\,, \quad \rho \to \Lambda^{-\beta/\nu_\perp} \rho\,, \tag{4.77}$$

where $\Delta = |p - p_c|$ and $Z$ is the dynamical exponent of DP. By a conformal mapping of the parabola to straight lines and using the mean-field approximation, they claimed that this boundary is a relevant perturbation for $\sigma > 1/Z$, irrelevant for $\sigma < 1/Z$ and marginal for $\sigma = 1/Z$ (see Fig. 4.9). The marginal case results in $C$-dependent non-universal power-law decay, while for the relevant case stretched exponential functions have been obtained. Kaiser and Turban have given support to these claims by numerical simulations.

### 4.2.7 Lévy flight anomalous diffusion in DP

Lévy flight anomalous diffusion generating long-range correlations has already been mentioned in Sects. 2.3.5 and 2.4.4 in the case of equilibrium models. In nonequilibrium systems [Grassberger (1986)], following the suggestion of [Mollison (1977)], introduced a variation of the epidemic processes with infection probability distribution $P(R)$, which decays with the distance $R$ as a power-law

$$P(R) \propto \frac{1}{R^{d+\sigma}}\,. \tag{4.78}$$

This formalism can model long-range epidemics mediated by flies, wind, etc. Grassberger claimed that the critical exponents should depend continuously on $\sigma$. This result was confirmed in one dimension by [Marques and Ferreira (1994)], who estimated $\beta$ based on GMF+CAM calculations.

The study of anomalous diffusion was extended for GEP processes (see Sect. 4.4) and for annihilating random walks (see Sect. 4.5.4) too. The

(a) (b) (c)

Fig. 4.9 The space-time evolution of the critical, 1 + 1-dimensional directed site percolation process confined by parabola. (a) $\sigma < 1/Z$, (b) $\sigma = 1/Z$, (c) $\sigma > 1/Z$. From [Kaiser and Turban (1994)].

effective action of the Lévy flight DP model is

$$S[\psi,\phi] = \int d^dx\, dt \left[\psi(\partial_t - \tau - D_N\nabla^2 - D_A\nabla^\sigma)\phi + \frac{g}{2}(\psi\phi^2 - \psi^2\phi)\right] , \tag{4.79}$$

where $D_N$ denotes the normal diffusion constant, $D_A$ the anomalous diffusion constant and $g$ the interaction coupling constant. Field theoretical RG method up to first order in an $\epsilon = 2\sigma - d$ expansion [Janssen *et al.* (1999)] gives:

$$\begin{aligned} \beta &= 1 - \frac{2\epsilon}{7\sigma} + O\left(\epsilon^2\right) , \\ \nu_\perp &= \frac{1}{\sigma} + \frac{2\epsilon}{7\sigma^2} + O\left(\epsilon^2\right) , \\ \nu_{||} &= 1 + \frac{\epsilon}{7\sigma} + O\left(\epsilon^2\right) , \\ Z = \frac{\nu_{||}}{\nu_\perp} &= \sigma - \epsilon/7 + O\left(\epsilon^2\right) . \end{aligned} \tag{4.80}$$

Moreover, the field theory indicated that the hyperscaling relation

$$\eta + 2\delta = d/Z \qquad (\delta = \beta/\nu_{||}) , \tag{4.81}$$

for the critical initial slip exponent $\eta$ and the scaling relation

$$\nu_{||} - \nu_{\perp}(\sigma - d) - 2\beta = 0\,, \tag{4.82}$$

hold exactly for arbitrary values of $\sigma$. Numerical simulations on 1 + 1-dimensional bond percolation confirmed these results except in the neighborhood of $\sigma = 2$ [Hinrichsen and Howard (1999)] (see Fig. 4.10).

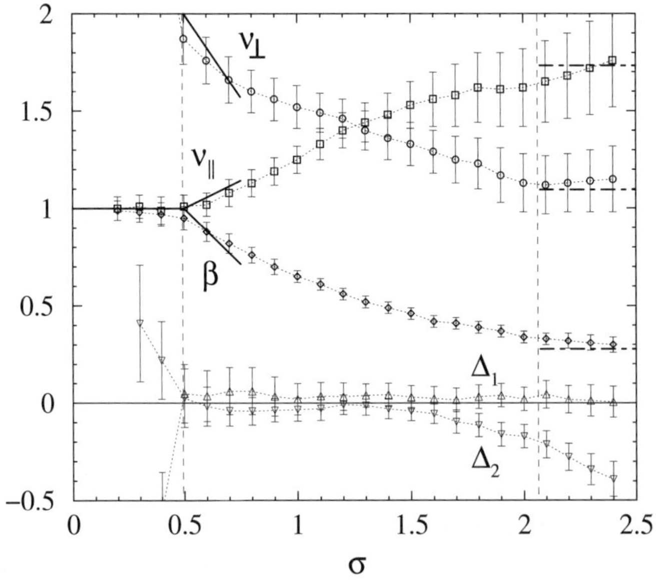


Fig. 4.10 Estimates for the exponent $\beta$ and the derived exponents $\nu_{\perp}$ and $\nu_{||}$ in comparison with the field-theoretic results (solid lines) and the DP exponents (dot-dashed lines). The quantities $\Delta_1$ and $\Delta_2$ represent deviations from the scaling relations (4.81) and (4.82), respectively. The vertical, dashed lines distinguish between regions of constant and continuously changing exponents. From [Hinrichsen and Howard (1999)].

### 4.2.8 *Long-range correlated initial conditions in DP*

It is well known that initial conditions influence the temporal evolution of nonequilibrium systems. The "memory" of systems for the initial state usually depends on the dynamical rules. For example, stochastic processes with a finite temporal correlation length relax to their stationary state in an exponentially short time. An interesting situation emerges when a system undergoes a nonequilibrium phase transition in which the temporal correlation length diverges. This setup raise the question whether it is possible construct initial states that affect the *entire* temporal evolution of such systems.

Monte-Carlo simulations of critical models with absorbing states usually employ two different types of initial conditions. On the one hand *uncorrelated random initial conditions* (Poisson distributions) are used to study the relaxation of an initial state with a finite particle density towards the absorbing state. In this case the particle density $\rho(t)$ *decreases* on the infinite lattice asymptotically as

$$\rho(t) \sim t^{-\beta/\nu_{||}} . \tag{4.83}$$

On the other hand, in spreading simulations [Grassberger and de la Torre (1979)], each run starts with a *single particle* as a localized active seed from which a cluster originates (this is a long-range correlated state). Although many of these clusters survive for only a short time, the number of particles $N(t)$ averaged over many independent runs *increases* as

$$\langle N(t) \rangle \sim t^{\eta} . \tag{4.84}$$

These two cases seem to represent extremal situations in which the average particle number either decreases or increases.

A *crossover* between these two extremal cases takes place in a critical spreading process that starts from a random initial condition of very low density. Here the particles are initially separated by empty intervals of a certain typical size, so that the the average particle number first increases according to Eq. (4.84). Later, when the growing clusters begin to interact with each other, the system crosses over to the algebraic decay of Eq. (4.83) – a phenomenon which is referred to as the "critical initial slip" of nonequilibrium systems [Janssen *et al.* (1989)].

[Hinrichsen and Ódor (1998)] and [Menyhárd and Ódor (2000)] investigated whether it is possible to interpolate *continuously* between the two extremal cases in 1 + 1-dimensional DP and PC processes. It was shown that one can in fact generate certain initial states in such a way that the particle density on the infinite lattice varies as

$$\rho(t) \sim t^{\kappa} \tag{4.85}$$

with a continuously adjustable exponent $\kappa$ in the range

$$-\beta/\nu_{||} \leq \kappa \leq +\eta . \tag{4.86}$$

To this end artificial initial configurations with algebraic long-range correlations of the form

$$C(r) = \langle s_i \, s_{i+r} \rangle \sim r^{-(d-\sigma)} , \tag{4.87}$$

were constructed, where $\langle\rangle$ denotes the average over many independent realizations and $s_i = 0, 1$ inactive and active sites. The exponent $\sigma$ is a free parameter and can be varied continuously between 0 and 1. This initial condition can be taken into account by adding the term

$$S_{ic} = \mu \int d^d x \psi(x, 0)\phi_0(x) \tag{4.88}$$

to the action, where $\phi_0(x)$ represents the initial particle distribution. The long-range correlations limit $\sigma \to d$ corresponds to a constant particle density and thus we expect Eq. (4.83) to hold ($\phi_0(x)$ = const. is irrelevant under rescaling). On the other hand, the short-range limit $\sigma \to 0$ represents an initial state in which active sites are separated by infinitely large intervals ($\phi_0(x) = \delta^d(x)$) so that the particle density should increase according to Eq. (4.84). In between we expect $\rho(t)$ to vary algebraically according to Eq. (4.85) with an exponent $\kappa$ depending continuously on $\sigma$.

In case of the 1 + 1-dimensional Domany-Kinzel SCA (see Sect. 4.2.3) field-theoretical renormalization group calculation and simulations have proved [Hinrichsen and Ódor (1998)] the exact functional dependence

$$\kappa(\sigma) = \begin{cases} \eta & \text{for } \sigma < \sigma_c \\ \frac{1}{z}(d - \sigma - \beta/\nu_\perp) & \text{for } \sigma > \sigma_c \end{cases} \tag{4.89}$$

with the critical threshold $\sigma_c = \beta/\nu_\perp$.

### 4.2.9 *Anisotropic DP systems*

The effect of diffusion drift in one dimensional DP was studied by simulation in [Park and Park (2005a)]. The BARWo model (see Sect. 4.2.4) with fully biased diffusion

$$A\emptyset \xrightarrow{p} \emptyset A, \quad AA \xrightarrow{p} \emptyset\emptyset, \quad A\emptyset \xrightarrow{1-p} AA \tag{4.90}$$

exhibited the same kind of density decay as the isotropic model at the critical point $\rho(t) \sim t^{-0.159(1)}$. Furthermore the BARWo with partial bias and other driven DP class models also show the same critical behavior. Therefore the Galilean invariance (4.5.5) holds for DP.

### 4.2.10 *Quench disordered DP classes*

Perhaps the lack of experimental observation of the robust DP class is owing to the fact that even weak disorder changes the critical behavior of such models. Therefore this section provides a view onto interesting generalizations of DP processes that may be observable in physical systems. First [Kinzel (1985)] showed using the Harris criterion (1.91) that **spatial quenched disorder** (frozen in space) changes the critical behavior of DP systems for $d \leq 3$. The numerical studies by [Noest (1986, 1988)] confirmed this. Then [Janssen (1997a)] studied the problem by field theory, taking into account the disorder in the action by adding the term

$$S \rightarrow S + \gamma \int d^d x \left[ \int dt\, \psi \phi \right]^2 . \tag{4.91}$$

This additional term causes **marginal perturbation** and the stable fixed point is shifted to an unphysical region, leading to runaway solutions of the flow equations in the physical region of interest, meaning that spatially quenched disorder changes the critical behavior of DP. This conclusion is supported by the simulation results of [Moreira and Dickman (1996)], who reported logarithmic spreading behavior in **two-dimensional contact process** at criticality. In the sub-critical region they found Griffiths phase, in which the time dependence is governed by non-universal power-laws, while in the active phase the relaxation of $P(t)$ is algebraic.

In $1+1$ dimension [Noest (1986, 1988)] predicted generic scale invariance. [Webman *et al.* (1998)] reported glassy phase with non-universal exponents in a $1+1$-dimensional DP process with quenched disorder. [Cafiero *et al.* (1998)] showed that DP with spatially quenched randomness in the large time limit can be mapped onto a non-Markovian spreading process with memory, in agreement with previous results. They also showed that the time reversal symmetry of the DP process (4.41) is not broken, therefore

$$\delta = \alpha . \tag{4.92}$$

Cafiero et al. were able to derive a hyperscaling law for the inactive phase:

$$\eta = dz/2 \tag{4.93}$$

and for the absorbing phase

$$\eta + \delta = dz/2 . \tag{4.94}$$

These relations were confirmed by simulations and it turned out that the dynamical exponents changed continuously as the function of the disorder probability.

A real space RG study by [Hooyberghs *et al.* (2003)] showed that in the case of strong enough disorder the critical behavior is controlled by an **infinite randomness** fixed point (IRFP), (related to the random-field Ising model) the static exponents of which in **one dimension** are

$$\beta = (3 - \sqrt{5})/2\ ,\ \nu_\perp = 2\ . \tag{4.95}$$

At the dirty critical point activated dynamical scaling of logarithmic forms (1.99), (1.100) emerge with the exponents

$$\psi = 0.5\ ,\ \bar{\alpha} = \bar{\delta} = 0.38197\ ,\ \bar{\eta} = 1.2360\ . \tag{4.96}$$

For disorder strengths outside the attractive region of the IRFP, disorder dependent, **continuously changing critical exponents** were found by numerical simulations [Cafiero *et al.* (1998)] and series expansion [Neugebauer and Taraskin (2007)].

Quenched, spatial disorder can be introduced into the contact process (see Sect. 4.2.1) by site or bond dilution of the underlying lattice. In **space dimensions** $d \geq 2$ the phase diagram of such system is shown on Fig. 4.11 Note that in $d = 1$, any dilution immediately destroys the active phase. For

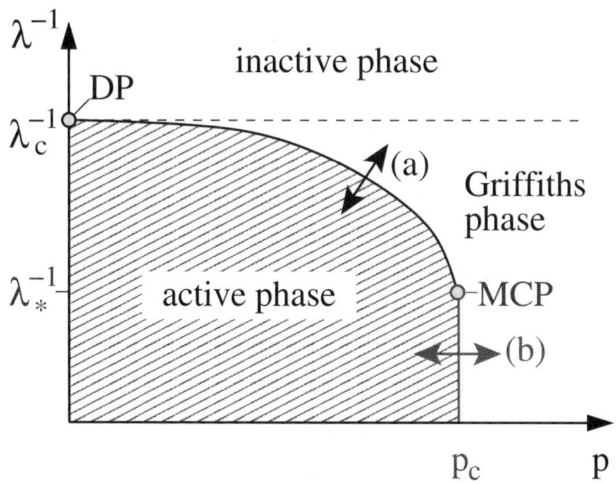


Fig. 4.11 Schematic phase diagram of the diluted contact process as a function of the dilution parameter $p$ and the birth rate $\lambda$, containing a multi-critical point (MCP). From [Dahmen *et al.* (2007)]

$p < p_c$ (below the isotropic percolation threshold of the lattice), the active phase survives, but the critical birth rate increases with $p$ (to compensate for the missing neighbors). At the percolation threshold the active phase survives on the infinite percolation cluster for $\lambda > \lambda_*$. For $p > p_c$, the lattice consists of disconnected clusters of finite size, hence no active phase can exist. Therefore there are two nonequilibrium phase transitions, separated by a multi-critical point (MCP) at $(p_c, \lambda_*)$. For $p < p_c$, the transition [(a) in Fig. 4.11] is expected to be in the universality class of the generic disordered contact process discussed above, while the phase transition (b) across the percolation threshold of the lattice is expected to show different critical behavior.

[Vojta and Lee (2006)] investigated this analytically, while [Dahmen *et al.* (2007)] performed simulations. The interplay of geometric (isotropic) criticality (see Sect. 4.4.1) and the dynamic fluctuations of the contact process leads to a novel universality class for this nonequilibrium phase transition. The transition lines (a) and (b) obey activated scaling (1.98), with different sets of critical exponents, the latter being related to the exactly known exponents of isotropic percolation in two dimensions (see Table 4.4). Along line (a) the model exhibits a continuous phase transition which belongs to the universality class of strongly disordered directed percolation.

Along line (b) activated dynamical scaling (1.98) with the critical exponent $\psi$ being equal to the fractal dimension of the critical percolation cluster $\psi = d_f$ was found. Here the supporting lattice is at the percolation threshold and thus decomposes into many clusters according to a scale-free distribution. As a consequence the critical behavior is very similar to that of the infinite-randomness critical point discussed. In particular the exponent $\bar{\delta} = \beta_c/(\nu_c f_f)$ is determined by $d_f$ together with the order parameter and correlation length exponents $\beta_c$ and $\nu_c$, of the isotropic percolation transition. On the other hand the static exponents are identical the corresponding lattice percolation exponents.

Sufficiently far away from the multi-critical point one finds a Griffiths-like power law decay

$$\rho(t) \propto (t/t_0)^{-d/Z'} \tag{4.97}$$

for $p > p_c$ and a stretched exponential of the form

$$\rho(t) - \rho_{\rm st} \propto \exp\left[\left(\frac{d}{Z''}\ln\left(\frac{t}{t_0}\right)\right)^{1-1/d}\right] \tag{4.98}$$

for $p < p_c$, where $\rho_{\rm st}$ denotes the stationary density. Here $Z'$ and $Z''$ are non-universal exponents with diverge as $p \to p_c$.

The MCP in case of bond dilution has been investigated by simulations [Dahmen *et al.* (2007)]. Its location is found at $p = 1/2$ and $\lambda = \lambda* = 3.55(1)$, and activated scaling with the exponents shown in Table 4.4 are reported.

Table 4.4 Numerical estimates of the critical exponents of the diluted CP in two dimensions.

| transition | $\beta$ | $\nu_\perp$ | $Z,\bar{Z}$ | $\delta,\bar{\delta}$ |
|---|---|---|---|---|
| line (a) | 0.93(5) | 1.00(9) | 1.98(12) | 0.47 |
| line (b) | 5/36 | 4/3 | 91/48 | 5/91 |
| MCP | 0.81(7) | 0.88(10) | 0.57(4) | 1.63(10) |

Kinzel [Kinzel (1985)] also considered the effect of **temporal disorder** and argued that a Harris type of criterion can be applied but with $d\nu_\perp$ replaced by $\nu_{||}$. Since $\nu_{||} = 1.733847(6)$ for (1+1)-dimensional DP this would make *temporal disorder a relevant perturbation.* The corresponding field theoretical action is

$$S \to S + \gamma \int dt \left[ \int d^d x \, \psi\phi \right]^2 . \qquad (4.99)$$

[Jensen (1996b)] investigated the $1+1$-dimensional directed bond percolation (see Sect. 4.2.3) with temporal disorder via series expansions and Monte Carlo simulations. The temporal disorder was introduced by allowing time slices to become fully deterministic ($p_1 = p_2 = 1$), with probability $\bar{\alpha}$. Jensen found $\bar{\alpha}$ dependent, **continuously changing critical point and critical exponent** values between those of the $1+1$-dimensional DP class and those of the deterministic percolation. This latter class is defined by the exponents:

$$\beta = 0, \quad \delta = 0, \quad \eta = 1, \quad Z = 1, \quad \nu_{||} = 2, \quad \nu_\perp = 2 . \qquad (4.100)$$

For small values of the disorder parameter he found $\nu_{||} < 2$ over a wide range of values, suggesting the **violation of the Harris criterion**. These findings have been strengthened by an other precise low-density series expansion analysis of the cluster spreading [Jensen (2005)].

If **quenched disorder takes place in both space and time** the corresponding term to action is

$$S \to S + \gamma \int dt d^d x \, [\psi\phi]^2 \qquad (4.101)$$

which is an **irrelevant** perturbation to the Reggeon field theory. This has the same properties as the intrinsic noise in the system and can be considered as being annealed.

## 4.3 Generalized, $n$-particle contact processes

As we learned in Sect. 4.2 the generic feature of DP class models is a corresponding triangular topology of the phase portrait (see Fig. 4.3), in which the zero energy trajectories intersect in a single point by tuning to criticality. This seems to be quite robust against RG transformations and makes the base of the DP hypothesis (see Fig. 4.12). For a long time it was be-

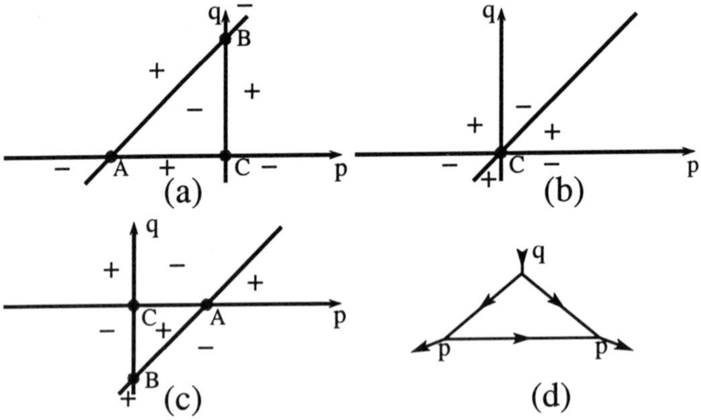


Fig. 4.12 Generic phase portrait of DP models in the vicinity of the phase transition [after the shift, Eq. (4.39)]. (a) Active phase, $m > 0$; (b) transition point, $m = 0$; (c) extinction phase, $m < 0$. The plus and minus signs show the sign of the Hamiltonian in each sector. (d) The one-loop diagram renormalizing $u$ vertex (vertices $m$ and $v$ are renormalized in a similar way) (from [Elgart and Kamenev (2006)]).

lieved that DP is the only class among nonequilibrium phase transitions to absorbing states. Historically the exploration of the PC class was the first exception, which is be discussed in Sect. 4.6, due to the additional parity conservation or $Z_2$ symmetry. However in the early times there were some numerical hints for other type of critical behavior in binary production, diffusive models [Grassberger (1982a)]. Since no other conventional symmetry corresponded to such models those numerical findings had been forgotten for a long time and rediscovered only by a field theoretical paper by [Howard and Täuber (1997)].

Let's see what happens if we change the DP phase space topology. This would be possible by the requirement of intersecting more than 3 lines for example. Tuning $k > 3$ nontrivial zero energy lines, which survive the RG transformation is not easy, and possible by multi-critical points (see Sect. 5.1).

Another possibility is when the $q = 0$ trajectory is $n$ times degenerate due to $n$ particle reactions, if no lower level reactions $j < n$ are generated under renormalization. Exactly this possibility was pointed out by the field theory [Howard and Täuber (1997)] for the annihilation-fission process

$$2A \xrightarrow{\lambda} \emptyset\,; \quad 2A \xrightarrow{\sigma} 3A\,; \quad 3A \xrightarrow{\mu} 2A\,, \tag{4.102}$$

where the last reaction is included (as usual in bosonic models) to prevent infinite proliferation of particles in the active phase. It is easy to construct the corresponding reaction Hamiltonian by the recipe given in Sect. 1.6.1.

$$\begin{aligned} H_R &= \frac{\lambda}{2}(1-p^2)\,q^2 + \frac{\sigma}{2}(p^3-p^2)\,q^2 + \frac{\mu}{6}(p^2-p^3)\,q^3 \\ &= \frac{1}{2}\left(\lambda\,(1+p) - \sigma\,p^2 + \frac{\mu}{3}p^2\,q\right)(1-p)\,q^2\,. \end{aligned} \tag{4.103}$$

The corresponding zero energy lines are shown on Fig. 4.13 with the $q = 0$ being double degenerated.

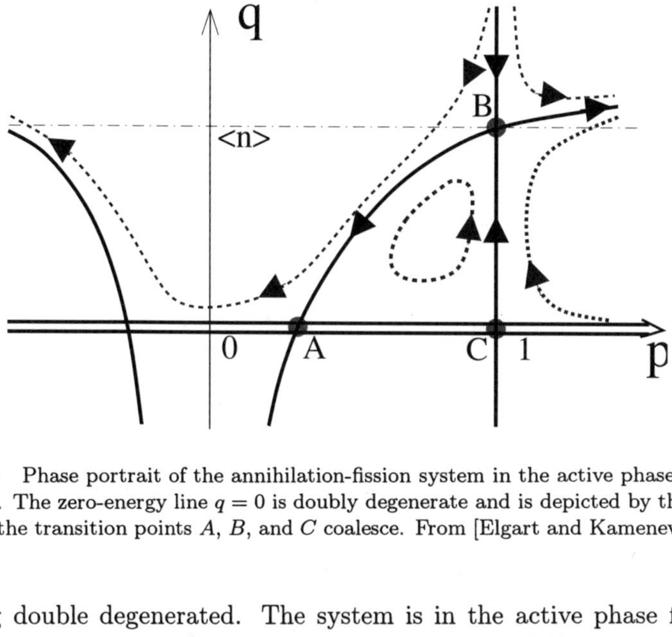


Fig. 4.13 Phase portrait of the annihilation-fission system in the active phase (cf. DP, Fig. 4.3). The zero-energy line $q = 0$ is doubly degenerate and is depicted by the double line. At the transition points $A$, $B$, and $C$ coalesce. From [Elgart and Kamenev (2006)].

The system is in the active phase for $\sigma > 2\lambda$ corresponding to point $B$ (and particle density $\langle n \rangle$). By by tuning the control parameter $m = \sigma/2 - \lambda$ to zero the model can be driven to extinction. The $\sigma/2 = \lambda$ condition corresponds to a continuous phase transition. At the transition point the *four* zero-energy lines are intersecting in the point $C$.

In the site restricted variant of this model we don't need the $3A \to 2A$ reaction to have a stable active phase, but an explicit diffusion of lonely particles is necessary to differentiate it from the pair contact process (see Sect. 6.6), in which pairs as reactants generate a phase transition, similar to that of the CP in many aspects (see Sect. 6.6.1). The diffusive version, the so called diffusive pair contact process (PCPD) exhibits novel critical behavior, but in $d = 1$ the difference from DP is rather small and generated a long history of contradictory numerical studies (for a review see [Henkel and Hinrichsen (2001)]. The PRG treatment has not found a stable fixed point [Janssen *et al.* (2004b)] and probably only a proper NPRG analysis capable accessing strong couplings can describe the true critical behavior. Since reacting pairs and diffusing monomers behave quite differently (shown by numerical simulations) it is more appropriate to consider this model as a coupled system, thus its more detailed discussion is postponed to Sect. 6.7.

Apart from Eq. (4.102), there are infinitely many other reaction sets, with the same topology of the phase portraits. The phase transition of these other models can be described by the same reaction Hamiltonian (4.103). Examples include $2A \to A$, $2A \to 4A$, $4A \to \emptyset$, etc. More generally one can consider higher level RD models in which at least $n$ particles are needed for a reaction

$$nA \to (n-l)A \quad nA \to (k+n)A \quad (n+1)A \to (n-j)A \tag{4.104}$$

or the set of PARWs models in terms of site restricted versions. In the phase portraits there are $n$ times degenerate $q = 0$ lines, a $p = 0$ line [after the shift, Eq. (4.39)], and the "accidental" $q = (m+up)/v$ diagonal line (see Fig. 4.14(b)). Thus these models can be described by a generic reaction Hamiltonian of the form

$$H_R = (m + up - vq)\, p\, q^n \,, \tag{4.105}$$

where $m$ is the control parameter of the transition, $u$ and $v$ are positive constants. The field theory of these models for $n > 2$ are even less known

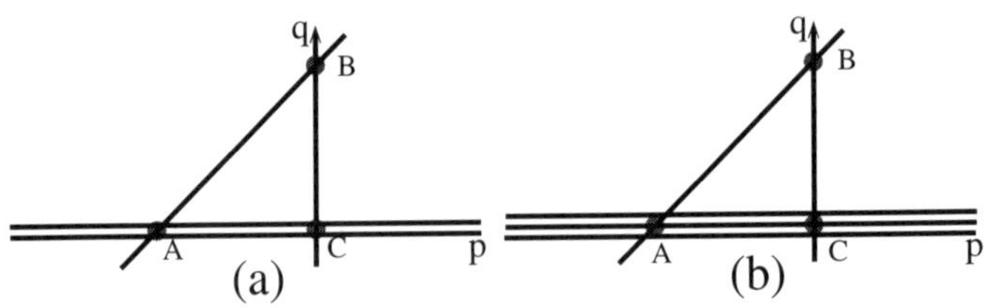


Fig. 4.14 Generic phase portraits of (a) binary reaction models and (b) trimer reaction models. From [Elgart and Kamenev (2006)].

as the that of the PCPD, but the naive power-counting analysis predicts low upper-critical dimensions (Eq. (4.30)), hence mean-field class behavior.

Nontrivial physical dimension may be expected only for $n = 3$, where $d_c = 4/3$ (3CPD model), but extensive simulations seem to exclude even this, suggesting a marginal mean-field behavior in one dimension [Ódor (2006b)].

## 4.4 Dynamical isotropic percolation (DIP) classes

In this section I depart from basic classes of RD models, which can be analyzed by the phase space topology method for historical reasons. If we allow memory in the unary DP spreading process (Sect. 4.2) such that the infected sites may have a different re-infection probability ($p$) than the virgin ones ($q$) we obtain different percolation behavior [Grassberger *et al.* (1997)] (see Fig. 4.15). The model in which the re-infection probability

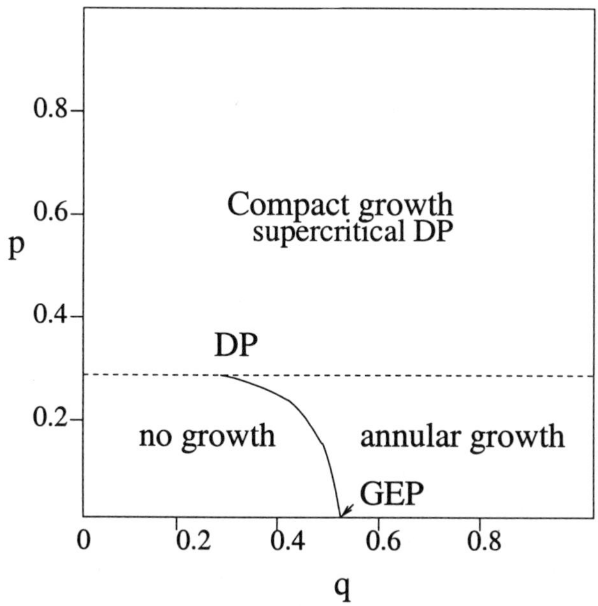


Fig. 4.15 Phase diagram of the GEP model in $d = 2$ dimensions. Along the curved line DIP, at the endpoint DP ($p = q$) criticality is observed. On the horizontal line re-infection processes are DP type, but first infections may be sub- or supercritical.

is zero is called the General Epidemic Model (GEP) [Mollison (1977)]. In this case the epidemic stops in finite systems but an infinite epidemic is possible in the form of a single wave of activity. When starting from a

Table 4.5 Dynamical critical exponents of DIP from [Muñoz *et al.* (1999)].

| $d$ | 2 | 3 | 6 |
|---|---|---|---|
| $Z$ | 1.1295 | 1.336 | 2 |
| $\eta$ | 0.586 | 0.536 | 0 |
| $\nu_{\|\|}$ | 1.506 | 1.169 | 1 |

single seed this leads to annular growth patterns. The transition between survival and extinction is a critical phenomenon called dynamical percolation (DIP) [Grassberger (1982c)]. Clusters generated at criticality are the **isotropic percolation clusters** of the lattice in question (see Sect. 4.4.1). Field theoretical treatments have been given by [Cardy (1983b); Cardy and Grassberger (1985); Janssen (1985); Muñoz *et al.* (1998); J.-Dalmaroni and Hinrichsen (2003); Janssen (2005)]. The action of the model is

$$S = \int d^d x dt \frac{D}{2}\psi^2\phi - \psi\left(\partial_t\phi - \nabla^2\phi - r\phi + w\phi\int_0^t ds\phi(s)\right) . \quad (4.106)$$

This functional is invariant under the nonlocal symmetry transformation

$$\phi(x,t) \leftrightarrow -\partial_t\psi(x,-t) , \quad (4.107)$$

which results in the hyperscaling relation [Muñoz *et al.* (1997)]:

$$\eta + 2\delta + 1 = \frac{dz}{2} . \quad (4.108)$$

As in case of DP the relations

$$\beta = \beta' , \quad \alpha = \delta \quad (4.109)$$

again hold. The upper critical dimension is $d_c = 6$ as for ordinary percolation (the statics of this model) (see Sect. 4.4.1). The dynamical critical exponents as well as spreading and avalanche exponents [Muñoz *et al.* (1999)] are summarized in Table 4.5. Dynamical percolation was observed in forest fire models [Drossel and Schwabl (1993); Albano (1994)] and in some Lotka-Volterra type lattice prey-predator models [Antal *et al.* (2001)] as well.

### 4.4.1 *Static isotropic percolation universality classes*

Ordinary percolation is a geometrical phenomenon that describes the occurrence of infinitely large connected clusters by **completely random** displacement of some variables (sites, bonds, etc) (with probability $p$) on lattices (see Fig. 4.16). For reviews see [Stauffer and Aharony (1994); Grimmett (1999)]. The dynamical percolation process is known to generate such percolating clusters (see Sect. 4.4). At the transition point moments of the $s$ cluster size distribution $n_s(p)$ show singular behavior. Ordinary percolation corresponds to the $q = 1$ limit of the Potts model (see Sect. 2.4). That means its generating functions can be expressed in terms of the free energy of the $q \to 1$ Potts model.

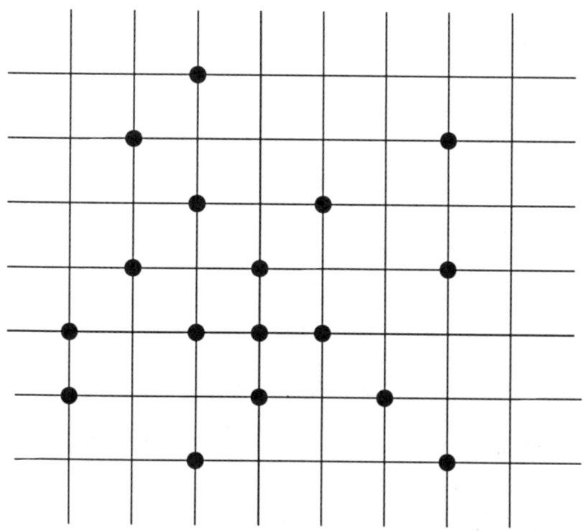

Fig. 4.16 Isotropic, ordinary site percolation in $d = 2$ dimensions.

In the low-temperature dilute Ising model (see Sect. 2.3) the occupation probability ($p$) driven magnetization transition is an ordinary percolation transition as well. As a consequence the critical exponents of the magnetization can be related to the cluster-size exponents. For example the susceptibility obeys simple homogeneity form with $p - p_c$ replacing $T - T_c$

$$\chi \propto |p - p_c|^{-\gamma} \tag{4.110}$$

Table 4.6 summarizes the known critical exponents of ordinary percolation. The exponents are from the overview of [Bunde and Havlin (1991)]. A field theoretical treatment [Benzoni and Cardy (1984)] provided an upper critical dimension $d_c = 6$. The $d = 1$ case is special : $p_c = 1$ and the order parameter jumps ($\beta = 0$). Furthermore here some exponents exhibit non-universal behavior by increasing the interaction length unless we redefine the scaling variable (see [Stauffer and Aharony (1994)]).

Table 4.6 Critical exponents of the isotropic (ordinary) percolation.

| d | $\beta = \beta'$ | $\gamma_P$ | $\nu_\perp$ | $\sigma$ | $\tau$ |
|---|---|---|---|---|---|
| 1 | 0 | 1 | 1 | 1 | 2 |
| 2 | 5/36 | 43/18 | 4/3 | 36/91 | 187/91 |
| 3 | 0.418(1) | 1.793(4) | 0.8765(17) | 0.452(1) | 2.189(1) |
| 4 | 0.64 | 1.44 | 0.68 | 0.48 | 2.31 |
| 5 | 0.84 | 1.18 | 0.57 | 0.49 | 2.41 |
| 6 | 1 | 1 | 1/2 | 1/2 | 5/2 |

### 4.4.2 *DIP with spatial boundary conditions*

There are very few numerical results exists for surface critical exponents of the dynamical percolation. The GEP process (see Sect. 4.4) in three dimensions was investigated numerically by [Grassberger (1992b)]. The surface and edge exponents (for angle $\pi/2$) were determined in case of IBC. Different measurements (density and cluster simulations) resulted in a single surface $\beta_1 = 0.848(6)$ and a single edge $\beta_e = 1.36(1)$ exponent.

### 4.4.3 *Lévy flight anomalous diffusion in DIP*

To model long-range epidemic spreading in a system with immunization the effect of Lévy flight diffusion [Eq. (4.78)] was investigated by [Janssen *et al.* (1999)]. The renormalization group analysis of the GEP with anomalous diffusion resulted in the following $\epsilon = 3\sigma - d$ expansion results for the critical exponents:

$$\eta = \frac{3\epsilon}{16\sigma} + O\left(\epsilon^2\right), \tag{4.111}$$

for the critical initial slip;

$$\beta = 1 - \frac{\epsilon}{4\sigma} + O\left(\epsilon^2\right), \tag{4.112}$$

for the order parameter (density of removed (immune) individuals);

$$\nu_\perp = \frac{1}{\sigma} + \frac{\epsilon}{4\sigma^2} + O\left(\bar{\varepsilon}^2\right), \tag{4.113}$$

for the spatial correlations;

$$\nu_{||} = 1 + \frac{\epsilon}{16\sigma} + O\left(\epsilon^2\right). \tag{4.114}$$

and for the temporal correlation length. As we can see the exponents are non-universal, they are continuous functions of the parameter $\sigma$ of the Lévy distribution, as in case of DP or the Ising model (see Sects. 4.2.7 and 2.3.5).

## 4.5 Voter model (VM) classes

Now we turn to models that can describe the spreading of voter opinion, arranged on regular lattices. These models exhibit **first-order transition** but dynamical scaling still can be observed in them. Therefore these classes should be discussed in Chapter 4. Since they occur very often in inactive phases of other models to be discussed as a dominant behavior I introduce them here. The voter model (VM) [Ligget (1985); Durret (1988)] is defined by the following spin-flip dynamics. A site is selected randomly which takes the "opinion" (or spin) of one of its nearest neighbors (with probability $p$). This rule ensures that the model has two homogeneous absorbing states (all spin up or down) and is invariant under the $Z_2$ symmetry. A general feature of these models is that dynamics takes place only at the boundaries. The action that describes this behavior proposed by [Dickman and Tretyakov (1995); Peliti (1986)]

$$S = \int d^d x dt \left[ \frac{D}{2} \phi(1-\phi)\psi^2 - \psi(\partial_t \phi - \lambda \nabla^2 \phi) \right] \tag{4.115}$$

is invariant under the symmetry transformation

$$\phi \leftrightarrow 1-\phi, \quad \psi \leftrightarrow -\psi \quad . \tag{4.116}$$

This results in the "hyperscaling" relation [Muñoz *et al.* (1997)]

$$\delta + \eta = dz/2 \ , \tag{4.117}$$

which is valid for all first order transitions ($\beta = 0$) with $d \leq 2$, hence $d_c = 2$ is the upper critical dimension. It is also valid for all **compact growth processes** (where "compact" means that the density in surviving colonies remains finite as $t \to \infty$).

In **one dimension** at the upper terminal point of the DK SCA (Fig.4.7 $p_1 = \frac{1}{2}, p_2 = 1$) an extra $Z_2$ symmetry exists between 1's and 0's, hence the scaling behavior is not DP class like but corresponds to the fixed point of the inactive phase of PC class models (see Sect. 4.6). As a consequence *compact domains* of 0's and 1's grow such as the domain walls follow annihilating random walks (ARW) (see Sect. 4.5.1) and belong to the one dimensional VM class. In one dimension the compact directed percolation (CDP) is also equivalent to the $T = 0$ Glauber Ising model (see Sects. 4.10 and 2.3). When a non-zero temperature is applied (corresponding to spin-flips inside the domains) or symmetry is broken (by changing $p_2$ or adding an external magnetic field) a discontinuous transition (in terms of magnetization) takes place ($\beta = 0$).

The **crossover from DP to CDP** in one dimensions was studied numerically first by [Mendes *et al.* (1996)]. More recently [Lübeck (2006a)] compared numerical results with a recently performed field theoretical approach [Janssen (2005)] and confirmed the value $\phi = 2$ for the crossover exponent. From CDP point of view the particle-hole symmetry is broken for any positive vertical distance

$$\kappa = 2(1 - p_2) \tag{4.118}$$

from the termination point in the phase diagram (4.7). The other distance from the critical line is $\tau = 2\delta p_1$ (see Eq. 4.63). At the crossover point the steady state order parameter obeys for any positive value of $\lambda$

$$\rho(\tau, \kappa) \sim \lambda^{-\beta_{\mathrm{CDP}}} \tilde{r}(\tau\,\lambda, \kappa\,\lambda^{\phi}) \tag{4.119}$$

with $\beta_{\mathrm{CDP}} = 0$ and where $\phi$ denotes the crossover exponent. Setting $\lambda = \kappa^{-1/\phi}$ we obtain $\rho(\tau, \kappa) = \tilde{r}(\tau\,\kappa^{-1/\phi}, 1)$. Comparing this result with the steady state solution (4.62)

$$\rho(\tau, \kappa) = \frac{\tau/\kappa}{1 + \tau/\kappa} \tag{4.120}$$

we can identify the crossover exponent as well as the crossover scaling function

$$\phi_{\mathrm{MF}} = 1\,, \qquad \tilde{r}_{\mathrm{MF}}(x, 1) = \frac{x}{1 + x}\,. \tag{4.121}$$

Performing a field theoretical treatment, Janssen derived the crossover exponent [Janssen (2005)]

$$\phi = \begin{cases} \frac{2}{d} \text{ if } d < d_c = 2 \\ 1 \text{ if } d \geq d_c = 2\,. \end{cases} \tag{4.122}$$

Remarkably, these field theoretical results are expected to be exact since the involved diagrams of PRG can be summed up exactly.

In **two (and higher) dimensions** the $p = 1$ situation corresponds to the $p_1 = 3/4$, $p_2 = 1$ point in the phase diagram of $Z_2$ symmetric models (see Fig. 2.1). This model has a "duality" symmetry with coalescing random walks: going backward in time, the successive ancestors of a given spin follow the trail of a simple random walk; comparing the values of several spins shows that their associated random walks necessarily merge when they meet [Ligget (1985)]. This correspondence permits us to solve many aspects of the kinetics. In particular, the calculation of the density of interfaces $\rho_m(t)$ (i.e. the fraction of $+-$ nearest neighbor (n.n.) pairs) starting from random initial conditions of magnetization $m$, is ultimately given by the

Table 4.7 Critical exponents of VM classes.

| d | $\beta$ | $\beta'$ | $\gamma$ | $\nu_{\|\|}$ | $\nu_\perp$ | $Z$ | $\delta$ | $\eta$ | $\theta_g$ |
|---|---|---|---|---|---|---|---|---|---|
| 1 | 0.0 | 1 | 2 | 2 | 1 | 2 | 1/2 | 0 | 0.25 |
| 2 | 0.0 | 1 | 1 | 1 | 1/2 | 2 | 1 | 0 | 0.37 |

probability that a random walk initially at unit distance from the origin, has not yet reached it at time $t$. Therefore, owing to the recurrence properties of random walks, the VM shows coarsening for $d \leq 2$ (i.e. $\rho_m(t) \to 0$ when $t \to \infty$). For the the 'marginal' $d = 2$ case one finds the slow logarithmic decay [Scheucher and Spohn (1988); Krapivsky (1992); Frachebourg and Krapivsky (1996)]:

$$\rho_m(t) = (1 - m^2)\left[\frac{2\pi D}{\ln t} + \mathcal{O}\left(\frac{1}{\ln^2 t}\right)\right] , \tag{4.123}$$

with $D$ being the diffusion constant of the underlying random walk ($D = 1/4$ for the standard case of n.n., square lattice walks, when each spin is updated on average once per unit of time).

Simulating general, $Z_2$ symmetric spin-flip rules in two dimensions [Dornic *et al.* (2001)] conjectured that all critical $Z_2$-symmetric rules without bulk noise form a co-dimension-1 'voter-like' manifold separating order from disorder, characterized by the logarithmic decay of both $\rho$ and $m$. The critical exponents for this class are summarized in Table 4.7. Furthermore [Dornic *et al.* (2001)] found that this $Z_2$ symmetry is not a necessary condition, VM behavior can also be observed in systems without bulk fluctuations, where the total magnetization is conserved. Field theoretical understanding of these results are still lacking.

### 4.5.1 *The $2A \to \emptyset$ (ARW) and the $2A \to A$ models*

As mentioned in Sect. 4.5 in one dimension the annihilating random walk (ARW) (also called diffusion limited pair-annihilation) and the voter model are equivalent. In higher dimensions this is not the case (see Sect. 4.7). The simplest reaction-diffusion model — in which identical particles follow random walk and annihilate on contact of a pair — is adequately described by mean-field-type equations in $d_c > 2$ dimensions

$$\rho(t) \propto t^{-1} , \tag{4.124}$$

which is reaction-limited, but in lower dimensions fluctuations become relevant. Omitting boundary and initial condition terms, the field theoretical action is

$$S = \int d^d x dt \left[\psi(\partial_t \phi - D\nabla^2 \phi) - \lambda(1-\psi^2)\phi^2\right] , \tag{4.125}$$

where $D$ denotes the diffusion coefficient and $\lambda$ is the annihilation rate.

For $d = d_c = 2$ the leading order decay of the ARW was derived exactly by [Lee (1994)] using field theoretical RG method:

$$\rho(t) = \frac{1}{8\pi D} \ln(t)/t + O(1/t) \quad . \tag{4.126}$$

For $d = 1$ [Rácz (1985); Lushnikov (1987)] predicted that the particle density decays as

$$\rho(t) = A_2 (Dt)^{-1/2} \quad . \tag{4.127}$$

This scaling law was confirmed by $\epsilon$ expansion and the universal amplitude $A_2$ was found to be

$$\frac{1}{4\pi\epsilon} + \frac{2\ln 8\pi - 5}{16\pi} + O(\epsilon) \quad . \tag{4.128}$$

The universal scaling behavior of the ARW has been shown to be equivalent to that of the $A + A \to A$ coagulation random walk process (also called diffusion limited pair-annihilation below $d_c = 2$) by [Peliti (1986)]. The renormalization-group approach provided universal decay amplitudes (different from those of the ARW) to all orders in epsilon expansion. It was also shown [Domany and Kinzel (1984)] that the motion of kinks in the compact version of directed percolation (CDP) [Essam (1989)] and the Glauber-Ising model [Glauber (1963)] at the $T = 0$ transition point are also described exactly by Eq. (4.127). These reactions also have intimate relationship to the EW interface growth model (see Sect. 7.2). The asymptotic density decay in one dimension $\sim t^{-1/2}$ has been experimentally observed (in an intermediate time window) in the fusion kinetics of laser-induced excitons in quasi one-dimensional $N(CH_3)_4MnCl_3$ (TMMC) polymer chains [Kroon *et al.* (1993)].

In case of addition of an **external particle source** ($\emptyset \xrightarrow{J} A$) to the reactions

$$2A \to \emptyset \ , \ 2A \to A \ , \tag{4.129}$$

a PRG study [Droz and Sasvári (1993)] found that the upper critical dimension is unchanged ($d_c = 2$) and the steady state density density scales

$$\rho \propto J^{1/\delta_H} \tag{4.130}$$

with the field exponent

$$\delta_H = 1 + 2/d \ , \tag{4.131}$$

to all orders of $\epsilon = d - 2$. Meanwhile the characteristic relaxation time $t_R$ behaves as

$$t_R \propto J^{-2/(d+2)} \ , \tag{4.132}$$

hence $\nu_\perp = -2/(d+2)$.

### 4.5.2 *Compact DP (CDP) with spatial boundary conditions*

By introducing a wall in CDP, one alters the survival probability and obtains surface critical exponents just as for DP. With **IBC**, the cluster is free to approach and leave the wall, but not to cross it. For $d = 1$, this gives rise to $\beta_1' = 2$. On the other hand, for **ABC**, the cluster is stuck to the wall and therefore described by a single random walker for $d = 1$. By reflection in the wall, this may be viewed as *symmetric* compact DP, which has the same $\beta'$ as normal compact DP, giving $\beta_1' = 1$ [Essam and TanlaKishani (1994); Essam and Guttmann (1995)].

### 4.5.3 *CDP with parabolic boundary conditions*

Space-time boundaries are also of interest in CDP. Cluster simulations in 1+1-dimension and MF approximations [Ódor and Menyhárd (2000); Dickman and ben Avraham (2001)] for CDP confined by repulsive parabolic boundary condition of the form $x = \pm Ct^\sigma$ resulted in $C$ dependent $\delta$ and $\eta$ exponents (see Fig. 4.17) similarly to the DP case (see Sect. 4.2.6) with the marginal condition: $\sigma = 1/2$. In the mean-field approximations [Ódor and Menyhárd (2000)] similar results were obtained as for the DP [Kaiser and Turban (1995)]. Analytical results can be obtained only in limiting cases. For narrow systems (small $C$) one obtains the following asymptotic behavior for the connectedness function to the origin:

$$P(t,x) \sim t^{-\pi^2/8C^2} \cos\left(\frac{\pi x}{2C\sqrt{t}}\right) \ . \tag{4.133}$$

Recently an analytical solution was derived for a related problem [Dickman and ben Avraham (2001)]. For a one-dimensional lattice random walk with an absorbing boundary at the origin and a movable partial reflector (with probability $r$) $\delta$ varies continuously between 1/2 and 1 as $r$ varies between 0 and 1.

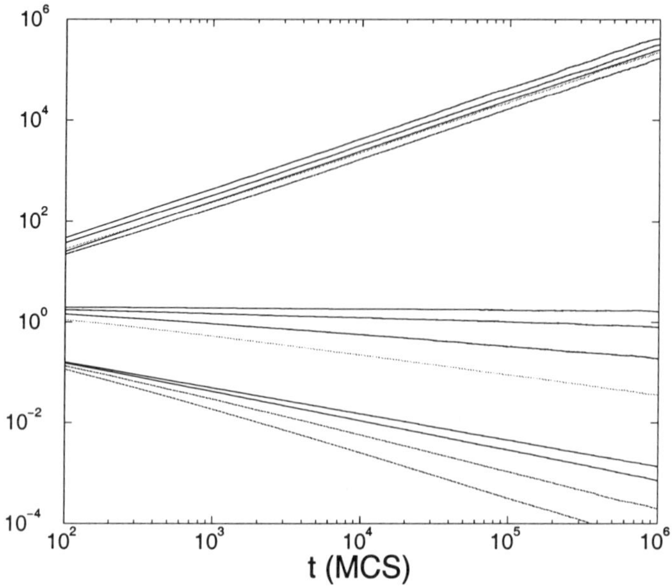


Fig. 4.17 Parabola boundary confinement cluster simulations for CDP [Ódor and Menyhárd (2000)]. Middle curves: number of active sites ($C = 2, 1.5, 1.2, 1$ top to bottom); Lower curves: survival probability ($C = 2, 1.5, 1.2, 1$ top to bottom); Upper curves: $R^2(t)$ ($C = 2, 1.5, 1.2, 1$ top to bottom). From [Ódor and Menyhárd (2000)].

### 4.5.4 *Lévy flight anomalous diffusion in ARW-s*

Long-range interactions generated by non-local diffusion in annihilating random walks result in the recovery of mean-field behavior. This has been studied by different approaches. Particles performing simple random walks subject to the reactions $A+B \to \emptyset$ (Sect. 4.5.1) and $A+A \to \emptyset$ (Sect. 6.1) in the presence of a quenched velocity field were investigated in [Zumofen and Klafter (1994)]. The quenched velocity field enhances the diffusion in such a way that the effective action of the velocity field is reproduced if Lévy flights are substituted for the simple random walk motion. In the above mentioned reactions the particle density decay is algebraic with an exponent related to the step length distribution of the Lévy flights defined in Eq. (4.78). These results have been confirmed by several renormalization-group calculations [Oerding (1996a); Deem and Park (1998); Park and Deem (1998)].

The $A + A \to \emptyset$ process with anomalous diffusion was investigated by field theory [Hinrichsen and Howard (1999)]. The action of this model is

$$S[\bar{\psi},\psi] = \int d^d x\, dt \Big\{ \bar{\psi}(\partial_t - D_N \nabla^2 - D_A \nabla^\sigma)\psi + 2\lambda\bar{\psi}\psi^2 + \lambda\bar{\psi}^2\psi^2 - n_0\bar{\psi}\delta(t) \Big\} , \qquad (4.134)$$

where $n_0$ is the initial (homogeneous) density at $t = 0$. The density decays for $\sigma < 2$ as

$$n(t) \sim \begin{cases} t^{-d/\sigma} & \text{for } d < \sigma , \\ t^{-1}\ln t & \text{for } d = d_c = \sigma , \\ t^{-1} & \text{for } d > \sigma . \end{cases} \qquad (4.135)$$

The simulation results of the corresponding $1+1$-dimensional lattice model [Hinrichsen and Howard (1999)] can be seen in Fig. 4.18. It was also shown

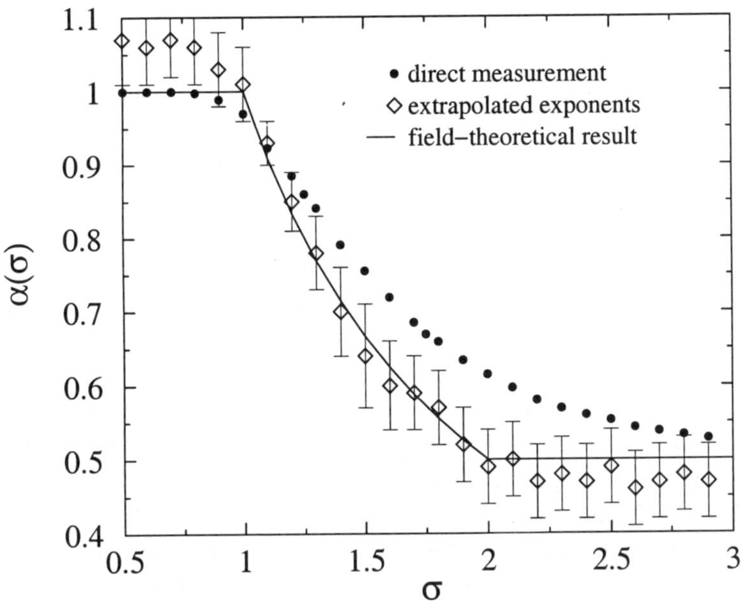


Fig. 4.18 The anomalous annihilation process: the graph shows direct estimates and extrapolations for the decay exponent $\alpha$, as a function of $\sigma$. The solid line represents the exact result (neglecting log corrections at $\sigma = 1$). From [Hinrichsen and Howard (1999)].

by [Hinrichsen and Howard (1999)] that Lévy flight **annihilation and coagulation processes** ($A + A \to A$) are in the same universality class.

Table 4.8 Cluster critical exponents of the Glauber Ising and it's different spin anisotropic versions. The anisotropy is increasing as: Glauber → SAGI → MDG. The hyperscaling law: $\eta + \delta = z/2$ is satisfied.

| Exponents | Glauber-Ising '+' | Glauber-Ising '-' | MDG '+' | MDG '-' | SAGI '+' | SAGI '-' |
|---|---|---|---|---|---|---|
| $\eta$ | 0 | 0 | 0.0 | 0.5 | 1 | 0 |
| $\delta$ | 1/2 | 1/2 | 0.5 | 0.0 | 0 | 0 |
| $z$ | 1 | 1 | 1.0 | 1.0 | 2 | 0 |

### 4.5.5 *ARW with anisotropy*

When we add biased diffusion (drift), to the the pair annihilation/coagulation model the corresponding action will be

$$S = \int dtdx \left[\psi(\partial_t - D\nabla^2 + \mathbf{v}\phi + \lambda_1\psi\phi^2 + \lambda_2\psi^2\phi^2\right] \ , \tag{4.136}$$

where $D$ is the diffusion constant, $\mathbf{v}$ is the drift velocity, and $\lambda_1$, $\lambda_2$ are properly scaled reaction parameters. Exact results for **one-dimensional** reaction-diffusion models $A + A \to \emptyset$ and $A + A \to A$ showed no changes in the universal properties as the result of the anisotropy [Privman *et al.* (1995)]. Therefore the Galilean invariance

$$\phi(t, \mathbf{x}) \to \phi(t, \mathbf{x} - \mathbf{v}t) \tag{4.137}$$

$$\psi(t, \mathbf{x}) \to \psi(t, \mathbf{x} - \mathbf{v}t) \tag{4.138}$$

holds in such simple one component RD models.

Since a one-dimensional ARW can be mapped onto the zero temperature Glauber-Ising model searching for the effects of **spin anisotropy** (anisotropy in the spin-flip rates) was proposed in [Menyhárd and Ódor (2002)]. This has been done within the framework of NEKIMA (see Sect. 4.6.9), since ARW is the dominant process of that model in the absorbing phase. Due to these anisotropies the behavior of '+' and '−' spins become decoupled, one obtains effectively two-component models, with novel critical scaling behavior in the cluster quantities [Menyhárd and Ódor (2003)]. Furthermore simulations showed, that this kind of anisotropy enhances the sensitivity to disorder [Menyhárd and Ódor (2007)]. The space-time evolution of such model can be seen on (Fig. 4.19). The clusters are compact as in case of CDP, but they exhibit different scaling exponents (see Table 4.8).

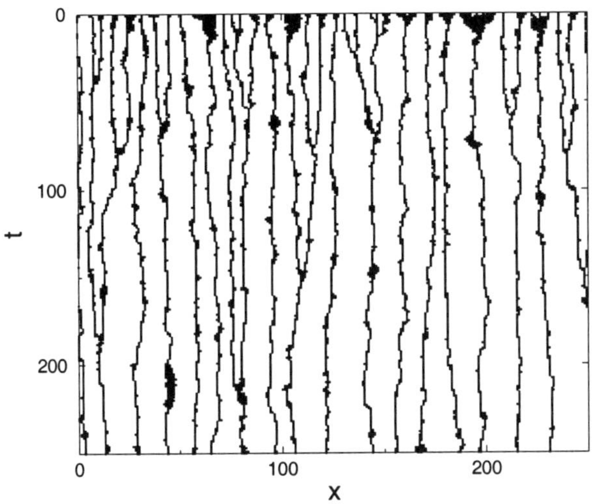


Fig. 4.19 Space-time development of '+' (white) and '−' (black) spins evolving from a random initial state for parameters $p_+ = 0.1$, $p_{ex} = 0$ of NEKIMA. From [Menyhárd and Ódor (2003)].

### 4.5.6 *ARW with quenched disorder*

In the one-dimensional ARW model ($\nu_\perp = 1$), thus Harris criterion (1.91) predicts relevant spatial disorder. The renormalization study of ARW with spatial **randomness in the reaction rates** found marginal perturbations to the fixed point of the pure system [Doussal and Monthus (1999)]. On the other hand an exact study of the infinite reaction rate ARW with **space-dependent hopping rates** obtained non-universal power-law decay of the density of $A$-s below a critical temperature [Schütz and Mussawisade (1998)].

A PRG study [Richardson and Cardy (1999)] arrived at conclusions for the bosonic ARW with **short-ranged, weakly quenched diffusion disorder**. Ray and Cardy The showed that for $d < d_c = 2$ the process becomes super-diffusive, with dynamic exponent

$$Z = 2 + 2\epsilon^2 + O(\epsilon^3) , \tag{4.139}$$

hence the density decay

$$\rho \propto \frac{1}{Dt^{d/Z}} \tag{4.140}$$

slows down in one dimension, in agreement with the numerical findings presented later. The effect of **long-ranged potential disorder** [Park and Deem (1998)] (when the random field velocity can be considered as a gradient of a random potential) is more drastic, it even changes the density decay exponent in **two dimensions** in a non-universal manner.

A simulation study applied quenched, power-law distributed hopping probabilities

$$p(y) = \nu y^{\nu-1} \tag{4.141}$$

with $0 < \nu \leq 1$ and $0 < y \leq 1$ in **one dimension** to the bosonic ARW model [Lee (2000)]. Similarly to the analytical results the density decay exponent was found to be a continuously changing function of the disorder

$$\alpha = \nu/(\nu + 1) \ . \tag{4.142}$$

Another recent, extensive simulation study [Menyhárd and Ódor (2006)] investigated the site-restricted ARW in **one dimension** (within the framework of the NEKIMCA model (see Sect. 4.6.3)) with uniformly distributed, quenched disorder $\chi(j)$ of the form $\chi(j) \in (-\epsilon, \epsilon)$ (with amplitude $\epsilon$), that is added to the reaction/diffusion rates. Contrary to the expectation following from the Harris criterion, an insensitivity to weak disorder was found. Only in case of very strong disorder — where diffusion could be blockaded at certain sites — did the simulations find logarithmic corrections to the scaling of the pure model.

Such blocking or **complete dilution** was studied in a one-dimensional toy model of random, quantum, ferromagnetic Ising model [Fisher (1999)], in which continuously variable power-laws were found at the phase transition point. The effect of disconnected domains in the reactions of ARW has been investigated in [Mandache and ben Avraham (2000)]. This study reported **stretched exponential decay** in case of exponential domain size distributions and continuously changing density decay for blocks distributed in a power-law manner.

## 4.6 Parity conserving (PC) classes

In an attempt to generalize DP and CDP like systems we turn to new models, in which a conservation law is relevant. A new universality class appears among 1+1-dimensional, single component, reaction-diffusion models. Although it is usually named parity conserving class (PC) examples have proved that the parity conservation alone is not a sufficient condition for the behavior of this class. For example [Inui *et al.* (1995)] showed that a one dimensional stochastic cellular automaton with *global parity conservation exhibited DP class transition.* The binary spreading process (see Sect. 6.7) in one [Park *et al.* (2001)] and two dimensions [Ódor *et al.* (2002)] were also found to be insensitive to the presence of parity conservation. Multi-component BARW2 models in one dimension (see Sect. 6.12) generate different, robust classes again [Cardy and Täuber (1996); Ódor (2001b); Hooyberghs *et al.* (2001)]. Today it is known that the BARW2 dynamics in single-component, single-absorbing-state systems (without inhomogeneities, long-range interactions or other symmetries) provide sufficient conditions for the PC class [Cardy and Täuber (1996)]. In single-component, multiabsorbing state systems, the $Z_2$ symmetry is a necessary but not sufficient condition — the BARW2 dynamics (Sect. 4.6.1) of domain walls is also a necessary condition. Some studies have shown [Park and Park (1995); Menyhárd and Ódor (1996); Hwang *et al.* (1998)] that an external field that destroys the $Z_2$ symmetry of absorbing states (but preserves the the BARW2 dynamics) yields a DP instead of a PC class transition in the system. Some other names for this class are also used, like the directed Ising (DI) class, the generalized VM, or the BARW class.

### 4.6.1 *Branching and annihilating random walks with even number of offspring (BARWe)*

Generic models of the PC class, for which field theoretical treatment exists are the branching and annihilating random walks — introduced in Sect. 4.2.4 — with $m = 2$ and even number ($k$) of offspring (BARWe) [Zhong and Avraham (1995); Lipowski (1996); Janssen (1997b); ben Avraham *et al.* (1994); Jensen (1993c, 1994)]. These conserve the particle number modulo 2 and hence have two distinct, an odd- and an even-parity one. In the even sector particles finally die out ($\delta \neq 0$, $\eta = 0$), while in the odd one at least one particle always remain alive ($\delta = 0$, $\eta \neq 0$). Considering the simplest possible PC reaction set

$$A \xrightarrow{\sigma} 3A\,, \qquad 2A \xrightarrow{\lambda} \emptyset\,, \tag{4.143}$$

one can construct the corresponding reaction Hamiltonian described in Sect. 1.6.1

$$H_R = (up - v\,q)\,(p^2 - 1)\,q\,, \tag{4.144}$$

where the bare values of the coupling constants are given by $u = \sigma$ and $v = \lambda/2$. This Hamiltonian (and the corresponding action) is invariant under the transformations

$$p \to -p\,, \qquad q \to -q\,. \tag{4.145}$$

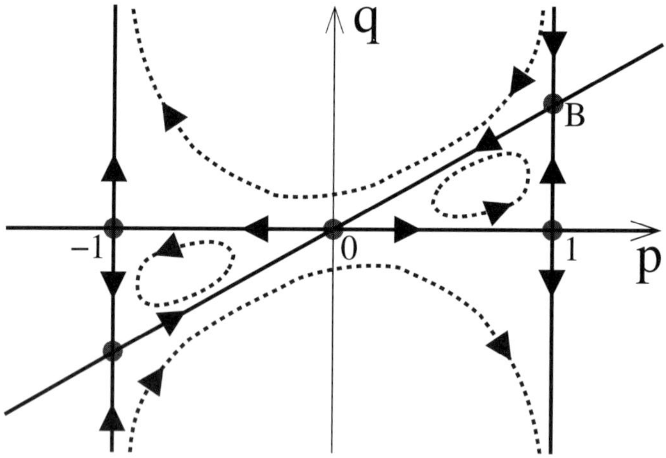


Fig. 4.20 Phase portrait of the parity conserving model, Eq. (4.143). Notice the reflection symmetry around the origin. From [Elgart and Kamenev (2006)].

As a result the phase portrait of (4.144) shown on Fig. 4.20, possesses a reflection point at the origin. One can show that this symmetry is preserved under RG transformations. Therefore the "accidental" zero-energy line [$q = up/v$, according to Eq. (4.144)] is an odd function and passes through the origin. Although its shape may change in the process of renormalization, the phase transition cannot be described by the coalescence of three points as in case of for BARWo (see lines in $A$, $B$, and $C$ in Fig. 4.3), hence one may expect that it is different from that of the DP class.

BARWe dynamics may also appear in spin or multi-component particle models exhibiting $Z_2$ symmetric absorbing states in terms of the kinks between ordered domains. Examples of such systems are the NEKIM (Sect. 4.6.2) and the GDK model (Sect. 4.6.3) for example. It has been conjectured [Menyhárd and Ódor (2000)] that in all models with $Z_2$ symmetric absorbing states an underlying BARWe process is a necessary condition for phase transitions with PC criticality. Sometimes it is not so easy to find the underlying BARWe process, which can bee seen on the coarse grained level only (in the GDK model for instance the kinks are spatially extended objects). This might have lead some studies to the conjecture that $Z_2$ symmetry is a sufficient condition for the PC class [Hwang and Park (1999)]. However the example of CDP (see Sect. 4.5) shows that this cannot be true.

According to the mean-field equation

$$\dot{\rho} = \sigma\rho - \lambda\rho^2 \tag{4.146}$$

the model is always in the active phase above the upper critical dimension (for any non-zero branching). The upper critical dimension and can be computed from Eq. (4.30):

$$d_c = 2 \ . \tag{4.147}$$

The mean-field exponents derived by power-counting (see Sect. 4.1.1) are tabulated in Table 4.9. The field theory of BARWe models was investigated by [Cardy and Täuber (1996, 1998)]. For this case the action

$$S = \int d^d x dt \ [\ \psi(\partial_t - D\nabla^2)\phi - \lambda(1-\psi^2)\phi^2 + \\ + \sigma(1-\psi^2)\psi\phi\ ] \tag{4.148}$$

is invariant under the simultaneous transformation of fields

$$\psi \leftrightarrow -\psi, \quad \phi \leftrightarrow -\phi \ . \tag{4.149}$$

Owing to the nonrecurrence of random walks in $d \geq 2$ the system is in the active phase for $\sigma > 0$ and mean-field transition occurs with

$$\beta_{MF} = 1 \ . \tag{4.150}$$

However the survival probability of a particle cluster is finite for any $\sigma > 0$ implying

$$\beta'_{MF} = 0 \ . \tag{4.151}$$

Hence in contrast with the DP class $\beta \neq \beta'$ for $d \geq 2$. At $d = d_c = 2$ random walks are barely recurrent and logarithmic corrections arise. In this case the generalized hyperscaling law [Mendes *et al.* (1994)] is valid among the cluster exponents

$$2\left(1+\frac{\beta}{\beta'}\right)\delta + 2\eta = dz. \tag{4.152}$$

In $d = 1$ however, owing to an exact duality mapping [Mussawisade *et al.* (1998)]

$$\beta = \beta' \tag{4.153}$$

holds, and the hyperscaling relation is the same as that of the DP (Eq. 4.43).

The perturbative RG analysis of BARWe for $d < 2$ run into difficulties. These stem from, the presence of another critical dimension

$$d'_c = 4/3 \tag{4.154}$$

above which the branching reaction is relevant at $\sigma = 0$, (and irrelevant for $d < d'_c$). Hence the $d = 1$ dimension cannot be accessed by controlled expansions from $d_c = 2$. The truncated one-loop expansions [Cardy and Täuber (1996)] for $d = 1$ resulted in the exponents

$$\beta = 4/7, \; \nu_\perp = 3/7, \; \nu_{||} = 6/7, \; Z = 2 \; , \tag{4.155}$$

which are quite far from the numerical values determined by Jensen's simulations [Jensen (1994)] (Table 4.9). Here the cluster exponents $\delta$ and $\eta$

Table 4.9 Critical exponents of PC classes, see also Table 4.10

| $d$ | $\beta$ | $\beta'$ | $\gamma$ | $\delta$ | $Z$ | $\nu_{\|\|}$ | $\nu_\perp$ | $\eta$ |
|---|---|---|---|---|---|---|---|---|
| 1 | 0.92(3) | 0.92(3) | 0.00(5) | 0.285(2) | 1.75 | 3.25(10) | 1.85(5) | 0.000(1) |
| 2 | 1 | 0 | 1 | 0 | 2 | 1 | 1/2 | -1/2 |

corresponding to the sector with even number of initial particles are shown. In case of odd number of initial particles they exchange values. Further dynamical exponents for the one-dimensional PC class are displayed in Table 4.10. The PRG study has pointed out that the whole class of parity conserving models

$$A \to (k+1)A \quad 2A \to \emptyset \tag{4.156}$$

with $k$ even, exhibits phase transitions belonging to the PC class. One can easily check that the phase portraits of such models are topologically identical to that of Fig. 4.20 indeed.

More recently the NPRG method (see Sect. 1.6) has been applied for BARWe and the fixed point structure has been determined in the local potential approximation (LPA) [Canet *et al.* (2005)]. The flow diagram can be seen on Fig. 4.21, which shows the appearance a non-perturbative $F^*$ fixed point corresponding to the PC class. By applying truncations in the local potential approximation the best estimate for the correlation length exponent $\nu_\perp = 2.0017$ was found, which is just less than 10% away from the simulation results [Jensen (1994)] 1.84(6) and the best analytic GMF+CAM estimates (based on $n = 12$ clusters) [Ódor and Szolnoki (2005)] : 1.85(3).

Another interesting result of the NPRG study is that due to the (4.149) symmetry, the effective potential $u(\phi, \psi)$ can be built up from three quadratic invariant quantities, $\phi^2$, $\psi^2$, and $\phi\psi$. The action (4.148) can be expressed in terms of these invariants and the potential is proportional to $1 - \psi^2$ and vanishes for $\phi = 0$. This structure is preserved by the renormalization flow, which further constrains the functional subspace in which the running potential evolves:

$$u(\phi, \psi) = (1 - \psi^2)\mathcal{F}(\phi^2, \phi\psi) \;\text{ with }\; \mathcal{F}(0,0) = 0 \; . \tag{4.157}$$

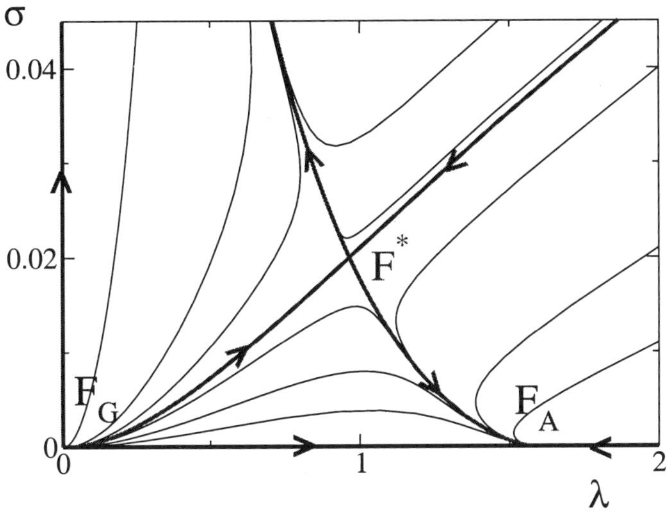


Fig. 4.21 Flow diagram in the lowest-order local potential approximation of NPRG for the $d = 1$ BARWe (as usual, arrows represent the RG evolution as (the cutoff scale) is decreased towards the "infra-red", macroscopic limit $k \to 0$). $F_A$ is the pure annihilation fixed point, $F_G$ is the Gaussian one, while $F^*$ denotes the PC class fixed point. From [Canet *et al.* (2005)].

Interestingly these invariants are just those of the topological phase space method of [Elgart and Kamenev (2006)] (using the $\phi \to q$ , $\psi \to p$ transcription), while the $1 - \psi^2 = 0$, $\phi = 0$ conditions correspond to the zero energy trajectories.

It was conjectured [Deloubriére and van Wijland (2002)] that in one dimension fermionic (single occupancy) and bosonic (multiple-occupancy) models may have different critical behaviors. Since only bosonic field theory exists, which gives rather inaccurate critical exponent estimates [Ódor and Menyhárd (2002)] performed bosonic simulations to investigate the density decay of BARW2 from random initial state. Figure 4.22 shows the local slopes of density decay ($\alpha_{eff}$ (4.50)) around the critical point for several branching rates ($\sigma$). The critical point is estimated at $\sigma_c = 0.04685(5)$, with the corresponding decay exponent $\alpha = 0.290(3)$. This value agrees with that of the PC class.

If there is no explicit diffusion of particles an implicit can still be generated via the reactions: $AA \to \emptyset$, $A \to 3A$ processes, which is called DBAP model by [Sudbury (1990)]. Using spatially asymmetric branching: $A\emptyset\emptyset \to AAA$ and $\emptyset\emptyset A \to AAA$ a diffusion process may go on by two lattice

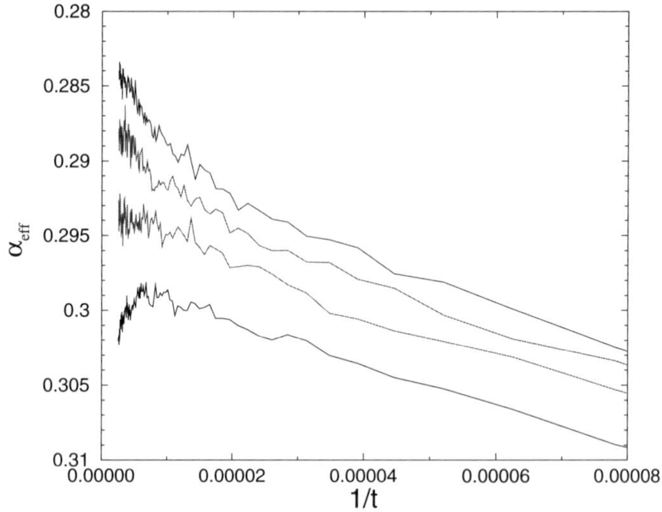


Fig. 4.22 Local slopes (4.50) of the density decay in a bosonic BARW2 model. Different curves correspond to $\sigma = 0.466$, 0.468 0.469, 0.47 (from bottom to top). From [Ódor and Menyhárd (2002)].

steps: $A\emptyset\emptyset \to AAA \to \emptyset\emptyset A$. As a consequence a single particle can not join a domain, the decay process slows down, and domain sizes exhibit a parity conservation. This results in additional **new sectors** (besides the existing two BARW2 sectors) that depend on the initial conditions. For example in case of random initial distribution $\delta \sim 0.13(1)$ was measured by simulations [Hinrichsen and Ódor (1999a)]. Similar sector decomposition has been observed in diffusion of $k$-mer models (see for example [Barma *et al.* (1993); Barma and Dhar (1994)]).

### 4.6.2 *The NEKIM model*

Another important representative of PC class appear among nonequilibrium Ising models, in which the steady state is generated by kinetic processes in connection with heat baths at different temperatures [DeMasi *et al.* (1985, 1986); Gonzaléz-Miranda *et al.* (1987); Wang and Lebowitz (1988); Droz *et al.* (1989)]. First it was shown that phase transition is possible even in one dimension under nonequilibrium conditions (for a review see [Rácz (1996)]). In short-ranged interaction models any nonzero temperature spin-flip dynamics cause disordered steady state. Menyhárd proposed a class of general nonequilibrium kinetic Ising models (NEKIM) with combined spin-flip dynamics at $T = 0$ and Kawasaki spin-exchange dynamics at $T = \infty$, in which, for a range of parameters, a PC-type transition takes place [Menyhárd (1994)].

A general form [Glauber (1963)] of the Glauber spin-flip transition rate in one-dimension for spin $s_i = \pm 1$ sitting at site $i$ is:

$$w_i = \frac{\Gamma}{2}(1 + \tilde{\delta} s_{i-1} s_{i+1}) \left(1 - \frac{\tilde{\gamma}}{2} s_i (s_{i-1} + s_{i+1})\right) . \tag{4.158}$$

Here $\tilde{\gamma} = \tanh(2J/kT)$, $J$ denotes the coupling constant in the ferromagnetic Ising Hamiltonian and $\Gamma$ and $\tilde{\delta}$ are further parameters, which can in general, also depend on temperature. The Glauber model is a special case corresponding to $\tilde{\delta} = 0$, $\Gamma = 1$. There are three independent rates:

$$\begin{aligned} w_{\uparrow\uparrow\uparrow} &= \frac{\Gamma}{2}(1 + \tilde{\delta})(1 - \tilde{\gamma}), \qquad w_{\downarrow\uparrow\downarrow} = \frac{\Gamma}{2}(1 + \tilde{\delta})(1 + \tilde{\gamma}) \\ w_{\uparrow\uparrow\downarrow} &= \frac{\Gamma}{2}(1 - \tilde{\delta}). \end{aligned} \tag{4.159}$$

In the NEKIM model $T = 0$ is taken, thus $\tilde{\gamma} = 1$, $w_{\uparrow\uparrow\uparrow} = 0$ and $\Gamma$, $\tilde{\delta}$ are the control parameters to be varied.

The Kawasaki spin-exchange rate of neighboring spins is:

$$w_{ii+1}(s_i, s_{i+1}) = \frac{p_{ex}}{2}(1 - s_i s_{i+1})[1 - \frac{\tilde{\gamma}}{2}(s_{i-1} s_i + s_{i+1} s_{i+2})]. \tag{4.160}$$

At $T = \infty$ ($\tilde{\gamma} = 0$) the above exchange is simply an unconditional nearest neighbor exchange:

$$w_{ii+1} = \frac{1}{2} p_{ex} [1 - s_i s_{i+1}] \tag{4.161}$$

where $p_{ex}$ is the probability of spin-exchange. The transition probabilities in Eqs. (4.158) and (4.161) are responsible for the basic elementary processes of kinks ($K$). Kinks separating two ferromagnetically ordered domains can carry out random walks with probability

$$p_{rw} \propto 2 w_{\uparrow\uparrow\downarrow} = \Gamma(1 - \tilde{\delta}) , \tag{4.162}$$

while two kinks getting into neighboring positions will annihilate with probability

$$p_{an} \propto w_{\downarrow\uparrow\downarrow} = \Gamma(1 + \tilde{\delta}) . \tag{4.163}$$

Here $w_{\uparrow\uparrow\uparrow}$ is responsible for creation of kink pairs inside of ordered domains at $T \neq 0$. In the case of the spin-exchanges, which act only at domain boundaries, the process of main importance here is that a kink can produce two offspring at the next time step with probability

$$p_{K\to 3K} \propto p_{ex}. \tag{4.164}$$

The above-mentioned three processes compete, and it depends on the values of the parameters $\Gamma$, $\tilde{\delta}$ and $p_{ex}$ what the result of this competition will be. It is important to realize that the process $K \to 3K$ can develop into propagation of offspring only if $p_{rw} > p_{an}$, i.e., the new kinks are able to travel some lattice points away from their place of birth and can thus avoid immediate annihilation. It is seen from the above definitions that $\tilde{\delta} < 0$ is necessary for this to happen. In the opposite case the only effect of the $K \to 3K$ process on the usual Ising kinetics is to soften domain walls. Note, that in the NEKIM model the normalization condition $p_{rw}+p_{an}+p_{k\to 3k} = 1$ was set.

The phase diagram determined by simulations and GMF calculations [Menyhárd and Ódor (1995, 2000)] is shown in Fig. 4.23. The line of phase

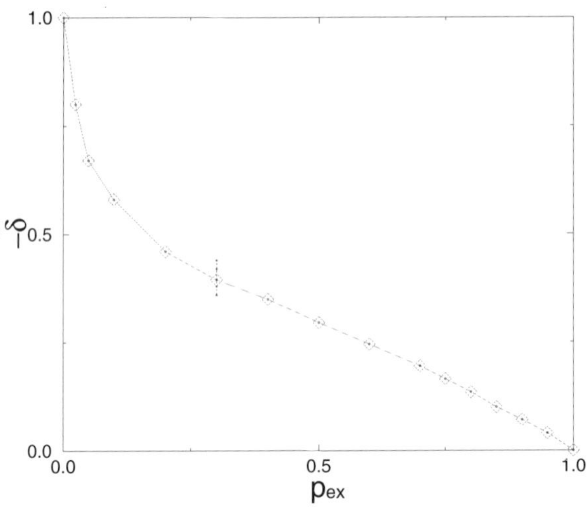


Fig. 4.23 Phase diagram of the two-parameter model. The transversal dotted line indicates the critical point that was investigated in more detail. From [Menyhárd and Ódor (2000)].

transitions separates two kinds of steady states reachable by the system for large times: in the Ising phase, supposing that an even number of kinks are present in the initial states, the system orders in one of the possible ferromagnetic states of all spins up or all spins down, while the active phase is disordered from the point of view of the underlying spins. The cause of disorder is the steadily growing number of kinks with time. While the low-level, $N = 1, 2$ GMF solutions for the SCA version of NEKIM exhibit first order transitions, for $N > 2$ this becomes continuous. GMF approximations (up to $N = 6$) with CAM extrapolation found $\beta \simeq 1$ [Menyhárd and Ódor (1995)]. Recent high precision Monte Carlo simulations [Menyhárd and Ódor (2000)] resulted in critical exponents $\beta = 0.95(2)$ (see Fig. 4.24) and $\delta = 0.280(5)$ at the dotted line of the phase diagram.

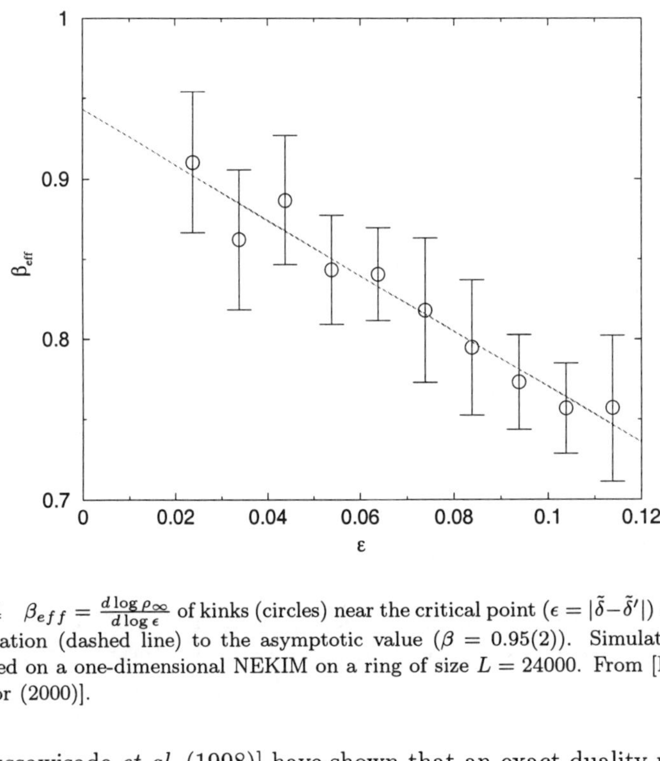


Fig. 4.24 $\beta_{eff} = \frac{d \log \rho_\infty}{d \log \epsilon}$ of kinks (circles) near the critical point ($\epsilon = |\tilde{\delta} - \tilde{\delta}'|$) and linear extrapolation (dashed line) to the asymptotic value ($\beta = 0.95(2)$). Simulations were performed on a one-dimensional NEKIM on a ring of size $L = 24000$. From [Menyhárd and Ódor (2000)].

[Mussawisade *et al.* (1998)] have shown that an exact duality mapping exist in the phase diagram of the NEKIM:

$$\begin{aligned} p'_{an} &= p_{an}, \\ p'_{rw} &= p_{an} + 2p_{ex}, \\ 2p'_{ex} &= p_{rw} - p_{an} \ . \end{aligned} \tag{4.165}$$

Table 4.10 Simulation data for static and dynamic critical spin exponents for NEKIM.

| | $\beta_s$ | $\gamma_s$ | $\nu_{\perp,s}$ | Z | $\theta_g$ | $\lambda_s$ |
|---|---|---|---|---|---|---|
| Glauber-Ising | 0 | 1/2 | 1/2 | 2 | 1/4 | 1 |
| PC | 0.00(1) | 0.444(2) | 0.444(2) | 1.75(1) | 0.67(1) | 1.50(2) |

The regions mapped onto each other have the same physical properties. In particular, the line $p_{ex} = 0$ maps onto the line $p_{rw} = p_{an}$ and the fast-diffusion limit to the limit $p_{ex} \to \infty$. There is a self-dual line at

$$\tilde{\delta} = \frac{-2p_{ex}}{1 - p_{ex}} \quad . \tag{4.166}$$

By various static and dynamical simulations spin and kink density critical exponents have been determined in [Menyhárd and Ódor (1996)] and as the consequence of the generalized hyperscaling relation for the structure factor

$$S(0,t) = L[< M^2 > - < M >^2] \propto t^x, \tag{4.167}$$

and kink density

$$n(t) = \frac{1}{L} < \sum_i \frac{1}{2}(1 - s_i s_{i+1}) > \propto t^{-y} \tag{4.168}$$

the exponent scaling relation

$$2y = x \tag{4.169}$$

is established. Spins-clusters at the PC point grow by compact domains as in the Glauber point, albeit with different exponents [Menyhárd and Ódor (1998)]. The spin-cluster critical exponents in a magnetic field are summarized in Table (5.1). By applying an external magnetic field $h$ that breaks the $Z_2$ symmetry the transition type of the model changes to DP type (see Table 4.12) [Menyhárd and Ódor (1996)]. The **global persistence** ($\theta_g$) and time autocorrelation exponents ($\lambda$) were determined both at the Glauber and at the PC critical points [Menyhárd and Ódor (1997)] and are shown in Table (4.10). While at the Glauber point the scaling relation (1.31) is satisfied by these exponents it is not the case at PC criticality, therefore **the magnetization is non-Markovian process here.**

An other possible variant of NEKIM was introduced by [Menyhárd and Ódor (2000)] in which the Kawasaki exchange Eq. (4.160) is considered at **finite temperature**. When the temperature is lowered the spin-exchange process acts against the kink production and a PC class transition occurs.

In this case the active phase part of the phase diagram shrinks. For more details see ([Ódor *et al.* (1999)]).

The **DS transition** of this model coincides with the critical point and the scaling behavior of spin and kink damage is the same as that of the corresponding NEKIM variables [Ódor and Menyhárd (1998)].

### 4.6.3 *Parity conserving, stochastic cellular automata*

The first models in which a non-DP class transition to absorbing state was firmly established were one dimensional stochastic cellular automata defined by [Grassberger *et al.* (1984)]. In these models the 00 and 11 pairs follow BARW2 dynamics and their density is the order parameter that vanishes at some critical point. The PC class critical exponents estimated by simulations for these models by [Grassberger (1989b)]. While in the **"A" model** the critical point coincides with the DS transition point and both of them are PC type, in the **"B" model** the DS transition occurs in the active phase — where the symmetry of replicas is broken — and therefore the DS exponents belong to the DP class [Ódor and Menyhárd (1998)].

Another stochastic CA, which may exhibit PC class transition and which is studied later from different directions is also introduced here. It points out that the underlying BARW2 dynamics of domain walls in $Z_2$ symmetric systems can sometimes be seen on the coarse grained level only. This generalization of the Domany-Kinzel stochastic cellular automaton [Domany and Kinzel (1984)] (see Sect. 4.2.3) called **(GDK)** was introduced by [Hinrichsen (1997)]. This model has $n+1$ states per site: one active state $Ac$ and $n$ different ('colored') inactive states $I_1, I_2, \ldots, I_n$. The conditional updating probabilities are given by $(k, l = 1, \ldots, n;\ k \neq l)$

$$P(I_k \mid I_k, I_k) = 1\,, \tag{4.170}$$

$$P(Ac|Ac, Ac) = 1 - n\, P(I_k|Ac, Ac) = q\,, \tag{4.171}$$

$$\begin{aligned} P(Ac|I_k, Ac) = P(Ac|Ac, I_k) &= p_k\,, \\ P(I_k|I_k, Ac) = P(I_k|Ac, I_k) &= 1 - p_k\,, \end{aligned} \tag{4.172}$$

$$P(Ac|I_k, I_l) = 1\,, \tag{4.173}$$

and the symmetric case $p_1, \ldots, p_n = p$ was explored. Equations (4.170)–(4.172) are straightforward generalizations of Eqs. (4.58)–(4.60). The only different process is the creation of active sites between two inactive domains of different colors in Eq. (4.173). For simplicity the probability of this process was chosen to be equal to 1.

For $n = 1$ the model defined above reduces to the original Domany-Kinzel model. For $n = 2$ it has two $Z_2$ symmetrical absorbing states. The phase diagram of this model is very similar to that of the DK model (see Fig. 4.7) except the transition line is of the PC type. If we call the regions separating inactive domains $I_1$ and $I_2$ as domain walls (denoted by $K$), they follow BARW2 process

$$K \to 3K \quad 2K \to \emptyset \quad K\emptyset \leftrightarrow \emptyset K \tag{4.174}$$

but for $0 < q < 1$ the size of active regions, hence the domain walls remains finite. Therefore a the observation of the BARW2 process is not so easy and can be done on coarse grained level only (except at the endpoint at $q = 0$, where active sites really look like kinks of the NEKIM model). Series expansions for the transition point and for the order parameter critical exponent resulted in $\beta = 1.00(5)$ [Jensen (1997)], which is slightly higher than the most precise simulation results [Menyhárd and Ódor (2000)] but agrees with the estimates of GMF+CAM analysis [Menyhárd and Ódor (1995)]. Similarly to the NEKIM the application of an external symmetry breaking field changes the PC class transition into a DP class [Hinrichsen (1997)]. The other symmetrical end point ($q = 1$, $p = \frac{1}{2}$) in the phase diagram again shows different scaling behavior. Here three types of compact domains grow in competition, and the boundaries perform annihilating random walks with exclusions (see Sect. 6.2).

Models with $n > 3$ symmetric absorbing states in one dimension do not show phase transitions (they are always active). In terms of domain walls as particles they are related to $N > 1$ component N-BARW2 processes, which exhibit a phase transition only for zero branching rate (see Sect. 6.12) [Cardy and Täuber (1996); Hooyberghs *et al.* (2001)].

GDK type models — exhibiting $n$ symmetric absorbing states — can be generalized to **higher dimensions**. In two dimensions Hinrichsen's spreading simulations for the $n = 2$ case yielded mean-field like behavior with $\delta = 1$, $\eta = 0$ and $z = 1$ leading to the conjecture that $1 < d_c < 2$. A similar model exhibiting Potts-like $Z_n$ symmetric absorbing states in $d = 2$ yielded similar spreading exponents but a first order transition ($\beta = \alpha = 0$) for $n = 2$ [Lipowski and Droz (2002a)]. In **three dimensions** the same model seems to exhibit mean-field like transition with $\beta = 1$. The verification of these findings would require further research.

The probabilistic cellular automaton version of NEKIM the **(NEKIMCA)** [Menyhárd and Ódor (1995, 2000)], which consists in keeping the spin-flip rates given in Eqs. (4.159) and prescribing *synchronous updating*. It is easy to show that the dual variables of spins ($\uparrow$) the kinks ($\bullet$) exhibit BARW2 dynamics via the synchronous spin-flip dynamics. In this SCA the parity conserving kink branching is also generated due to the synchronous spin-update of neighboring sites (without the need of an additional, explicit spin-exchange reaction as in the NEKIM model). The reactions are the followings:

- Random walk : $\uparrow\ \ \uparrow\bullet\downarrow \xrightarrow{w_i} \uparrow\bullet\downarrow\ \ \downarrow$
- Annihilation : $\uparrow\bullet\downarrow\bullet\uparrow \xrightarrow{w_o} \uparrow\ \ \uparrow\ \ \uparrow$
- Branching : $\uparrow\ \ \uparrow\bullet\downarrow\ \ \downarrow \xrightarrow{w_i^2} \uparrow\bullet\downarrow\bullet\uparrow\bullet\downarrow$

In the NEKIMCA there are two independent variables, parameterized as

$$w_i = \Gamma(1-\tilde{\delta})/2 \tag{4.175}$$

$$w_o = \Gamma(1+\tilde{\delta})\,. \tag{4.176}$$

The phase boundary of NEKIMCA in the $(\Gamma, -\tilde{\delta})$ plane is similar to that shown in Fig. 4.23, except that the highest value of $\Gamma = 1$, $\tilde{\delta}_c = 0$ cannot be reached by simulations.

### 4.6.4 *PC class surface catalytic models*

In this section I discuss some one-dimensional, surface catalytic-type reaction-diffusion models exhibiting PC class transition. Strictly speaking they are multi-component models, but I show that the symmetries among species enable us to interpret the domain-wall dynamics as a simple BARW2 process.

The two-species monomer-monomer (MM) model was first introduced by [Zhuo *et al.* (1993)]. Two monomers, called $A$ and $B$, adsorb at the vacant sites of a one-dimensional lattice with probabilities $p$ and $q$, respectively, where $p + q = 1$. The adsorption of a monomer at a vacant site is affected by monomers present on neighboring sites. If either neighboring site is occupied by the same species as that trying to adsorb, the adsorption probability is reduced by a factor $r < 1$, mimicking the effect of a nearest-neighbor repulsive interaction. Unlike monomers on adjacent sites react immediately and leave the lattice, leading to a process limited by adsorption only. The basic reactions are

$$\emptyset \to A, \quad \emptyset \to B, \quad AB \to \emptyset\,. \tag{4.177}$$

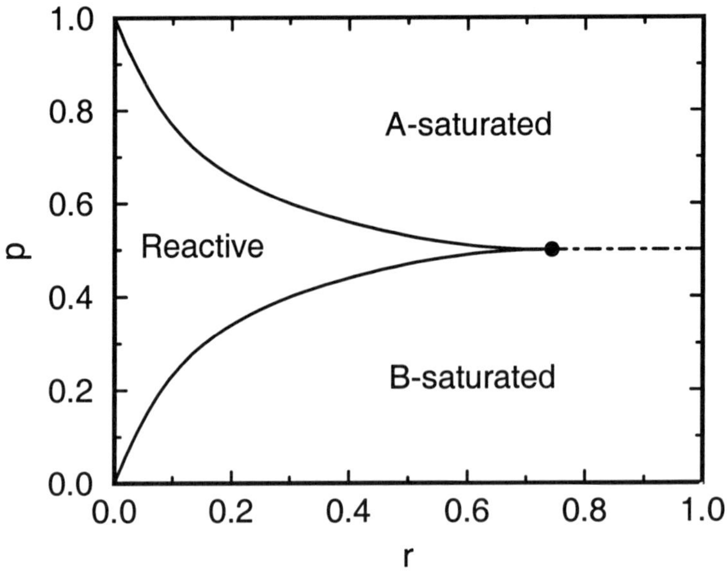


Fig. 4.25 Phase diagram of the two-species monomer-monomer model. From [Brown *et al.* (1997)].

The phase diagram, displayed in Fig. 4.25 with $p$ plotted *vs.* $r$, shows a reactive steady state containing vacancies bordered by two equivalent saturated phases (labeled A and B). The transitions from the reactive phase to either of the saturated phases are continuous, while the transition between the saturated phases is first-order discontinuous. The two saturated phases meet the reactive phase at a *bi-critical* point at a critical value of $r = r_c$. In the case of $r = 1$, the reactive region no longer exists and the only transition is a first-order discontinuous line between the saturated phases. Considering the density of vacancies between unlike species as the order parameter (which can also be called a species "C") the model is the so called "three species monomer-monomer model". Simulations and cluster mean-field approximations were applied to investigate the phase transitions of these models [Bassler and Browne (1996, 1997); Brown *et al.* (1997); Bassler and Browne (1998)]. As Fig. 4.26 shows if we call the extended objects filled with vacancies between different species as domain-walls ($C$) we can observe $C \to 3C$ and $2C \to \emptyset$ BARW2 processes in terms of them. These $C$ parity conserving processes arise as the combination of the elementary reaction steps (4.177). The reactions always take place at domain

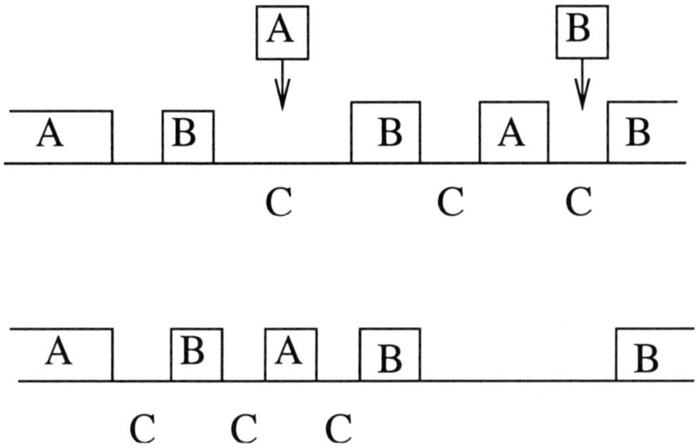


Fig. 4.26 Domain-wall dynamics in the interacting monomer-monomer model. Left part: branching; right part: annihilation

boundaries, hence the $Z_2$ symmetric $A$ and $B$ saturated phases are absorbing.

The interacting monomer-dimer model (IMD) [Kim and Park (1994)] is a generalization of the simple monomer-dimer model [Ziff *et al.* (1986)], in which particles of the same species have nearest-neighbor repulsive interactions. The IMD is parameterized by specifying that a monomer ($A$) can adsorb at a nearest-neighbor site of an already-adsorbed monomer (restricted vacancy) at a rate $r_A k_A$ with $0 \leq r_A \leq 1$, where $k_A$ is an adsorption rate of a monomer at a free vacant site with no adjacent monomer-occupied sites. Similarly, a dimer ($B_2$) can adsorb at a pair of restricted vacancies ($B$ in nearest-neighbor sites) at a rate $r_B k_B$ with $0 \leq r_B \leq 1$, where $k_B$ is an adsorption rate of a dimer at a pair of free vacancies. There are no nearest-neighbor restrictions in adsorbing particles of different species and the $AB \to \emptyset$ desorption reaction happens with probability 1. The case $r_A = r_B = 1$ corresponds to the ordinary noninteracting monomer-dimer model, which exhibits a first-order phase transition between two saturated phases in one dimension. In the other limiting case $r_A = r_B = 0$, there exists no fully saturated phase of monomers or dimers. However, this does not mean that this model no longer has any absorbing states. In fact, there are two equivalent ($Z_2$ symmetric) absorbing states in this model. These states comprise of only the monomers at the odd- or even-numbered lattice sites. A pair of adjacent vacancies is required for a dimer to adsorb, so a state with alternating sites occupied by monomers can be identified with an absorbing state. The PC class phase transition of the $r_A = r_B = 0$ infinite repulsive case has been thoroughly investigated by [Kim and Park (1994); Park *et al.* (1995); Park and Park (1995); Kwon and Park (1995); Hwang *et al.* (1998)]. As one can see the basic reactions are similar to those of the MM model (Eq. (4.177)) but the order parameter here is the density of dimers ($K$) that may appear between ordered domains of alternating sequences: '0A0..A0A.' and 'A0A..0A0', where monomers are on even or odd sites only. The recognition of an underlying BARW2 process (4.174) is not so easy in this case. Still considering regions between odd and even filled ordered domains one can identify domain-wall random-walk, annihilation and branching processes through the reactions with dimers as one can see on the examples below.

```
t      A 0 A 0 A 0 A 0 A 0 0 A 0 A 0 A          K
t+1    A 0 A 0 A 0 A 0 A B B A 0 A 0 A          K
t+2    A 0 A 0 A 0 A 0 0 0 B A 0 A 0 A      K K K

t      A 0 A 0 A 0 A 0 A 0 0 0 A 0 A 0        K K
t+1    A 0 A 0 A 0 A 0 A 0 A 0 A 0 A 0
```

The introduction of a $Z_2$ symmetry-breaking field, which makes the system prefer one absorbing state to the other, was shown to change that transition type from PC to DP [Park and Park (1995)].

### 4.6.5 *Long-range correlated initial conditions*

The effect of initially long-range correlations has already been discussed in two-dimensional Ising models (Sect. 2.3.5) and for one-dimensional bond percolating systems (Sect. 4.2.8). In both cases continuously changing decay exponents have been found. In case of the NEKIM (see Sect. 4.6.2) simulations [Ódor *et al.* (1999); Menyhárd and Ódor (2000)] have shown that the density of kinks $\rho_k(t)$ changes as

$$\rho_k(t) \propto t^{\kappa(\sigma)} \tag{4.178}$$

when the system begins with two-point correlated kink distributions of the form of Eq. (4.87). The $\kappa(\sigma)$ changes linearly between the two extremes $\beta/\nu_{||} = \pm 0.285$, as shown in Fig. 4.27. This behavior is similar to that of the DP model (Sect. 4.2.8), but here one can observe a symmetry:

$$\sigma \leftrightarrow 1 - \sigma, \qquad \kappa \leftrightarrow -\kappa\ , \tag{4.179}$$

which is related to the duality symmetry of the NEKIM (4.165).

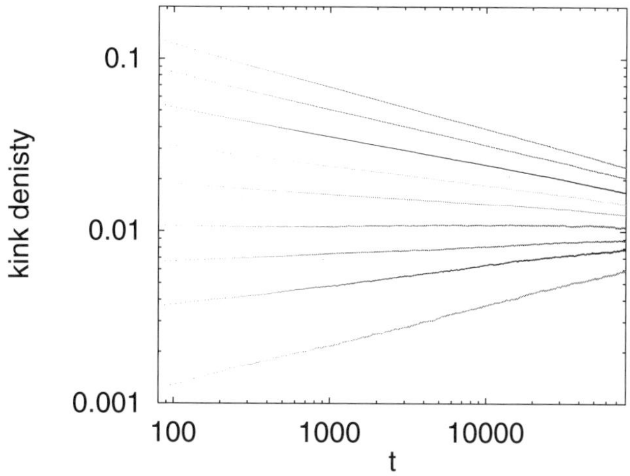


Fig. 4.27 Kink density $\log(\rho_k(t))$ vs $\ln(t)$ in NEKIM simulations for $\sigma = 0, 0.1, 0.2..., 1$ initial conditions (from bottom to top curves). From [Menyhárd and Ódor (2000)].

### 4.6.6 *Spatial boundary conditions*

The surface critical behavior of the PC class has been explored through the study of the GDK model (Sect. 4.6.3) [Lauritsen *et al.* (1998); Howard *et al.* (2000); Fröjdh *et al.* (2001)]. The basic idea is that on the surface one may include not only the usual BARW2 reactions (4.174) but potentially also a parity symmetry breaking $A \to \emptyset$ reaction. Depending on whether or not the $A \to \emptyset$ reaction is actually present, we may then expect different boundary universality classes. Since the rapidity-reversal symmetry (4.41) is broken for BARW2 processes two independent exponents ($\beta_{1,\text{seed}}$, $\beta_{1,\text{dens}}$) characterize the surface critical behavior.

The surface phase diagram for the mean field theory of BARW (valid for $d > d_c = 2$) is shown in Fig. 4.28. Here $\sigma_m$, $\sigma_{m_s}$ are the rates for the branching processes: $A \to (m+1)A$ in the bulk and at the surface, respectively, and $\mu_s$ is the rate for the surface spontaneous annihilation reaction $A \to \emptyset$. Otherwise, the labeling is the same as that for the DP phase diagram (see Fig. 4.8). The $\mu_s > 0$ corresponds to the parity symmetry-breaking RBC.

For the $\mu_s = 0$ (IBC) parity conserving case, the surface action is of the form

$$S_s = \int d^{d-1}x_\parallel \int_0^\tau dt \sum_{l=1}^{m/2} \sigma_{2l_s} \left(1 - \psi_s^{2l} \psi_s \phi_s\right) , \qquad (4.180)$$

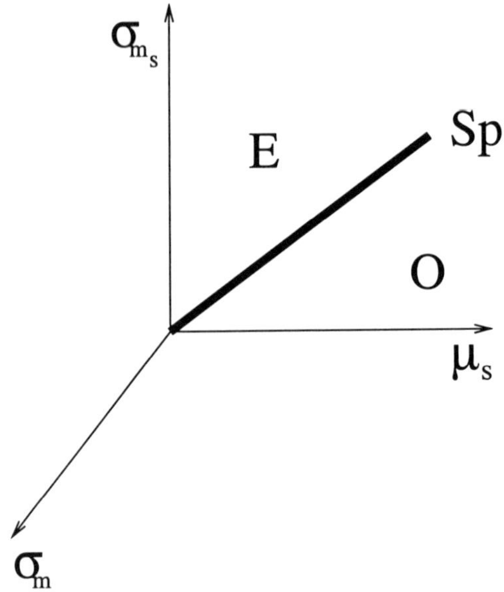


Fig. 4.28 Schematic mean field boundary phase diagram for BARW. See text for an explanation of the labeling. From [Fröjdh *et al.* (2001)].

where $\psi_s = \psi(\mathbf{x}_\parallel, x_\perp = 0, t)$ and $\phi_s = \phi(\mathbf{x}_\parallel, x_\perp = 0, t)$. In $d = 1$ the boundary and bulk transitions are inaccessible to controlled perturbative expansions, but scaling analysis shows that surface branching is *irrelevant*, leading to the $Sp*$ and $Sp$ special transitions. For the case of $\mu_s > 0$ (RBC) parity symmetry breaking, the surface action is

$$S_2 = \int d^{d-1}x_\parallel \int_0^\tau dt \left[ \sum_{l=1}^{m} \sigma_{l_s}(1 - \psi_s^l)\psi_s\phi_s \right. \\ \left. + \mu_s(\psi_s - 1)\phi_s \right] \qquad (4.181)$$

and the RG procedure shows that the stable fixed point corresponds to an ordinary transition. Therefore in one dimension the phase diagram looks very differently from the mean-field case (Fig. 4.29). One can differentiate two cases corresponding to (a) the annihilation fixed point of the bulk and (b) the PC critical point of the bulk. As on can see in both cases the ordinary transition ($O$,$O^*$) corresponds to $\mu_s > 0$, the RBC and the special transitions ($Sp$, $Sp^*$) to $\mu_s = 0$, the IBC. The ABC condition obviously behaves as if there existed a surface reaction equivalent to $\emptyset \to A$,

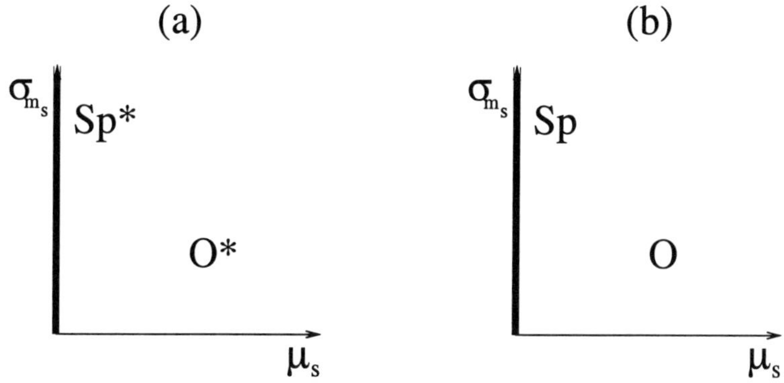


Fig. 4.29 Schematic surface phase diagrams for BARW in $d = 1$ for (a) $\sigma_m < \sigma_{m,\text{critical}}$, and (b) $\sigma_m = \sigma_{m,\text{critical}}$. See text for an explanation of the labeling. From [Fröjdh *et al.* (2001)].

and thus it belongs to the normal transition universality class. By scaling considerations the following exponent relations can be derived:

$$\tau_1 = \nu_{||} - \beta_{1,\text{dens}} \quad , \tag{4.182}$$

$$\nu_{||} + d\nu_{\perp} = \beta_{1,\text{seed}} + \beta_{\text{dens}} + \gamma_1 . \tag{4.183}$$

[Howard *et al.* (2000)] showed that on the self-dual line of the one-dimensional BARWe model (see Sect. 4.6.2 and [Mussawisade *et al.* (1998)]) the scaling relations between exponents of ordinary and special transitions

$$\beta^{\text{O}}_{1,\text{seed}} = \beta^{\text{Sp}}_{1,\text{dens}} \tag{4.184}$$

and

$$\beta^{\text{Sp}}_{1,\text{seed}} = \beta^{\text{O}}_{1,\text{dens}} \tag{4.185}$$

hold. Relying on universality Howard et al. claim that they should be valid elsewhere close to the transition line. Numerical simulations support this hypothesis as shown in Table 4.11.

Table 4.11 Critical boundary exponents of the PC class in $d = 1, 2$ for ordinary and special cases.

| | 1$d$ (IBC) | 1$d$ (RBC) | 2$d$ (O) | 2$d$ (Sp) |
|---|---|---|---|---|
| $\beta_{1,\text{seed}}$ | 2.06(2) | 1.37(2) | 0 | 0 |
| $\beta_{1,\text{dens}}$ | 1.34(2) | 2.04(2) | 3/2 | 1 |
| $\tau_1$ | 1.16(4) | 1.85(4) | 1 | 1 |
| $\gamma_1$ | 2.08(4) | 2.77(4) | 1/2 | 1/2 |

### 4.6.7 *BARWe with long-range interactions*

The effect of long-range interaction in BARWe models via Lévy flights (i.e. when particle hopping probability to distance $r$ decays as $r^{-d-\sigma}$) was investigated by field theory and simulations in one dimension [Vernon and Howard (2001)]. The following field theoretical action – the modification of Eq. (4.67) – describing both normal ($D\nabla^2$) and anomalous ($D_A\nabla^\sigma$) diffusion was proposed

$$S = \int d^dxdt[\psi(\partial_t - D\nabla^2 - D_A\nabla^\sigma)\phi - \lambda(1-\psi^k)\phi^k + \mu(1-\psi^m)\psi\phi] \ . \tag{4.186}$$

Simple power counting reveals that the upper critical dimension, at which the fluctuations become important, is $d_c = \sigma$, for $\sigma < 2$. This field theory was analyzed by PRG in the neighborhood of $d_c$. For $\sigma < 2$ the normal diffusion term can be dropped, since it is less relevant than the anomalous one. According to the one-loop approximation, the branching process is relevant at the annihilation fixed point for

$$\sigma < \sigma_c'(d) = 3d/2 \ . \tag{4.187}$$

Hence, as in the mean field case of BARWe, we expect an active phase for all nonzero values of the branching rate $\mu$, for sufficiently small $\sigma$ (or high $d$). For $\sigma = 1$ in $d = 1$, the critical point for the transition from an absorbing to an active phase occurs at zero branching. However, for $\sigma$ bigger than about $3/2$ (in $d = 1$), the critical branching rate moves smoothly away from zero with increasing $\sigma$, and the transition lies in a different universality class, inaccessible by controlled perturbative expansions.

This theory was supplemented by simulations in one dimension and the following complex phase structure has been found:

(1) For $\sigma < 1$ the process exhibits mean-field type of behavior with the density decaying as $\rho \sim 1/t$.
(2) In case of $\sigma < 1.5$ the branching process is irrelevant and one finds the same behavior as for pair annihilation with Lévy flights as described in Sect. 4.5.4 and the density decays according to Eq. (4.135).
(3) In the range $1.5 < \sigma < 2.5$ one observes a nontrivial phase, with continuously varying exponents.
(4) For $\sigma > 2.5$ the process is short-ranged and the transition belongs to the usual $d = 1$ PC class.

The observed phenomenology in 1+1 dimensions suggests the general phase structure shown in Fig. 4.30 ([Hinrichsen (2007a)]). The precise form of

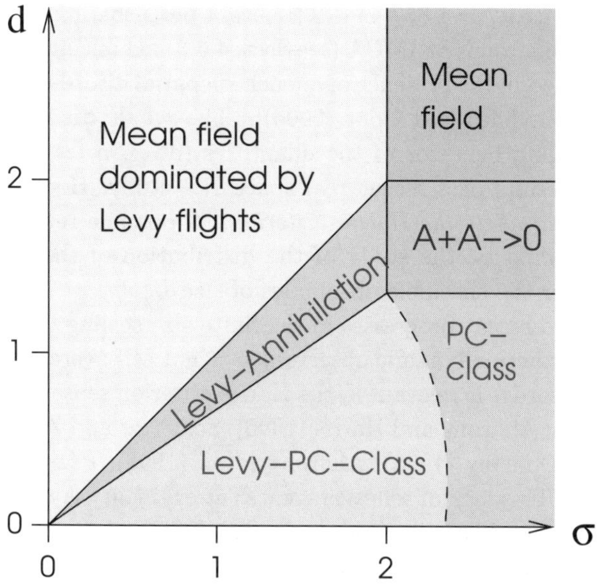


Fig. 4.30 General phase structure of parity-conserving branching annihilating random walks with Lévy-distributed long range interactions. From [Hinrichsen (2007a)].

the phase boundary between short- and long-ranged branching annihilating random walks (the dashed line in Fig. 4.30) is not yet known. However, taking the field-theoretic action (4.186) and assuming that the fractional derivative is not renormalized, Hinrichsen conjectured the following scaling form ([Hinrichsen (2007a)]) in the phases (2) and (3), where the exponents vary continuously

$$\sigma = d + Z(1 - \alpha) \ . \tag{4.188}$$

If this relation were is correct, it should give the phase boundary of phase (3) (the dashed line). One can check by inserting the numerically known exponents of the PC class and solving the equation for $\sigma$ that in one dimension this gives the threshold $\sigma_c \approx 2.50$, which is in qualitative agreement with the numerical simulations [Vernon and Howard (2001)].

### 4.6.8 *Parity conserving NEKIMCA with quenched disorder*

According to the Harris criterion (1.91) the pure critical point of the PC class transition in *one dimension should be instable* against quenched, spatial disorder, since $\nu_\perp = 1.857(1) < 2$ (relevant perturbation). The stochastic cellular automaton NEKIMCA (see Sect. 4.6.3) exhibiting PC class transition was studied in the presence of quenched spatial disorder by large scale simulations [Menyhárd and Ódor (2006)]. The **weak disorder** does not change the scaling behavior of the quantities ($\alpha$, $\beta$, $\eta$, $z$) studied. This supports the results of a recent real space RG study [Hooyberghs *et al.* (2004)] but *contradicts the Harris criterion.* A possible resolution of this contradiction could be the study of the distribution of the critical point coordinates over the ensemble of samples of size $L$.

According to recent progresses in the finite size scaling theory of disordered systems, thermodynamic observable are not self-averaging at critical points it the disorder is relevant in the Harris criterion sense [Wiseman and Domany (1995); Aharony and Harris (1996); Scalettar and Zimányi (1997); Wiseman and Domany (1998); Aharony *et al.* (1998); Pázmándi and Batrouni (2000)]. This lack of self-averageness at criticality is directly related to the distribution of pseudo-critical temperatures $T_c(i, L)$ over the ensemble of samples ($i$) of size $L$. This has been shown numerically in case of a wetting transition of a polymer chain with quenched disorder [Monthus and Garel (2005)]. An interesting further direction of research would be the study of those distributions for nonequilibrium systems like the NEKIMCA.

For **stronger disorder** continuously changing density decay exponent ($\alpha$) were observed. This corroborates a former RG study [Doussal and Monthus (1999)] for ARW with site dependent reaction rates and an exact calculation for the infinite reaction rate ARW with disordered diffusion traps [Schütz and Mussawisade (1998)].

**Very strong disorder** introduces complete blocking of reactions or diffusion with an exponential domain size distribution. If the system becomes segmented by diffusion walls a **blocking transition** to fluctuating states occur. In this case in odd parity blocks residual particles remain active. Large scale simulations suggest that this kind of disorder is marginal, the density decay slows down from the $\rho \propto t^{-1/2}$ power-law by a logarithmic factor or changes continuously if the annihilation is blockaded too. In the frozen phase the concentration of residual kinks increases logarithmically, with $\beta_{eff} \to 1$ asymptotically. This is in agreement with the results of [Mandache and ben Avraham (2000)], who considered the ARW model with complete blockades. However one can't see a crossover to stretched exponential decay reported in [Mandache and ben Avraham (2000)] for long times, but rather the decay slows down. This is due to the fact that in NEKIMCA, in the frozen state not only blockades exist, but reaction and

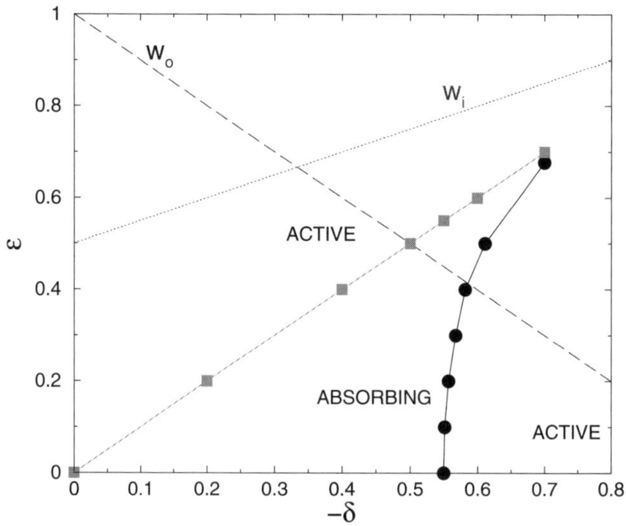


Fig. 4.31 Phase diagram of the disordered NEKIMCA for negative $\delta$. Bullets correspond to the disordered PC class transition points, squares to the freezing (blocking) transition (lines are for the guide of eye only). The dashed line shows $w_o$, while dotted line denotes $w_i$. From [Menyhárd and Ódor (2006)].

diffusion probabilities are randomized, which can introduce slower power-laws [Noest (1986); Schütz and Mussawisade (1998); Doussal and Monthus (1999)].

In the inactive phase and at the critical point kink depletion process dominates, via the $AA \to \emptyset$ annihilation. However for strong enough disorder active domains of arbitrary sizes may also emerge due to the exponential distribution of such events. The contribution of these large regions can be estimated as in [Noest (1986)] with the difference that for PC class even the absorbing phase decays algebraically i.e. one does not have exponential decay, that could slow down to stretched exponential at the clean critical point as for DP. The probability $p_a$ for finding a rare region of size $l_a$ is

$$p_a \propto \exp(-q l_a) \ , \tag{4.189}$$

where $q$ is the probability that a site is active. The long-time kink decay density is dominated by these rare regions, when any finite region decays exponentially, hence

$$\rho(t) \propto \int dl_a \, l_a \, p_a \, \exp(-t/\tau(l_a)) \ , \tag{4.190}$$

Table 4.12 CAM estimates for the kink density and its fluctuation exponents.

| h | 0.0 | 0.01 | 0.05 | 0.08 | 0.1 | DP |
|---|---|---|---|---|---|---|
| $\beta$ | 1.0 | 0.281 | 0.270 | 0.258 | 0.285 | 0.2767(4) |
| $\gamma$ | | 0.674 | 0.428 | 0.622 | 0.551 | 0.5438(13) |

where $\tau(l_a)$ is the characteristic decay time of a region of size $l_a$. The average lifetime of such active regions grows as

$$\tau(l_a) \propto \exp(al_a) \tag{4.191}$$

because a coordinated fluctuation of the entire region is required to take it to the absorbing state (the constant $a$ vanishes at the clean critical point and increases with increasing $q$ (see, e.g., [Goldenfeld (1992)])).

The saddle point analysis of (4.190) as in [Noest (1986)] results in power-law decay

$$\rho(t) \propto t^{-p_a/a} \ . \tag{4.192}$$

Hence one can expect continuously changing decay exponents either as the effect of rare active regions or as the effect of high diffusion barriers [Schütz and Mussawisade (1998)].

Since in one dimension $\nu_{||} = 3.25(10) > 2$ for the PC class (see Table 4.9), the Harris criterion predicts irrelevant temporal disorder. This has not been checked yet.

### 4.6.9 *Anisotropic PC systems*

One can introduce anisotropy into PC class models in different ways. The effect of **diffusion drift** in a one dimensional BARWe (see Sect. 4.6.1) has been studied by simulations by [Park and Park (2005a)]. This model with a fully biased diffusion

$$A\emptyset \xrightarrow{p} \emptyset A, \quad AA \xrightarrow{p} \emptyset\emptyset, \quad A\emptyset \xrightarrow{1-p} AAA \tag{4.193}$$

exhibited the same kind of density decay at the critical point as the isotropic one ($\rho(t) \propto t^{-0.285(1)}$). Furthermore such driven PC system with a partial bias also show this scaling behavior. Therefore the Galilean invariance (4.5.5) holds for the PC class.

When we apply an **external magnetic field** $h$ to NEKIM, which breaks the $Z_2$ symmetry the transition **changes from PC to DP** type (see Table 4.12). [Menyhárd and Ódor (1996)] This is in agreement with the simulation results of the interacting monomer-dimer model (see Sect. 4.6.4) [Park and Park (1995)].

Another kind of anisotropy can be introduced in the NEKIM by **breaking the symmetry of spin-updates** [Menyhárd and Ódor (2002)] in such a way that two different types of domain walls emerge. The following changes to the Glauber spin-slip rates (4.159) (with $\Gamma = 1$, $\tilde{\delta} = 0$) were introduced in the so-called NEKIMA:

$$w_{\uparrow\downarrow\uparrow} = 0, \tag{4.194}$$

$$w_{\uparrow\uparrow\downarrow} = w_{\downarrow\uparrow\uparrow} = p_+ < 1/2\,. \tag{4.195}$$

In the terminology of domain walls as particles the following reaction-diffusion picture arises. Owing to the symmetry breaking there are two kinds of domain walls $\downarrow\uparrow \equiv A$ and $\uparrow\downarrow \equiv B$, which can only occur alternately (...B..A..B..A..B...A...) owing to the spin background. Upon meeting $AB \to \emptyset$ happens, while in the opposite sequence, $BA$, the two domain-walls are repulsive due to (4.194). The spin-exchange leads to $A \leftrightarrow ABA$ and $B \leftrightarrow BAB$ type of kink reactions, which together with the diffusion of $A$-s and $B$-s leads to a kind of two-component, coupled branching and annihilating random walk (see Sect. 6.12).

There are two control parameters in this model: $p_{ex}$, which regulates the kink production-annihilation and $p_+$, responsible for the local-symmetry breaking (4.195). Simulations show that for $p_{ex} \to 0$, $p_+ < 0.5$ an absorbing phase emerges with N-BARW2 class static exponents (Sect. 6.12). Owing to the pairwise order of kinks the hard-core effects cannot play a role. The transition on the $p_{ex} > 0$ line belongs to the 1+1 dimensional DP class. The phase diagram in the plane of parameters $p_{ex}$ and $p_+ < 1/2$, as obtained by computer simulations, is shown on Fig. 4.32.

Since the $AB \to \emptyset$ reaction breaks the parity conservation of species (but preserves the global parity conservation) the necessary conditions for N-BARW2 class can be eased. On the other hand the appearance of the DP transition suggests a zero branching rate condition for N-BARW2 universal behavior.[4] This study and the results for the generalized contact process (Sect. 6.12.1) reveal that the conditions for the N-BARW2 class should be further investigated.

The **cluster spreading behavior of NEKIMA** is rather interesting, due to the spin anisotropy (anisotropy in the spin-flip rates) the '+' and '−' spins decouple and follow different scaling laws (see Table 4.13) [Menyhárd and Ódor (2003)]. Since spin clusters are compact at the N-BARW2 line

[4]Similar phenomenon arises in other RD systems (Sect. 4.7), i.e. the phase transition of a model at zero branching rate is different from the one, which occurs at finite offspring production probability.

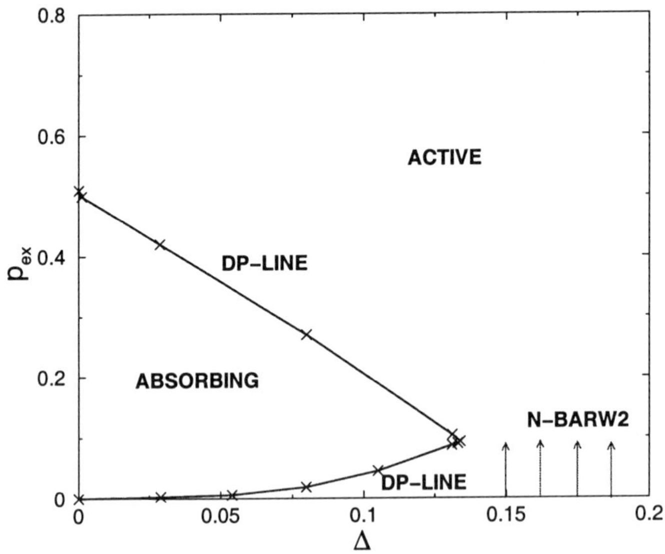


Fig. 4.32 The reentrant phase diagram of the NEKIMA model for parameters $\tilde{\delta} = -0.565$, $\Gamma = 0.5$. The absorbing phase is fully '−'. From [Menyhárd and Ódor (2002)]

Table 4.13 Cluster critical exponents of the NEKIMA model at the N-BARW2 transition and along the DP line. The hyperscaling relations among exponents of columns are satisfied.

| Exponents | $p_{ex} = 0$ '+' | $p_{ex} = 0$ '-' | On DP-line '+' | On DP-line '-' |
|---|---|---|---|---|
| $\eta$ | 1. 0 | 0.0 | 0.31 | 1.0 |
| $\delta$ | 0.0 | 0.0 | 0.16 | 0.0 |
| $z$ | 2.0 | 0.0 | 1.26 | 2.0 |

($p_{ex} = 0$) the spreading exponents of them satisfy the hyperscaling relation (4.117). Along the DP line they satisfy the hyperscaling of DP (4.43). The (1.50) scaling relation, connecting different cluster and the global anisotropy (dynamical) exponent is broken.

The effect of **quenched disorder** for decoupled spins of NEKIMA was also studied by [Menyhárd and Ódor (2007)]. The simulations pointed out that by increasing the spin anisotropy the sensitivity to disorder (which is very low in case of NEKIM and ARW) increases and novel type of fixed points/scaling behaviors arise. In the extreme case one type of spins can follow DP dynamics (independently from the other) and as the consequence Griffiths phase, with disorder dependent continuously changing dynamical exponents and activated scaling can also be observed in this model.

## 4.7 Classes in models with $n < m$ production and $m$ particle annihilation at $\sigma_c = 0$

In Sect. 4.5.1 the $2A \to \emptyset$ annihilating random walk (ARW) has already been discussed. When we add simple branching processes $A \to (1+k)A$ to ARW we arrive to BARW models, exhibiting continuous phase transitions and different critical behaviors (see Sects. 4.2.4 and 4.6.1). Now we generalize this construction to models with

$$nA \xrightarrow{\sigma} (n+k)A, \qquad mA \xrightarrow{\lambda} \emptyset, \tag{4.196}$$

production ($n \geq 1$) and annihilation ($n < m$) (PARWa).

Mean-filed classes of such models have been discussed in Sect. 4.1.5. In order to take into account the effect of fluctuations, relevant in low dimensions, we formulate the field theoretical action

$$S = \int d^dxdt[\psi(\partial_t - D\nabla^2)\phi - \lambda(1-\psi^m)\phi^m \\ +\sigma(1-\psi^k)\psi^n\phi^n] \, . \tag{4.197}$$

A dimensional analysis shows that these system exhibit a phase transition at $\sigma_c = 0$ with an upper critical dimension [Lee (1994); Cardy and Täuber (1998)]

$$d_c = 2/(m-1) \, , \tag{4.198}$$

and with mean-field exponents

$$\alpha = 1/(m-1), \quad \beta = 1/(m-n), \quad Z = 2 \, . \tag{4.199}$$

At $d = d_c$ **(which falls below physical dimensions for** $m > 3$) the decay has logarithmic corrections

$$\rho(t) = A_k(ln(t)/t)^{1/(k-1)} \, . \tag{4.200}$$

In the one-dimensional $AAA \to \emptyset$ process we find the density decay behavior [Lee (1994)]

$$\rho(t) = (\frac{1}{4\pi\sqrt{3}D})^{\frac{1}{2}}(\ln(t)/t)^{\frac{1}{2}} + O(t^{\frac{1}{2}}) \, , \tag{4.201}$$

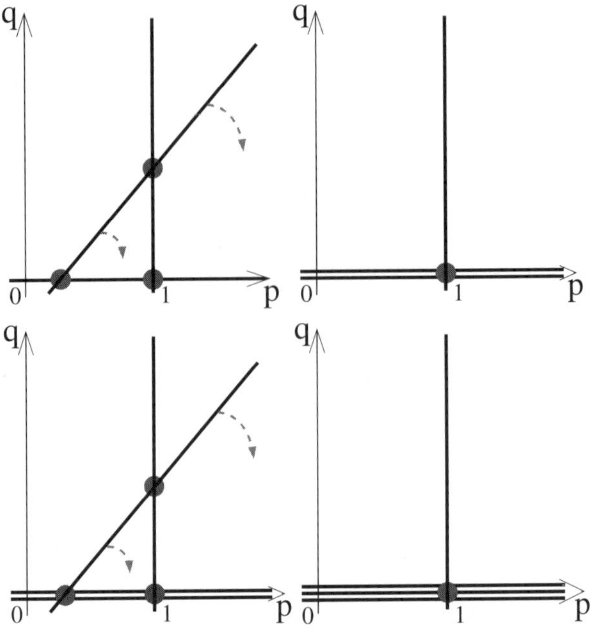


Fig. 4.33 Phase portrait evolution of PARWa model phase transitions: (a) $n = 1$ and $\sigma > 0$. Decreasing the rate $\sigma$ is schematically shown by the dashed arrows. (b) $n = 1$ and $\sigma = 0$, the "accidental" line becomes horizontal and the local topology of the phase portrait is that of the binary annihilation reaction; Fig. 1.3. (c) $n = 2$ and $\sigma > 0$, (d) $n = 2$ and $\sigma = 0$, the topology is that of the tripplet annihilation reaction, $3A \to \emptyset$. From [Elgart and Kamenev (2006)]

which is also the dominant behavior of the bosonic PCPD model in $d = 1$, at the transition point (see Sect. 6.7).

The evolution of the phase portrait corresponding to such transitions is depicted in Fig. 4.33. As one can see the change of reaction probabilities does not cause the collapse of $A, B, C$ points of the zero energy trajectories as in case of the BARWo or PARWs models (see Figs. 4.3 and 4.12) and the critical behavior is different too.

The field theory by [Cardy and Täuber (1996)] has shown that in models with $n = 1$, below $d_c$ the $A \to \emptyset$ reaction is also generated via the sequence $A \to 2A \to \emptyset$ and the transition point is shifted to $\sigma_c > 0$ with BARW type of critical behavior. However a non-perturbative GMF analysis [Ódor (2004d)] has found diffusion dependence for $m > 2$, i.e the effective $A \to \emptyset$ **may or may not be generated** depending on the diffusion/annihilation

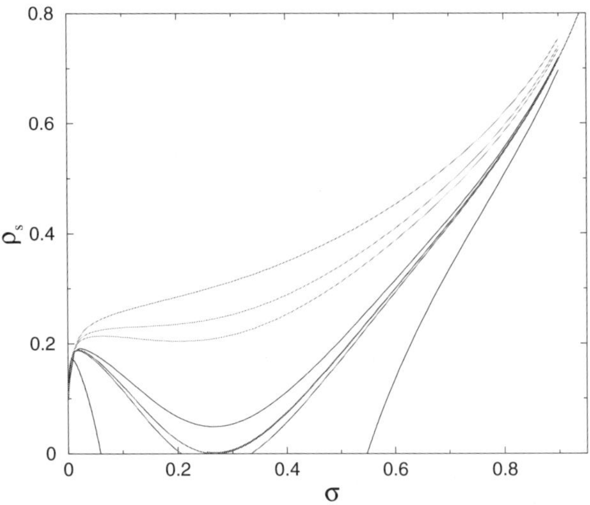


Fig. 4.34 Steady state density results in $N = 7$ GMF approximations of the $A \to 2A$, $4A \to \emptyset$ model. Different curves correspond to: $D = 1$, 0.7, 0.6, 0.4, 0.371, 0.36, 0.2 (top-to bottom). The $\sigma_c > 0$ critical point disappears for $D^* > 0.37083$. From [Ódor (2004c)].

ratio. This means that for strong diffusions the particles can escape each other before making a reaction (see Fig. 4.34). Similarly above $d_c$ the effective annihilation reaction generation may occur if the diffusion is slow enough. This has been proven by non-perturbative field theory and simulations [Canet *et al.* (2004); Canet and Delamotte (2004)] in case of the $A \to 2A$, $2A \to \emptyset$ BARWo models (Fig. 4.35), for the $A \to 2A$, $3A \to \emptyset$ triplet [Dickman (1990)] and for the $A \to 2A$, $4A \to \emptyset$ quadruplet annihilation models [Ódor (2004d)].

For $n = 2$ production the lower order annihilation terms $2A \to \emptyset$ may or may not be generated (like in the $2A \to 3A$, $4A \to \emptyset$ models [Ódor (2004c)], [Ódor (2004b)]), depending of the diffusion/reaction ratio. However the fluctuations here cause a PCPD transition for small diffusions (see Sect. 6.7) and a reentrant phase diagram can be observed (see Fig. 4.36). Note that similar diffusion dependent, reentrant phase diagram has also been observed in a NEKIM variant [Menyhárd and Ódor (2003)]. It is reasonable to assume that this kind of diffusion dependence is rather general and occurs in all $n > 2$ PARWa models.

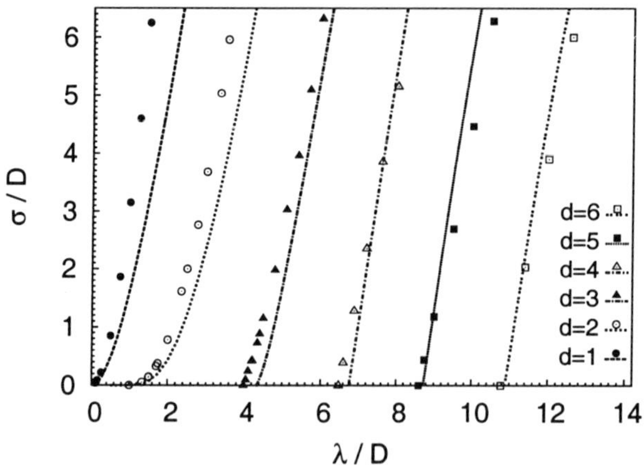


Fig. 4.35 Phase diagrams of BARWo $A \xrightarrow{\sigma} 2A$, $2A \xrightarrow{\lambda} \varnothing$ in dimensions 1 to 6. Lines present NPRG results. Symbols follow from numerical simulations. For each dimension, the active phase lies on the left of the transition line, the absorbing phase is on the right. From [Canet and Delamotte (2004)].

## 4.8 Classes in models with $n < m$ production and $m$ particle coagulation at $\sigma_c = 0$; reversible reactions (1R)

In this section I discuss the universal behavior of models, which are very similar to those of Sect. 4.7 except the vacuum state is not accessible. The reactions are

$$nA \xrightarrow{\sigma} (n+k)A, \qquad mA \xrightarrow{\lambda} (m-l)A, \tag{4.202}$$

with $m > l$. Earlier I pointed out the equivalence (from asymptotic scaling point of view) of the $2 \to \emptyset$ annihilation and the $2A \to A$ coagulation models (see Sect. 4.5.1). However if a $nA \to (n+k)A$ branching reaction is also present, field theory and simulations [Hammal *et al.* (2007)] have provided evidences for different behaviors in models with coagulation and annihilation.

It was shown in Sect. 4.2.4 the in the $A \to 2A$, $2A \to \emptyset$ model an effective $A \to \emptyset$ single particle decay is generated (for any diffusion rate below $d = 3$) and as a consequence DP type of criticality can be observed. In the coagulation model $nA \to (n+k)A$, $2A \to A$ this is not possible, which results in different, non-DP scaling behavior.[5]

[5]It is shown in Sects. 4.7 and 4.102, that the accessibility of the vacuum state alters the critical behavior.

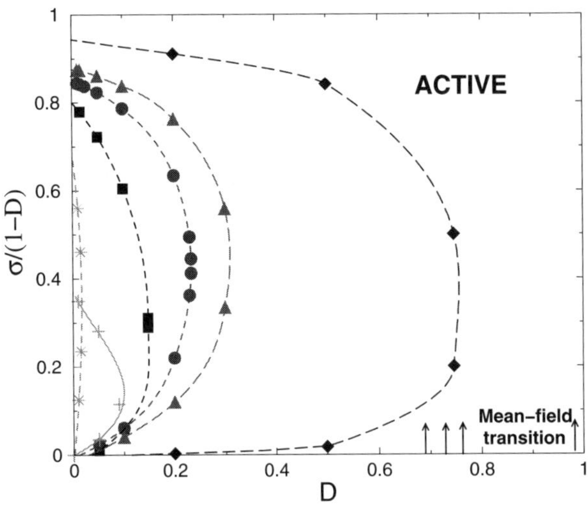


Fig. 4.36 Phase diagram of the $2A \to 3A \to 4A \to \emptyset$ model. Stars correspond to $N = 2$, boxes to $N = 3$, bullets to $N = 4$ and triangles to $N = 5$ cluster mean-field approximations. Diamonds denote one-, + signs two-dimensional simulation data, where PCPD class transitions are found. The lines serve to guide the eye. At the $\sigma = 0$ line PARWa type of mean-field transition occurs (from [Ódor (2004b)]).

The **reversible** $A \to 2A$, $2A \to A$ model (without branching is also called diffusion limited pair-coagulation) satisfies the detailed balance condition and is exactly solvable in one dimension [Burschka *et al.* (1989); Jack *et al.* (2006); Mayer *et al.* (2006)]. This model can be mapped on the Fredrickson-Anderson (FA) model [Fredrickson and Andersen (1984)] and exhibits LSI symmetry, at least for the global response [Mayer and Sollich (2007)]. Furthermore it exhibits a phase transition at $\sigma_c = 0$ characterized by the exponents shown in Table 4.14.

Table 4.14 Critical exponents of the $A \to 2A$, $2A \to \emptyset$ model.

| d | $\beta$ | $\alpha$ | $\nu_\perp$ | $\nu_{\|\|}$ | $Z$ | $a$ | $a'$ | $\lambda_R$ |
|---|---|---|---|---|---|---|---|---|
| 1 | 1 | 1/2 | 1 | 2 | 2 | 1 | 5/2 | 1 |
| 2 | 1 | 1 | 1/2 | 1 | 2 | 2 | 4 | 3 |

The diagrammatic technique [Elgart and Kamenev (2006)] pointed out different phase space topologies corresponding to these models, suggesting different class behavior. The reaction Hamiltonian is

$$H_R = \sigma(p^2 - p)\, q + \frac{\lambda}{2}(p - p^2)\, q^2$$
$$= (p^2 - p)(u\, q - v\, q^2)\,, \tag{4.203}$$

where $u = \sigma$ and $v = \lambda/2$. The phase portrait (shown in Fig. 4.37(b)) exhibits a characteristic rectangular shape comprised by the generic lines $p = 1$ and $q = 0$, along with the two "accidental" ones $p = 0$ and $q = u/v$. The mean-field approximation predicts the average density to be $\rho = u/v$. The control parameter is $u = \sigma$ with the critical value $u_c = 0$. This is very different from the triangular DP topology (Fig. 4.37(a)), where three lines coalesce at the transition. One can show that the latter is generated by adding the $A \to \emptyset$ reaction, which converts the absorbing state accessible. The phase portrait of the reversible reaction is stable

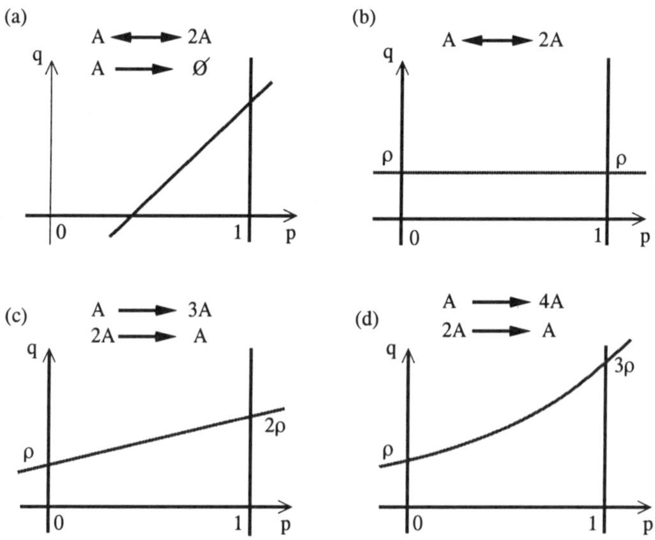


Fig. 4.37 Schematic zero-energy manifolds for different reactions. The non-trivial manifold lines, depending on the control parameter ($\sigma$) move downwards upon approaching the critical point in all cases. While DP is characterized by a triangular structure as in (a), models without an accessible vacuum state have a different form, being rectangular (b), trapezoidal (c), or more complicated geometries (d), for different reactions. $\rho = \sigma/\lambda$. Following [Hammal *et al.* (2007)].

against renormalization i.e. no terms violating the rectangular structure are generated. This can be confirmed by realizing that the action possesses the symmetry

$$p \to \frac{v}{u} q \, , \qquad q \to \frac{u}{v} p \, , \qquad t \to -t \, . \tag{4.204}$$

Very recently [Hammal *et al.* (2007)] have shown that the reversibility is just a **sufficient but not a necessary condition** for this class. In unary, branching models models (with $n \geq 2$)

$$A \to nA \; , 2A \to \emptyset \; , \tag{4.205}$$

**the vacuum state is not accessible**, i. e. no $A \to \emptyset$ reaction is generated by fluctuations at the $\sigma_c = 0$ transition point, which could make it DP — the same critical behavior was found in one dimension as for the $n = 2$ reversible case. Mean-filed classes of such models are discussed in Sect. 4.1.5 and differ from that of DP. Taking into account fluctuations the field theoretical action looks as

$$S = \int d^d x dt [\psi(\partial_t - D\nabla^2)\phi - \lambda(1 - \psi^l)\phi^m \psi^{m-l} \\ + \sigma(1 - \psi^k)\psi^n \phi^n] \tag{4.206}$$

and the engineering, upper-critical dimension is (like for all PARWa model)

$$d_c = 2/(m-1) \; . \tag{4.207}$$

Higher order branching terms ($A \to nA$) ($n > 2$) replace the rectangular phase space picture to more complex ones (see Fig. 4.37(b-c)), but the overall topology is unchanged and approaching the critical point they become negligible. This means that the "accidental" line merges the $q = 0$ line as $\sigma \to 0$ and do not intersect as in case of DP. The PRG analysis has also shown that renormalization generates the lower order branching processes, especially the $A \to 2A$ one, but for $\sigma_c = 0$ the reversible class scaling behavior is preserved.

For higher order production terms $n > 1$, the number of coagulating particles is $m > 2$, thus the upper-critical dimension shits below $d_c = 1$. Indeed simulations of the $2A \to 4A$, $4A \to 2A$ process in one and two dimension resulted in purely PARWa mean-field type of behavior (4.1.5) with $\beta = 1/2$ and $\alpha = 1/3$ at the $\sigma_c = 0$ critical point [Ódor (2003b)].

Summing up for production and coagulation models there is one non-mean-field class, the representative of it is the reversible $A \to 2A$, $2A \to A$ model, therefore I call it **1R** class.

## 4.9 Generalized PC models

Following the phase space description of classes of RD systems one may try to invent other general models exhibiting parity conservation. In Sect. 4.6.1 I have already mentioned that the class of parity conserving models

$$A \to (k+1)A\ , \quad 2A \to \emptyset \tag{4.208}$$

with $k$ even, exhibit phase transitions belonging to the PC class.

Another generalization is the replacement of the unary branching $A \to (k+1)A$ with a more general, $n$-particle production

$$nA \xrightarrow{\sigma} (n+2)A\ , \tag{4.209}$$

and the replacement of the simple annihilation $2A \to \emptyset$ with the

$$(n+1)A \xrightarrow{\lambda} (n-1)A \tag{4.210}$$

many particle process. The reaction Hamiltonian is

$$H_R = \left[up^{n-1} - v\, h_n(p)\, q\right] p(p^2-1)\, q^n\ , \tag{4.211}$$

where $h_n(p)$ is an even polynomial of the degree $n-2$, $u = \sigma/n!$, and $v = \lambda/(n+1)!$. The corresponding topologies are shown on Fig. 4.38 and make up a stable (under RG) and distinct set of classes. One can call them $n$-particle parity conserving classes (n-PC). In Sect. 4.7 I have already considered such (PARWa) models and found that $d_c < 2$ for $n > 2$, and even the $n = 2$ (2-PC model) has a mean-field type of phase transition with logarithmic corrections.

One can show by RG that other generalizations do not lead to different phase-space topologies. For example the reaction Hamiltonian of the $2A \to 4A$, $4A \to \emptyset$ model (Fig. 4.38(c)) exhibits the

$$q \to -q \tag{4.212}$$

symmetry besides the PC class symmetry (Eq. 4.145), but that is not the symmetry of the full action, hence unstable under renormalization, and the the phase portrait drifts towards Fig. 4.38(a) of the 2-PC class.

If one adds a competing annihilation to the reaction set (4.209), (4.210), for example:

$$2A \dashrightarrow 4A, \quad 3A \to \emptyset, \quad 2A \to \emptyset\ , \tag{4.213}$$

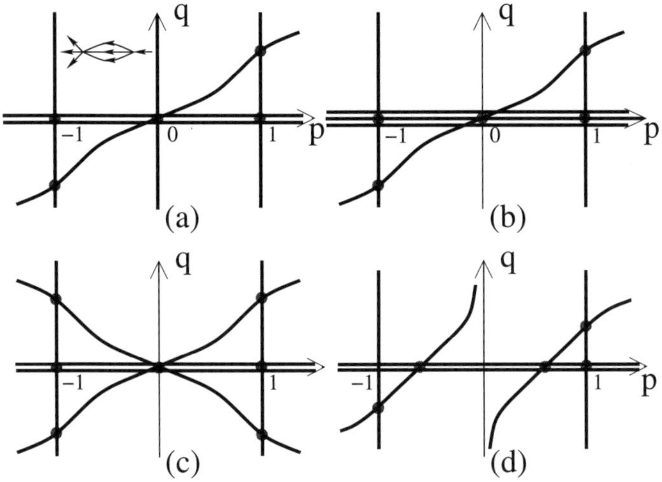


Fig. 4.38 Phase diagrams of n-PC models. (a) 2-PC model, the inset shows the diagram leading to the logarithmic corrections in $d = 1$. (b) 3-PC model. (c) 2-PC model with all reactions including only an even number of particles. This topology is not stable and evolves towards (a). (d) 2-PC with $2A \to \emptyset$, the corresponding topology is essentially equivalent to the PCPD model, Fig. 4.13. From [Elgart and Kamenev (2006)].

the reaction Hamiltonian

$$H_R = [up^2 - v(1+p^2)q^2](p^2-1)q^2 \tag{4.214}$$

exhibits PC symmetry (Eq. 4.145) and the corresponding phase topology looks as Fig. 4.38(d). However near the transition point the topology is the same as in Fig. 4.13, hence it is expected to belong to the PCPD class. Indeed numerical simulations have confirmed this in one [Park *et al.* (2001)] and two dimensions [Ódor *et al.* (2002)].

One can show that the other n-PC processes with competition in the $n$-particle channel belong to the same universality classes as the corresponding generalized n-particle diffusive contact models (PCPD, 3CPD,... etc.). Therefore the **parity conservation is irrelevant in generalized n-particle diffusive contact models**. This suggests that parity conservation is not a sufficient condition to alter the universal behavior in general (except the $n = 1$ unary, branching case), but rather the topology of the phase portrait in the vicinity of the transition discriminates various universality classes.

## 4.10 Multiplicative noise classes

In the hope of classifying nonequilibrium phase transition classes according to their noise terms [G. Grinstein and Tu (1996); Tu *et al.* (1997)] introduced and studied systems via the Langevin equation

$$\partial_t \phi(x,t) = -a\phi(x,t) - b\phi^p(x,t) + D\nabla^2\phi(x,t) + c\phi(x,t)\eta(x,t) \ , \quad (4.215)$$

where $\eta(x,t)$ is a Gaussian white noise, which is proportional to the field ($\phi$)

$$\langle \eta(x,t) \rangle = 0 \ , \quad \langle \eta(x,t)\eta(x',t') \rangle = 2\nu\, \delta^d(x-x')\delta(t-t') \ , \quad (4.216)$$

$a, b > 0, D > 0$ and $c > 0$ are parameters and $p$ is an integer constant obeying $p \geq 2$ (typically $p = 2$ or 3).

The applicability of such **multiplicative noise (MN)** equations in RD models has been proposed. Since for the DP class the noise is square root proportional to the field; for ARW and PC systems the noise is imaginary one may expect distinct universal behavior for such systems. [Howard and Täuber (1997)] on the other hand argued on field theoretical basis that such 'naive' Langevin equations fail to accurately describe systems controlled by *particle pair-reaction* processes, where the noise is in fact 'imaginary' and one should derive a proper Langevin equation by starting from the master equation of particles. Howard and Taüber have investigated the simplest RD systems, in which both real and imaginary noise is present and compete

- $2A \to \emptyset$, $2A \to 2B$, $2B \to 2A$, $2B \to \emptyset$
- $2A \to \emptyset$ and $2A \to (n+2)A$ (see Sect. 6.7)

They have set up the master equation of particles first and derived the corresponding field theoretical action. In neither case did they recover the MN critical behavior reported by [G. Grinstein and Tu (1996); Tu *et al.* (1997)]. Therefore they suspected that there **might not be real RD system possessing the MN behavior**.

On the other hand Langevin equation with multiplicative noise can well represent (bosonic) surface-interface models, exhibiting pinning-depinning, nonequilibrium wetting and synchronization transitions. In particular [G. Grinstein and Tu (1996); Tu *et al.* (1997)] have established a connection between MN systems and the Kardar-Parisi-Zhang (KPZ) theory (see Sect. 7.4)

$$\partial_t h(x,t) = -a - b\, e^{(p-1)\ h} + D\nabla^2 h + D(\nabla h)^2 + c\eta(x,t) \quad (4.217)$$

via the Cole-Hopf transformation

$$\phi(x,t) = e^{h(x,t)} . \tag{4.218}$$

Equation (4.217) is the famous KPZ equation for surface heights ($h$), with an additional bounding term, $-b\, e^{[(p-1)h]}$, pushing regions where $h(x,t) > 0$ towards negative height values (*an upper wall*). Had we defined the transformation as $\phi = \exp(-h)$, the sign of the emerging KPZ-non-linearity would be negative, inducing a *lower wall.* The KPZ interface with positive non-linearity in the presence of an upper wall is completely equivalent to a KPZ with negative non-linearity and a lower wall and the corresponding universality classes are called **MN1** classes.

Grinstein et al. have shown that the phase diagram and the critical exponents $Z$, $\nu_\perp$ and $\beta$ of KPZ and MN classes agree within numerical accuracy. They have found diverging susceptibility (with continuously changing exponent as the function of $a$) for the entire range of $a$.

Other cases (like positive KPZ non-linearity in the presence of a lower wall) are connected to

$$\partial_t\phi(x,t) = -a\phi - b\phi^p + D\nabla^2\phi(x,t) - 2D\frac{(\nabla\phi(x,t))^2}{\phi(x,t)} + c\phi(x,t)\eta(x,t) \tag{4.219}$$

which is identical to Eq. (4.215) except for an extra singular term. Eq. (4.219) is a *modified* MN equation and called (**MN2**) equation. The extra term introduces a relevant perturbation, making MN2 different from MN1.

Finding a well defined **mean-field approximation** of Eq. (4.215) has not been proven to be a trivial task [Genovese and Muñoz (1999)]. In case of MN1, for the order parameter exponent the following non-trivial, noise amplitude dependent result has been derived [Birner *et al.* (2002)]

$$\beta_{MF} = \sup\left(\frac{1}{p-1}, \frac{c^2}{2D}\right) . \tag{4.220}$$

This has been obtained by approximating the Laplacian in the Langevin equation by its mean-field analog. On the other hand in case of MN2 only a depinned (absorbing) phase exists in the mean-field limit.

The field theoretical action of MN1 depending on the order parameter field $\phi(x,t)$ and response field $\psi(x,t)$

$$S = \int d^dx dt[(\partial_t - D\nabla^2 + a + b\phi^{p-1})\phi - \frac{c^2}{2}\phi^2\psi^2] \tag{4.221}$$

can be obtained by a standard Gaussian transformation. Naive power-counting shows that the upper critical dimension is $d_c = 2$ and following standard perturbative techniques, one can perform an $\epsilon$-expansion below $d_c$. The RG flow-diagram is very similar to that of KPZ, in particular, for $d < 2$ the flow runs towards infinity for any noise amplitude, (no stable fixed point can be find by PRG), while for $d \geq 2$ there is a separatrix between the trivial fixed-point basin of attraction and the runaway-trajectories, between the weak- and the strong-coupling fixed points.

Since at the critical point the interface falls continuously to minus infinity, one can prove [Tu *et al.* (1997)] that the dynamical exponent of MN1 coincides with that of KPZ, the correlation length exponent, $\nu_\perp$ can be related to the $Z$ exponent of KPZ by $\nu_\perp = 1/(2Z-2)$ (also for MN2) and the critical exponent of the order-parameter ($\phi$) is $\beta > 1$. Numerical

Table 4.15 Critical exponents for the MN1 in $d = 3$, for weak and strong coupling regimes. $\beta_{MF}$ is from Eq. (4.220).

| | $\beta$ | $\nu_\perp$ | $Z$ | $\alpha$ | $\eta$ |
|---|---|---|---|---|---|
| Weak coupling | 0.97(05) | 0.50(5) | 2.00(0.5) | 1.0(1) | −0.10(05) |
| Strong coupling | 2.5(1) | 0.75(3) | 1.67(03) | 2.0(1) | −0.5(1) |
| Mean Field | $\beta_{MF}$ | 1/2 | 2 | $\beta_{MF}$ | 0 |

estimates **in one dimension** for the MN1 (bKPZ+) and MN2 (bKPZ-) class exponents are shown in Table 7.4.

In **three dimensions** two different universality classes — predicted by PRG calculation — can be recognized: for small $c$ values, very close to mean-field expectations have been reported, while for a larger $c$-s non-trivial exponents have been found [Genovese and Muñoz (1999)] (see Table 4.15).

# Chapter 5

# Scaling at First-Order Phase Transitions

At first-order transitions the correlation length is finite, the relaxation time decays exponentially and one can observe phase co-existence. In equilibrium systems several thermodynamical quantities, such as the internal energy and the order parameter (magnetization) exhibit discontinuity. As a consequence one cannot expect static universal scaling behavior. In nonequilibrium systems thermodynamical potentials cannot be defined in general, but there are phase transitions in which the order parameter changes discontinuously if we vary a control parameter. The definition of the order parameter is somewhat arbitrary, and there exist systems (see for example CDP in Sect. 4.5.1 or the NEKIM model Sect. 4.6.2), in which this quantity jumps at the transition point (the density in case of CDP) and some other static quantity (the percolation probability) changes continuously. This provides a possibility for scaling and universal behavior at first-order transitions in nonequilibrium models.

Dynamical quantities at first-order transitions may exhibit scaling behavior even if the static ones do not. This is due to the fact that correlations of an initial state can change very slowly (power-law manner) by short-ranged reactions and diffusion. For example in CDP (without external field) the ordered (absorbing) state can be reached by a power-law decay of density if we start from a homogeneous state with finite particle concentration. Other dynamical quantities, like the autocorrelation shows this behavior too.

There is a hypothesis based on the phase transition classification scheme advanced by [Elgart and Kamenev (2006)] that *bosonic, one-component RD systems with $n$-particle creation and $m$-particle annihilation always exhibit a first-order transition if $n > m$.* This is indeed sure above the upper critical dimension (Sect. 4.1.4). In lower-dimensional, site-restricted system several studies reported continuous transitions (at least for slow diffusion) [Dickman and Tomé (1991); Kockelkoren and Chaté (2003a); Cardozo and Fontanari (2006b)].

Interestingly first-order phase transitions have rarely been seen in **one dimension**. This is due to the fact that in lower dimensions fluctuations are relevant and can destabilize the ordered phase. Therefore fluctuation induced second ordered phase transitions are likely to appear. Some years ago [Hinrichsen (2000c)] advanced a hypothesis, that *first-order transitions do not exist in one-component* $(1+1)$*-dimensional systems without extra symmetries, conservation laws, special boundary conditions or long-range interactions (which can be generated by macroscopic currents or anomalous diffusion in nonequilibrium systems for instance).* This hypothesis is based on the observation that the "surface tension" of domains in such system does not depend on their size.

However simulations of diffusive RD models like the $3A \to 4A$, $A \to \emptyset$ triplet-creation model in **one dimension** found first-order transition for strong diffusion [Dickman and Tomé (1991); Cardozo and Fontanari (2006b)]. A possible resolution of this controversy could be that such diffusive models can be considered effectively multi-component system (see Sects. 6.7 and 6.9) as the PCPD model, since the solitary particles exhibit a different explicit diffusion than the implicit diffusion of reacting triplets. The reason for such behavior might be due to a 'declustering transition' as we increase the diffusion rate.

Examples for first-order transitions in **one dimension** can be found in the Glauber and the NEKIM Ising spin systems (see Sects. 2.3 and 4.6.2), **possessing** $Z_2$ symmetry [Glauber (1963); Menyhárd and Ódor (1998)]. In these models the introduction of a "temperature like" spin-flip inside an ordered domain, or an external field ($h$) cause discontinuous jump in the magnetization order parameter ($m$). Interestingly, the correlation length diverges at the transition point and static magnetization

$$m \propto \xi^{-\beta_s/\nu_\perp} g(h\xi^{\Delta/\nu_\perp}) \tag{5.1}$$

as well as cluster critical exponents can be defined:

$$P_s(t,h) \propto t^{-\delta_s} \tag{5.2}$$

$$R_s^2(t,h) \propto t^{z_s} \tag{5.3}$$

$$|m(t,h) - m(0)| \propto t^{\eta_s} \tag{5.4}$$

$$\lim_{t\to\infty} P_s(t,h) \propto h^{\beta_s'} \; . \tag{5.5}$$

Table 5.1 Critical, one-dimensional Ising spin exponents at the Glauber and NEKIM transition points. From [Menyhárd and Ódor (1998)].

| | $\beta_s$ | $\nu_\perp$ | $\beta_s{}'$ | $\Delta$ | $\eta_s$ | $\delta_s$ | $z_s$ |
|---|---|---|---|---|---|---|---|
| Glauber | 0 | 1/2 | .99(2) | 1/2 | .0006(4) | .500(5) | 1 |
| NEKIM | .00(2) | .444 | .45(1) | .49(1) | .288(4) | .287(3) | 1.14 |

Here $s$ refers to the spin variables. Table 5.1 summarizes the numerical results obtained for these transitions. Note that at these transitions the exponent $\beta'$ is non-zero, meaning that if we define the order parameter by the cluster survival probability instead of the density of active sites we may also call them continuous ones [Lübeck (2006a)].

Other examples of first-order transition in one dimension are known in **driven diffusive systems** [Janssen and Schmittman (1986)], in ASEP [Derrida (1998)] and TASEP [Reichenbach *et al.* (2007)] models (see Sect. 3.1.2), in asymmetric triplet models [Dickman and Tomé (1991)] (Sect. 4.1.4) and in the so-called bridge model [Godrèche *et al.* (1995)]. Discontinuous transitions are also reported in some **one-dimensional, multi-component RD model** such as the DCF models [Wijland *et al.* (1998); Maia and Dickman (2007)] (see Sect. 6.9), the Ziff-Gulari-Barshad model [Ziff *et al.* (1986)] and in case of **particle-number-conserving** RD system [Fiore and de Oliveira (2004)].

It is quite difficult to decide by simulations whether a transition is really discontinuous. The order parameter of weak first-order transitions — where the jump is small — may look very similar to continuous transitions. Measuring the the hysteresis of the order parameter that is considered to be the indication of a first-order transition is a demanding task. For examples of debates over the order of the transition see [Dickman and Tomé (1991); Hinrichsen (2000c); Tomé and de Oliviera (1989); Szolnoki (2000); Lipowski and Lopata (1999); Lipowski (1999); Hinrichsen (2001c)]).

In **two and higher dimensions** one-component RD models with $n > m > 1$ (see Sect. 4.1.4) exhibit first-order transition. This has been confirmed by simulations [Grassberger (1982a); Bidaux *et al.* (1989); Ódor (2003a); Cardozo and Fontanari (2006a); Windus and Jensen (2007)] and cluster mean-field technique [Ódor and Szolnoki (1996)] in case of SCA models (see Table 5.2). In nonequilibrium models with **competing model-A heath bathes** and $Z_2$ symmetry, first-order transitions were found in case of the lack of bulk noise (the noisy voter model [Oliveira *et al.* (1993)] see Fig. 2.1) and in a segregation model [Ódor (2008)] for example. In case of **competing Glauber and Kawasaki dynamics** [Tomé and de Oliviera

Table 5.2 Convergence of the critical point estimates in various ($y = 1, 2, 3$) two dimensional SCA calculated by n-cluster (GMF) approximation (see Sect. 4.2.3).

| $n$ | $y = 1$ | $y = 2$ | | $y = 3$ | |
|---|---|---|---|---|---|
| | $p_c$ | $p_c$ | $\rho(p_c)$ | $p_c$ | $\rho(p_c)$ |
| 1 | 0.111 | **0.354** | **0.216** | **0.534** | **0.372** |
| 2 | 0.113 | **0.326** | **0.240** | **0.455** | **0.400** |
| 4 | 0.131 | **0.388** | **0.244** | **0.647** | **0.410** |
| simulation | 0.163 | **0.404** | **0.245** | **0.661** | **0.418** |

(1989)] the ferromagnetic-paramagnetic phase transition changes from second to first-order by increasing the spin exchange [Szolnoki (2000)].

The initial condition can also be regarded a special boundary condition. It is known from equilibrium statistical mechanics [Lipowsky (1982)] that there exist systems with first-order bulk transition (at $T^*$) but continuous surface transition (at $T_c \neq T^*$). Such transitions have been analyzed by Landau-Ginzburg mean-field theory [Lipowsky (1982)] and DMRG technique in case of $q$-state Potts models [Iglói and Carlon (1999)]. This phenomenon is known as **"surface induced disorder"** and is closely related to wetting phenomena (see Sects. 7.3.1 and 7.4.4), since as $T_c$ is approached from below a layer of disordered phase intervenes between the surface and the bulk. At $T_c$ this interface becomes delocalized as in wetting. So in a nonequilibrium models infinite correlation length in the time direction can appear and various quantities like the magnetization, or the autocorrelation can exhibit power-law time dependences. First-order transitions are also observed in nonequilibrium wetting models, where a hight dependent potential term gives attractiveness to the wall [Hinrichsen *et al.* (1997a)].

Very recently the first-order transition has been investigated in a **two-dimensional** model-A Ising system of mixed two-spin and four-spin interactions following a quench from the high-temperature phase to the transition point [Pleimling and Iglói (2007)]. The magnetization starting with an uncorrelated initial state ($m_i$) was found to decay by a power-law asymptotically towards either to zero (for $m_i < 0.5$) or to some finite limiting value in case of $m_i > 0.5$ as

$$m(t) \propto t^{\alpha} \tag{5.6}$$

with $\alpha \approx -1$. This exponent was related to the critical initial slip of a second-order transition point (1.20) using a relation for the anomalous dimension of the initial magnetization ($x_0$) to the dimension of the interface $x_0 = d - 1$. Pleimling and Iglói assumed this exponent to be universal. On the other hand the connected autocorrelation function $(A(t,0))$ followed a stretched exponential decay

$$A(t,0) \propto \exp(-bt^a) \tag{5.7}$$

with an exponent $a = 0.197(4)$. The above universality hypothesis should be tested in other spin models exhibiting first-order phase transition.

## 5.1 Tricritical directed percolation classes (TDP)

In a model where both first and second order transition is possible one can a find a tricritical point with distinct scaling behavior where the first and the second order transition lines (or surfaces) meet in the phase diagram. More generally, multi-critical behavior appears when global, lower order reactions vanish in a system with competing reactions. It is well known from equilibrium statistical mechanics within the framework of Landau-Ginzburg theory that by adding a $\phi^6$ term to the $\phi^4$ model (the simplest model with second order transition)

$$F = \tau\phi^2 + g_2\phi^4 + g_4\phi^6 \tag{5.8}$$

one finds first-order transition if $g_4 < 0$ and a crossover if $\tau = g_4 = 0$.

The possibility of such behavior in dynamical RD systems was pointed out long time ago [Grassberger (1982a)]. If we supplement pair reactions with a triplet coagulation

$$\begin{aligned} 2A &\xrightarrow{k_1} 3A\ , \\ 2A &\xrightarrow{k_2} \emptyset\ , \\ 3A &\xrightarrow{k_3} 2A\ , \end{aligned} \tag{5.9}$$

the mean-field rate equation is

$$\dot{\rho} = b\rho^2 - c\rho^3 \tag{5.10}$$

with $b = k_1 - 2k_2, c = k_3$, which exhibits a tricritical point at

$$b = 0, \quad c > 0. \tag{5.11}$$

This model is in principle the PCP (Sect. 6.6) or with diffusive particles the PCPD model (Sect. 6.7.1), thus one may find tricritical behavior (at least for the mean-field case) at some values of the parameter space, albeit different from TDP.[1] Note that in case of particle number conservation

[1]As discussed by [Grassberger (2006)] one can find different noise in the Langevin formalism than in case of TDP.

such non-trivial tricritical point has already been found by field theory (Sect. 6.9).

One can also extend the unary reaction mean-field theory of DP (see Eq. (4.32)) with a higher order term

$$\partial_t \rho = \tau \rho - g\rho^2 - c\rho^3 \ , \tag{5.12}$$

where $\tau$ is the control parameter of the transition. Finite particle density occurs above the transition point ($\tau > 0$), whereas the absorbing state ($\rho = 0$) is approached from below the transition point. Negative values of $g$ give rise to a first-order phase transition, whereas $g = 0$ is associated with a tricritical point [Lübeck (2006b)]

$$\rho = 0 \quad , \quad \rho = -\frac{g}{2c} \pm \sqrt{\frac{\tau}{c} + \left(\frac{g}{2c}\right)^2} \ . \tag{5.13}$$

The first solution corresponds to the absorbing phase, which is stable for $\tau < 0$ and unstable for ($\tau > 0$). The solution with the $-$ sign is unphysical and unstable, while the $+$sign solution for $g < 0$ is stable if $\tau > -g^2/4c$, otherwise it is unstable (see Fig. 5.1). If the system is in the active phase ($\rho > 0$) the order parameter jumps at $\tau = -g^2/4c$ from $\rho = |g|/2c$ to zero. Hence the absorbing and the active phases coexist between the spinodal $\tau = -g^2/4c$ and the lines $\tau = 0$ for $g < 0$. The tricritical behavior (TDP) is obtained for $g = 0$. The density for $\tau > 0$ behaves as

$$\rho = (\tau/c)^{\beta_t} \tag{5.14}$$

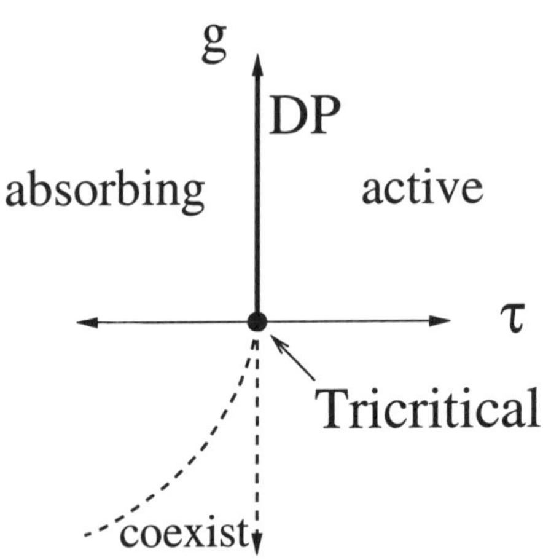


Fig. 5.1 Phase diagram of the mean-field TDP. Dashed lines correspond to the first-order transition limiting the phase coexistence, solid thin line denotes the critical DP transition.

with the tricritical order parameter exponent

$$\beta_t = 1/2. \tag{5.15}$$

For $g > 0$ one can observe a critical DP transition, with the usual $\beta_{MF} = 1$ mean-field exponent as $\tau \to 0$. In the active phase for $g \geq 0$ the order parameter obeys the scaling form

$$\rho = \lambda^{-\beta_t}\, \tilde{r}_{tDP}(\lambda\tau,\, g\lambda^{\phi}) \tag{5.16}$$

for all positive $\lambda$ with the crossover exponent

$$\phi = 1/2\ . \tag{5.17}$$

[Ohtsuki and Keyes (1987)] has set up general, field theoretical formalism for bosonic, multi-critical RD system with competing reactions. In a system composed of $2m - 1$ creation(+) and coagulation(−) reactions

$$lA \xrightarrow{k_{l-}} l'A \quad (l > l')\ , \tag{5.18}$$

$$lA \xrightarrow{k_{l+}} l''A \quad (l < l'')\ , \tag{5.19}$$

$$mA \xrightarrow{k_{m-}} m'A \quad (m > m')\ , \tag{5.20}$$

where $l = 1, 2, ...m - 1$, with $m$ being the regulating, highest order coagulation process, the relevant action is given by

$$S = \int dt d^d x [\psi(\partial_t - D\Delta)\phi + \sum_{l=1}^{m-1} D s_l \psi\phi^l + Du(\psi\phi^m - \psi^2\phi)]\ . \tag{5.21}$$

Here $D$ denotes the diffusion coefficient, as usual, and the multi-critical point is located at

$$s_1 = s_2 = ...s_{n-1} = 0\ , \tag{5.22}$$

meaning that the $l$-th order global reactions vanish due to the balance between creation and annihilation (this does not necessarily mean the absence of $l$-th order microscopic reactions). The engineering dimensions can be obtained as in Sect. 4.1.6 with $n = 1$ substitution:

$$[\psi] = \frac{m-1}{m} d\ , \quad [\phi] = \frac{1}{m} d\ . \tag{5.23}$$

The couplings can be expressed keeping the action dimensionless

$$[D] = \kappa^{-2} t^{-1}\ , \quad [s_l] = 2 - \frac{l-1}{m} d\ , \quad [u] = 2 - \frac{m-1}{m} d\ . \tag{5.24}$$

The upper critical dimension $d_c$ of the process can be obtained from the marginality condition $d_u(d_c) = 0$

$$d_c = 2m/(m-1)\ . \tag{5.25}$$

In case of TDP we need $m = 3$, hence $d_c = 3$. The perturbative RG analysis for $m = 3$ resulted in critical exponents in linear order of $\epsilon = d_c - d$ as shown in Table 5.3. These values satisfy up to $O(\epsilon^2)$ order the generalized hyperscaling relation (4.152), but since the rapidity symmetry (4.41) of DP is broken

$$\beta \neq \beta' \,. \tag{5.26}$$

[Lübeck (2006b)] have performed simulations in **two dimensions** for the following site restricted RD model. A contact process, exhibiting continuous DP type of transition

$$A \xrightarrow{p} 0\,, \quad A \xrightarrow{1-p} 2A \tag{5.27}$$

is supplemented with a binary creation

$$2A \xrightarrow{q} 3A\,. \tag{5.28}$$

We know from the previous section that a binary creation model with single particle annihilation possesses a first-order phase transition in two dimensions. Hence by tuning the relative strength $p/q$ we should see a crossover between continuous and discontinuous transitions. The critical exponents obtained by simulations are shown in Table 5.3.

Almost parallel to this study [Grassberger (2006)] investigated a generalized version of the Domany-Kinzel SCA (see Sect. 4.2.3) with extensive simulations. This SCA also exhibits a tricritical point in its two-dimensional parameter space. The obtained critical exponents show rather poor agreement with those of [Lübeck (2006b)] and with the field theory (see Table 5.3). The latter is surprising since at $d = 2$ we are close to the upper critical dimension, thus terms linear in $\epsilon$ should give rather good description in two dimensions. One has to admit that analyzing a tricritical point by simulations is quite hard, since two control parameters have to be fine tuned. If the scaling corrections are unknown a small deviation in the critical point coordinates causes a large error in the exponent estimates.

In **one dimension** the existence of a nontrivial TDP point is an open question. For unary creation one finds the trivial CDP point, as the end point of the critical line in Domany-Kinzel SCA (see Fig. 4.6), which exhibits a discontinuous phase transition. In case of higher order particle production processes (see [Dickman and Tomé (1991); Kockelkoren and Chaté (2003a); Cardozo and Fontanari (2006b)]) the behavior of the tricritical point has not been clarified yet (even its existence is debated [Hinrichsen (2000c)]).

Table 5.3 Critical exponents of TDP in 2+1 dimensions

| | [Grassberger (2006)] | $\epsilon$−expansion | [Lübeck (2006b)] |
|---|---|---|---|
| $\eta$ | −0.353(9) | $-0.0193\epsilon = -0.019$ | - |
| $\alpha$ | 0.087(3) | $1/2 - 0.4677\epsilon = 0.032$ | - |
| $\delta$ | 1.218(7) | $1 - 0.0193\epsilon = 0.981$ | - |
| $Z$ | 2.110(6) | $2 + 0.0086\epsilon = 2.009$ | - |
| $\beta$ | 0.100(4) | $1/2 - 0.4580\epsilon = 0.042$ | 0.14(2) |
| $\beta'$ | 1.408(10) | $1 + \mathcal{O}(\epsilon^2)$ | - |
| $\nu_\parallel$ | 1.156(4) | $1 + 0.0193\epsilon = 1.019$ | - |
| $\nu_\perp$ | 0.547(3) | $1/2 + 0.0075\epsilon = 0.507$ | 0.59(8) |
| $d_f$ | 1.817(8) | $2 - 0.0690\epsilon = 1.931$ | 1.76(5) |
| $\eta_\perp$ | 0.366(16) | $1 - 0.8620\epsilon = 0.138$ | 0.42(24) |
| $\phi$ | 0.36(4) | $1/2 - 0.0121\epsilon = 0.488$ | 0.55(3) |

## 5.2 Tricritical DIP classes

The dynamical isotropic percolation (Sect. 4.4) line followed by a first-order transition has been investigated by field theory by [Janssen *et al.* (2004a); Janssen (2005)]. One can generalize the GEP by introducing individuals as a further species. This species is weakened by an infection in its neighbor and not made immune. Hence a second infection in its neighborhood makes the weakened individual more susceptible. By introducing this instability the transition may become discontinuous. In the field theoretical action of DIP (4.106) this causes a higher order term with the coupling $w'$

$$S = \int d^dx dt \frac{D}{2}\psi^2\phi - \psi\left[\partial_t - \nabla^2 - r + w\int_0^t ds\phi(s) + w'\left(\int_0^t \phi(s)ds\right)^2\right]\phi\,. \tag{5.29}$$

The tricritical point of the mean-field theory is at $w = r = 0$ and the upper critical dimension is $d_c = 5$. This new term breaks the symmetry of DIP (4.107), thus the hyperscaling (4.108) is not valid either (in this case the generalized hyperscaling relation (4.152) is fulfilled) and

$$\beta \neq \beta' \tag{5.30}$$

as in case of DIP. The critical exponents have been obtained up to one loop order in $\epsilon = 5 - d$:

$$\nu_\perp = \frac{1}{2} + \frac{11}{450}\epsilon\ ,\ \beta = \frac{1}{2} - \frac{17}{45}\epsilon\ ,\ \beta' = 1 - \frac{2}{45}\epsilon\ \ \phi = \frac{1}{2} - \frac{1}{45}\epsilon\,. \tag{5.31}$$

The percolation cluster scaling exponents coincide with those of DIP in the mean-field limit, otherwise

$$\sigma = \frac{1}{2} + O(\epsilon^2)\ ,\ \tau = \frac{5}{2} - \frac{1}{45}\epsilon\ ,\ \gamma = 1 + \frac{2}{45}\epsilon\ \ \delta_p = 3 + \frac{8}{5}\epsilon\,. \tag{5.32}$$

Models of this type were studied some time ago by [Cieplak and Robbins (1988); Koiller and Robbins (2000)].

# Chapter 6

# Universality Classes of Multi-Component Systems

These genuine nonequilibrium classes emerge as the combination of the simple models discussed in previous chapters by coupling. An interesting feature is that the Galilean symmetry is broken in them in contrast with the simple RD models. They are also related to surface growth classes, which I overview in Chapter 7. First I recall some well known results for multi-component, reaction-diffusion systems without particle creation (for a detailed review see [Privman (1996)]) From the viewpoint of phase transitions these describe the sub-critical behavior generally. In models, where the transition occurs at zero offspring production rate (like in the $N$-component BARW models) this is just the dynamical critical behavior. Then I shall discuss the effect of particle exclusions in one dimension. Later in the chapter I shall present universal behavior of more complex, coupled multi-component systems. This field is under development and some of the results are still under debate.

## 6.1 The $A + B \to \emptyset$ classes

The simplest two-component reaction-diffusion model involves two types of particles undergoing diffusive random walks and reacting upon contact to form an inert particle. The action of this model is:

$$S = \int d^d x dt [\psi_A(\partial_t - D_A\nabla^2)\phi_A + \psi_B(\partial_t - D_B\nabla^2)\phi_B - \lambda(1 - \psi_A\psi_B)\phi_A\phi_B] - \rho(0)(\psi_A(0) + \psi_B(0)) \tag{6.1}$$

where $D_A$ and $D_B$ denotes the diffusion constants of species $A$ and $B$. In $d < d_c = 4$ dimensions and for **homogeneous, initially equal density**

**of $A$ and $B$ particles** ($\rho_0$) the density decays asymptotically as

$$\rho_A(t) = \rho_B(t) \propto C\sqrt{\Delta}t^{-d/4} , \tag{6.2}$$

where $\Delta = \rho(0) - C'\rho^{d/2}(0) + ...$, $C$ is a universal $C'$ is a non-universal constant [Burlatskii and Ovchinnikov (1978); Kang and Redner (1985); Bramson and Lebowitz (1988); Lee and Cardy (1995)]. This slow decay behavior is due to the fact that in the course of the reaction, local fluctuations in the initial distribution of reactants lead to the formation of clusters of like particles that do not react and they will be asymptotically **segregated** for $d < 4$. The asymptotically dominant process is the diffusive decay of the fluctuations of the initial conditions. Since this is a short-ranged process, the system has a long-time memory — appearing in the amplitude dependence — for the initial density $\rho(0)$.

For **segregated or non-equal initial $A$ and $B$ concentrations** the upper critical dimension is $d_c = 2$. For $d \leq 2$ a controlled RG calculation is not possible, but the result Eq. (6.2) gives the leading order term in $\epsilon = 2 - d$ expansion. For the case of $D_A \neq D_B$ a RG study [Lee and Cardy (1995)] found a new amplitude but the same exponents. If $\rho_A(0) > \rho_B(0)$, the asymptotic dynamics will approach a steady concentration of $\rho_B(0) - \rho_A(0)$ particles, with very few isolated $A$ particles surviving. In this situation exact results indicate an exponentially decaying $A$ particle density for $d > 2$, logarithmic corrections to an exponential in $d = 2$ and a stretched exponential

$$\rho \propto \exp(-c\sqrt{t}) \tag{6.3}$$

form for $d = 1$. In case of segregated initial state, where a $(d-1)$-dimensional surface separates the two species at time $t = 0$ reaction zone forms [Gálfi and Rácz (1988)] and the particle densities in the reaction zone scale as

$$\rho \propto t^{-1/3} . \tag{6.4}$$

Furthermore it is easy to see that in case of pairwise ordered initial conditions (..AA..BBBB.AA...) in one dimension the density decay of both species is the same as of ARW

$$\rho \propto t^{-1/2} \tag{6.5}$$

(see also Sect. 6.2).

The study of the $A + B \rightarrow \emptyset$ model was motivated originally by the cosmological problem of matter-antimatter annihilation [Ovchinnikov and

Zel'dovich (1978)]. Independently a few years later [Toussaint and Wilczek (1983)], asked whether in such simple annihilation model it would be possible that locally in space only matter would be left. Now it is well known that the $A + B \to \emptyset$ system exhibits segregation. The non-classical decay: $\rho \sim t^{-3/4}$ has experimentally been confirmed in **three dimensions** in a calcium / fluorophore reaction system [Monson and Kopelman (2004)].

The **persistence** behavior in one dimension with homogeneous, equal initial density of particles ($\rho_0 = \rho_A(0)+\rho_B(0)$) was studied by [O'Donoghue and Bray (2001)]. The probability $p(t)$, that an annihilation process has not occurred at a given site ("type I persistence") has the asymptotic form

$$p(t) \sim \text{const.} + t^{-\theta_l} \ . \tag{6.6}$$

For a density of particles $\rho \gg 1$, $\theta_l$ is identical to that governing the persistence properties of the one-dimensional diffusion equation, where $\theta_l \approx 0.1207$. In the case of an initially low density, $\rho_0 << 1$, $\theta_l \approx 1/4$ was found asymptotically. The probability that a site remains unvisited by any random walker ("type II persistence") decays in a stretched exponential way:

$$p(t) \sim \exp(-\text{const}\times\rho_0^{1/2} t^{1/4}) \tag{6.7}$$

provided $\rho_0 \ll 1$.

### 6.1.1 *Reversible $A + A \rightleftharpoons C$ and $A + B \rightleftharpoons C$ class*

[Zel'dovich and Ovchinnikov (1977)] introduced an extension of the simple pair annihilation models (Sect. 6.1), and the PRG method was applied for it [Rey and Cardy (1998)] to see what happens if a backward combination with some probability is also allowed

$$A + A \rightleftharpoons C, \quad A + B \rightleftharpoons C \, . \tag{6.8}$$

In most physically interesting systems it is unlikely that this possibility is totally forbidden.[1] Rey and Cardy found that even in case of extremely small backward probability fundamental change occurs in the system. Instead of decaying to a nonequilibrium steady state the system eventually reach an equilibrium state. Moreover a conserved quantity can be constructed $\rho_A(t) + 2\rho_C(t)$ for the one-species, $\rho_A(t) + \rho_B(t) + 2\rho_C(t)$ for the two-species case. Starting from random initial conditions, the approach of the $C$ species to its equilibrium density takes the form

$$\rho_C(t) - \rho_C(\infty) \propto t^{-d/2} \tag{6.9}$$

[1] These models are also similar to those discussed in Sect. 4.8.

in both cases and in all dimensions [Kang and Redner (1985)]. The exponent follows directly from the conservation laws and is universal, whereas the amplitude of the decay turns out to be model-dependent. In case of correlated initial conditions or unequal diffusion constants, the system exhibits more complicated behavior, including a non-monotonic approach to equilibrium.

### 6.1.2 *Anisotropic* $A + B \to \emptyset$

Based on a Burgers equation solution [I. Ispolatov (1995)] argued that in the two-species annihilation, in which all particles undergo strictly biased motion in the same direction and with an excluded volume repulsion between same species particles should exhibit super-diffusive behavior and density decay speeds up. For this system, one might naively expect that, by performing a Galilean transformation (4.5.5) to a zero momentum reference frame, the long-time kinetics should be the same as that of the isotropic model. Such a Galilean universality has been derived for the single species ARW (and for special type of two-species reactions) (see Sect. 4.5.5) [Privman *et al.* (1995)]. In the driven $A + B \to \emptyset$ model extensive numerical simulations [Janowsky (1995)] in **one dimension** showed a faster decay than in the isotropic case

$$\rho \propto t^{-1/3} , \tag{6.10}$$

in agreement with the analytical results [I. Ispolatov (1995)]. The drift is irrelevant for $d \geq 2$ dimensions and the reaction kinetics is the same as two-species annihilation with isotropic diffusion. This means that the Galilean invariance is broken in one dimension.

One can also add a linear shear flow $\mathbf{v} = v_0 y \hat{\mathbf{x}}$, where $\hat{\mathbf{x}}$ is a unit vector in the $x$-direction and investigate the effect of mixing of the reactants. We may expect that this weakens the appearance of segregation. Indeed PRG and simulations [Howard and Barkema (1996)] confirmed that the upper critical dimension is lowered to $d_c = 2$ and the asymptotic density decay speeds up to

$$\rho \propto t^{-(d+2)/4} , \tag{6.11}$$

in $d < 2$ dimensions. In the mean-field regime $d \geq 2$ the decay exponent remains $\alpha = 1$.

### 6.1.3 *Disordered $A + B \to \emptyset$ models*

The $A + B \to \emptyset$ model in quenched random velocity field was considered first by [Oerding (1996b)] by bosonic PRG. In this case (similarly to impure ARW Sect. 4.5.6) the process becomes super-diffusive below $d_c = 3$ and the asymptotic density decay process accelerates to

$$\rho = t^{-(d+3)/8} \tag{6.12}$$

up to first order in $\epsilon = 3 - d$. A simulation study applied power-law distributed hopping probabilities

$$p(y) = \nu y^{\nu-1} \tag{6.13}$$

with $0 < \nu \leq 1$ and $0 < y \leq 1$ in **one dimension** to the bosonic model [Lee (2000)]. The density decay exponent has been found to be unchanged ($\alpha = 1/4$) for $\nu > 0.4$ but decreases monotonically in case of stronger disorder ($\nu < 0.4$). This is in agreement with the results of an other recent Monte Carlo study of the site restricted model with uniformly distributed, quenched disorder with amplitude $\epsilon_D$ [Menyhárd and Ódor (2007)]. In this case the weak disorder ($\epsilon_D \leq 1$) did not make any effect. On the other hand for *very strong disorder in the diffusion* (hopping or velocity) probabilities logarithmic corrections to the scaling of the impure case can be seen (see Fig. 6.1). The same kind of disorder in the reaction rates did not have any effect even for $\epsilon_A \to 1$.

## 6.2 $AA \to \emptyset$, $BB \to \emptyset$ with hard-core exclusion

The next simplest two-component model in which particle blocking may be effective in low dimensions was investigated first in the context of a stochastic cellular automaton model. At the symmetric point of the GDK model (Sect. 4.6.3) compact domains of $I1$ and $I2$ grow separated by $A = Ac - I1$ and $B = Ac - I2$ kinks that cannot penetrate each other. In particle language this system is a reaction-diffusion model of two types $A + A \to \emptyset$, $B + B \to \emptyset$ with exclusion $AB \not\leftrightarrow BA$ and special **pairwise initial conditions** (because the domains are bounded by kinks of the same type):

```
....A...A...B.B..B.....B..A..A..
```

In the case of homogeneous, pairwise initial conditions simulations by [Ódor and Menyhárd (2000)] showed a density decay of kinks $\rho \propto t^{-\alpha}$ characterized by a power-law with an exponent somewhat larger than $\alpha = 0.5$. The

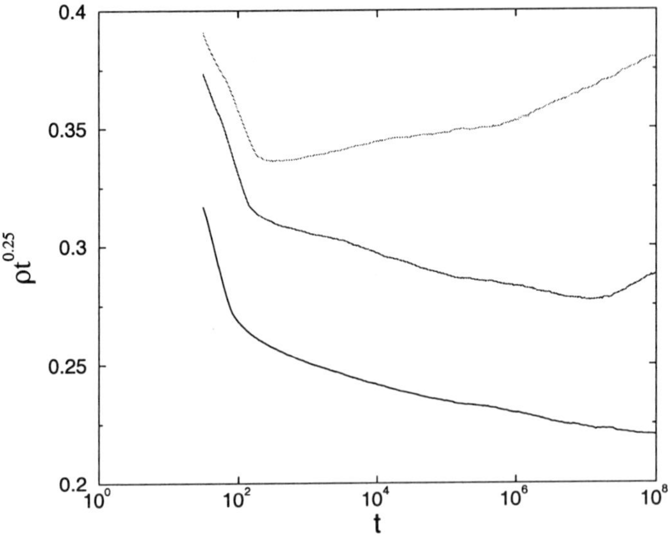


Fig. 6.1 Density decay in the diffusion disordered $AB \to \emptyset$ model for random initial conditions. Different curves correspond to disorder amplitudes $\epsilon_D = 1, 0.95, 0.5$ (top to bottom). From [Menyhárd and Ódor (2007)].

$\alpha = 1/2$ would have been expected in case of two copies of ARW systems that do not exclude each other. Furthermore the deviation of $\alpha$ from 1/2 showed an initial density dependence. [Ódor and Menyhárd (2000)] provided a possible explanation based on permutation symmetry between types, according to which hard-core interactions cause **marginal perturbation** resulting in non-universal scaling. The situation is similar to that of the compact directed percolation that is confined by parabolic boundary conditions (see Sect. 4.5.3) if we assume that $AB$ and $BA$ pairs exert parabolic space-time confinement on coarsening domains. Non-universal scaling can also be observed at surface critical phenomena. Similarly here $AB$, $BA$ pairs produce 'multisurfaces' in the bulk. However, simulations and independent interval approximations in a similar model predict logarithmic corrections to the single component decay with the form $\rho \sim t^{-1/2}/\ln(t)$ [Majumdar *et al.* (2001)]. Note that both kinds of behavior may occur in case of marginal perturbations. Cluster simulations [Ódor and Menyhárd (2000)] also showed initial density ($\rho_{I1}(0)$) dependent survival probability of $I2$-s in the sea of $I1$-s:

$$P_{I2}(t) \propto t^{-\delta(\rho_{I1}(0))} . \tag{6.14}$$

However this reaction-diffusion model with homogeneous, **random initial distribution** of $A$-s and $B$-s exhibits a much slower density decay. An exact duality mapping helps to understand the coarsening behavior. Consider the leftmost particle, which may be either $A$ or $B$, and arbitrarily relabel it as a particle of species $X$. For the second particle, we relabel it as $Y$ if it is the same species as the initial particle; otherwise we relabel the second particle as $X$. We continue to relabel each subsequent particle according to this prescription until all particles are relabeled from $\{A, B\}$ to $\{X, Y\}$. For example, the string

$$AABABBBA\cdots$$

translates to

$$XYYYYXYY\cdots.$$

The diffusion of the original $A$ and $B$ particles at equal rates translates into diffusion of the $X$ and $Y$ particles. Furthermore, the parallel single-species reactions, $A + A \to \emptyset$ and $B + B \to \emptyset$, translate directly to two-species annihilation $X + Y \to \emptyset$ (see Sect. 6.1) in the dual system. The interesting point is that in the $X + Y \to \emptyset$ model blockades do not exist, because $XY$ pairs annihilate, and there is no blockade between $XX$ and $YY$ pairs. Therefore the density decay should be proportional to $t^{-1/4}$. Simulations confirmed this for the $A + A \to \emptyset$, $B + B \to \emptyset$ [Ódor (2001c)] model. Nevertheless corrections to scaling were also observed. The pairwise initial condition transforms in the dual system to domains of $..XYXY..$ separated by $YY$ and $XX$ pairs, which do not allow $X$ and $Y$ particles to escape each other.

## 6.3 Symmetrical, multi-species $A_i + A_j \to \emptyset$ (q-MAM) classes

One can generalize the diffusion-limited reactions $AB \to \emptyset$ (Sect. 6.1) for $q > 2$ species [ben Avraham and Redner (1986)]

$$A_i + A_j \stackrel{k_{ij}}{\to} \emptyset , \tag{6.15}$$

where $1 \leq i < j \leq q$, thus no self-annihilation is allowed (q-MAM model) and the permutation symmetry of particle species is prescribed. For **unequal initial densities or different reaction rates** between the species, we generically expect the same scaling as in the $AB \to \emptyset$ model asymptotically, when only the two most numerous, or least reactive, species remain. In $d \geq 2$ dimensions with **equal rates and initial densities** particle **species segregation cannot occur**, hence the asymptotic density decay is the same as for ARW (Sect. 4.5.1). The q-MAM model has the upper critical dimension $d_c = 2$ and below this the recurrence properties of random walkers lead to depletion zones around each surviving particle. The annihilation kinetics induce particle anti-correlations.

In $d = 1$, however, particles of the same types can not pass each other and a segregation occurs for all $q < \infty$. The total density decays according to a $q$ dependent power law, $\rho \propto t^{-\alpha(q)}$ with

$$\alpha = (q-1)/2q \tag{6.16}$$

exactly [Deloubriére *et al.* (2002)]. For $q = 2$ this recovers the $AB \to \emptyset$ density decay law, while in the $q \to \infty$ limit we see the decay of ARW, since the effect of blockades vanishes. These findings were also supported through Monte Carlo simulations and the results are summarized in Table 6.1. Note

Table 6.1 Summary of the density decay exponent $\alpha$ of the q-MAM model. The asterisk (*) indicates segregation.

| | $q = 2$ $A + B \to 0$ | $q = 3, 4, \ldots$ | $q = \infty$ $A + A \to 0$ |
|---|---|---|---|
| $d \geq 4$ | 1 | 1 | 1 |
| $2 < d < 4$ | $d/4$ (*) | | |
| $d = 2$ | | 1 (log) | 1 (log) |
| $1 < d < 2$ | | | $d/2$ |
| $d = 1$ | 1/4 (*) | $(q-1)/2q$ (*) | 1/2 |

that **special initial conditions** such as ...$ABCDABCD$... prevent the segregation and lead to decay law of the $2A \to \emptyset$ model (Sect. 4.5.1).

In the $AB \to \emptyset$ system the segregation is usually explained heuristically through arguments, invoking the **local conservation** of the difference between the number of $A$ and $B$ particles. One may also construct models, in which **segregation occurs in dimensions** $d > 2$ despite the absence of any microscopic conservation law [Hilhorst *et al.* (2004)]. Such special situations arise when the $k_{ij}$ reaction rates are chosen such that certain subsets of the $A_i$ are equivalent under a symmetry operation (but lower than the permutation symmetry of q-MAM). A case of lower symmetry occurs, for example, when the particle species are ordered cyclically and annihilation, with rate $k_1$, is possible only between particles of two neighboring species along the cycle. Conditions for the matrix $k_{ij}$ have been determined under which a critical segregation dimension $d_{seg}$ exists. Below $d_{seg}$ the spatial uniformity condition (necessary for q-MAM decay) is violated; the symmetry between the species is then locally broken. In those cases the density decay slows down to

$$\rho(t) \propto t^{-d/d_{seg}} \tag{6.17}$$

for $2 < d < d_{seg}$. When $d_{seg}$ exists, its value can be expressed in terms of the ratio of the smallest to the largest eigenvalue of $k_{ij}$. It means that the **conservation law is sufficient but not a necessary** condition for segregation.

## 6.4 Heterogeneous, multi-species $A_i + A_j \to \emptyset$ system

One can further generalize the $A_i + A_j \to \emptyset$ multi-species models (Sect. 6.1) by allowing self-annihilation reactions [Krapivsky *et al.* (1994)] and different diffusion coefficients. Such a situation arises naturally when the reactants have different masses. To simplify the problem let's consider a two component systems where one of the two species is greatly in the majority (as is almost always the case asymptotically). For example, if the $A$ species is predominant, then we can safely neglect the reaction $B + B \to \emptyset$. This problem in one dimension is equivalent to the persistence of the zero temperature Glauber Ising model (see Sect. 2.4.3). In case of **immobile impurities** a PRG study by [Cardy (1995)] showed that the density of the minority species decays away with a universal power law:

$$\rho \propto t^{-\alpha} \tag{6.18}$$

for $d < 2$, where $\alpha = \frac{1}{2} + O(\epsilon)$ and $\epsilon = 2 - d$. induced by the annihilation kinetics. [Grassberger and Procaccia (1982)] have analyzed the model, in which the **majority species is immobile**. In this case the decay rate of the minority species is dominated by minority impurity particles existing in regions, where there is only a few of the majority particles present. Since these majority particles are strictly stationary, this situation is not describable using a rate equation approach, and it turns out that the minority species decays away with the stretched exponential law

$$\rho \propto e^{-t^{d/(d+2)}} . \tag{6.19}$$

This result is not accessible by PRG method either.

Typically the density of the species with the smallest diffusion coefficient has the same time dependence as that of in ARW, but the more mobile species decays with non-universal exponents, which depends on the ratios of diffusivities. Smoluchowski type approximations [Krapivsky *et al.* (1994)] and PRG [Howard (1996)] have been applied for the simple system

$$A + A \to \emptyset \qquad B + B \to \emptyset \qquad A + B \to \emptyset. \tag{6.20}$$

In particular for $\rho_A(0) \gg \rho_B(0)$ initial conditions and $\delta = (D_B/D_A) \leq 1$ ratio of diffusion constants the minority type for intermediate time decays with the exponent

$$\alpha \approx \frac{d}{2}\left(\frac{\delta+1}{2}\right)^{d/2} . \tag{6.21}$$

The majority type follows ARW law, which for $\delta = 0$ gives $\alpha = \frac{1}{2} + O(\epsilon)$ in agreement with [Cardy (1995)]. [Howard (1996)] has gone beyond the tree level calculation by including one loop diagrams and obtained an improved value for the decay exponent, which is in good agreement with simulations [Krapivsky *et al.* (1994)] and exact results [Derrida (1995)] in $d = 1$.

However, for $\delta < 1$, the system crosses over to a second regime, where $\rho_B \gg \rho_A$ (similarly as if we began with $\rho_B(0) \gg \rho_A(0)$). Here the majority ($B$) decays ARW type and the minority ($A$) exhibits a power-law with the exponent

$$\alpha' \approx \frac{d}{2}\left(\frac{1+\delta^{-1}}{2}\right)^{d/2} . \tag{6.22}$$

In **two dimensions** one finds extra logarithmic factors multiplying the power law decay rates. In case of **equal diffusion constants** (as a model for a steric RD system) for $d \leq d_c = 2$ the densities of both species always decay at the same rate asymptotically, in contrast with the predictions of the mean-field theory [Howard (1996); Konkoli *et al.* (1999)]. This result follows from the indistinguishability of the two species at large times. This means that below the upper critical dimension, the reaction rates run to identical fixed points. Since the diffusion constants are also equal, there is then no way to asymptotically distinguish between the two species, whose densities must therefore decay at the same rate.

[Howard and Täuber (1997)] have analyzed the **mixed annihilation/'scattering' process**

$$2A \to \emptyset, \quad 2A \to 2B, \quad 2B \to 2A, \quad 2B \to \emptyset\ . \tag{6.23}$$

They found that in $d < 2$ dimensions in all orders of PRG this problem reduces to the ARW. This is due to the re-entrance property of random walks (in $d < 2$ dimensions). As soon as two like particles approach each other, they annihilate regardless of the competing 'scattering' processes, which only produce particle pairs in a close proximity. Therefore with a large probability an immediate subsequent annihilation occurs.

## 6.5 Unidirectionally coupled ARW classes

Besides the symmetrically coupled ARW systems discussed before now let us turn to unidirectionally coupled ARW (Sect. 4.5.1) models:

$$\begin{array}{ll} A+A \to \emptyset & A \to A+B \\ B+B \to \emptyset & B \to B+C \\ C+C \to \emptyset & C \to C+D \\ \dots & \end{array} \tag{6.24}$$

These models were introduced and analyzed with the RG technique and simulations by [Goldschmidt (1998)]. This kind of coupling was chosen because $A \to B$ would constitute a spontaneous death process of $A$ particles leading to exponential density decay. On the other hand quadratic coupling of the form $A + A \to B + B$ leads to asymptotically decoupled systems [Howard and Täuber (1997)]. The mean-field theory is described by the rate equation for the density $\rho_i(x,t)$ at level $i$:

$$\frac{\partial \rho_i(x,t)}{\partial t} = D\nabla^2 \rho_i(x,t) - 2\lambda_i \rho_i(x,t)^2 + \sigma_{i,i-1}\rho_{i-1}(x,t) \ , \tag{6.25}$$

which relates the long time behavior of level $i$ to level $i-1$:

$$n_i(t) \propto n_{i-1}^{1/2} \ . \tag{6.26}$$

By inserting into this the exact solution for ARW (Sect. 4.5.1) one gets:

$$\rho_i(t) \sim \begin{cases} t^{-d/2^i} & \text{for } d < 2 \ , \\ \left(t^{-1}\ln t\right)^{1/2^{i-1}} & \text{for } d = d_c = 2 \ , \\ t^{-1/2^{i-1}} & \text{for } d > 2 \ . \end{cases} \tag{6.27}$$

The action of a two-component system with fields $a$,$\hat{a}$, $b$, $\hat{b}$ for equal annihilation rates ($\lambda$) takes the form

$$S = \int d^dx \int dt \Big[ \hat{a}(\partial_t - D\nabla^2)a - \lambda(1-\hat{a}^2)a^2 + \hat{b}(\partial_t - D\nabla^2)b - \lambda(1-\hat{b}^2)b^2 + \sigma(1-\hat{b})\hat{a}a \Big] \ . \tag{6.28}$$

The RG solution is plagued by infrared-divergent diagrams similarly to UCDP (see Sect. 7.6), which can be interpreted as evidence of an eventual *nonuniversal* crossover to the decoupled regime. Simulation results — exhibiting finite particle numbers and coupling strengths — really show the breakdown of scaling Eq. (6.27), but the asymptotic behavior could not be determined. Therefore the results (6.27) are valid for an intermediate time region.

## 6.6 DP coupled to frozen field classes

One of the first generalizations of absorbing phase transition models were such systems, in which percolation is coupled to frozen fields. These models exhibit many absorbing states, hence not fulfilling the conditions of the DP hypothesis [Janssen (1981); Grassberger (1982a)] (Sect. 3.1.4). Several variants of models with infinitely many absorbing states containing frozen particle configurations have been introduced. In these models non-diffusive (slave) particles are coupled to a DP-like (order parameter) process. In case of homogeneous, uncorrelated initial conditions DP-class exponents have been found, whereas in cluster simulations — which involve a correlated initial state of the order parameter particles — initial density dependent scaling exponents ($\eta$ and $\delta$) arise. These cluster exponents take the DP-class values only if the initial density of the slave particles agrees with the "natural density" that occurs in the steady state. The first such models introduced by [Jensen and Dickman (1993a); Jensen (1993a)] were the so-called pair contact process (PCP) (see Sect. 6.6.1) and dimer the reaction model. These systems seem to be single component ones when the rules for the pairs are defined, but the isolated, frozen particles behave as a second component. In the threshold transfer process (TTP) [Mendes *et al.* (1994)] the two components defined as the 2'-s, which follow the DP process, and the 1'-s, which decay or reappear as: $\emptyset \xrightarrow{r} 1 \xrightarrow{1-r} \emptyset$.

For the PCP model defined by the simple processes (6.31), a set of coupled Langevin equations were set up [Muñoz *et al.* (1998, 1996)] for the fields $n_1(\mathbf{x},t)$ and $n_2(\mathbf{x},t)$ (order parameter):

$$\frac{\partial n_2}{\partial t} = [r_2 + D_2\nabla^2 - u_2 n_2 - w_2 n_1]n_2 + \sqrt{n_2}\eta_2$$

$$\frac{\partial n_1}{\partial t} = [r_1 + D_1\nabla^2 - u_1 n_2 - w_1 n_1] + \sqrt{n_2}\eta_1 \qquad (6.29)$$

where $D_i, r_i, u_i$, and $w_i$ are constants and $\eta_1(\mathbf{x},t)$ and $\eta_2(\mathbf{x},t)$ are Gaussian white noise terms. Owing to the multiple absorbing states and the lack of the time reversal symmetry (4.41) a generalized hyperscaling law (1.45) was derived by [Mendes *et al.* (1994)]. As discussed by [Muñoz *et al.* (1998)] this set of equations can be simplified by dropping the $D_1$, $u_1$, and noise terms in the $n_1$ equation, and then solving that equation for $n_1$ in terms of $n_2$. Substituting this result for $n_1$ into Eq. 6.29 yields

$$\begin{aligned}\frac{\partial n_2(\mathbf{x},t)}{\partial t} &= D_2\nabla^2 n_2(\mathbf{x},t) + m_2 n_2(\mathbf{x},t) - u_2 n_2^2(\mathbf{x},t) + \\ &+ w_2(r_1/w_1 - n_1(\mathbf{x},0))n_2(\mathbf{x},t)e^{-w_1\int_0^t n_2(\mathbf{x},s)ds} \\ &+ \sqrt{n_2(\mathbf{x},t)}\eta_2(\mathbf{x},t) \ , \end{aligned} \tag{6.30}$$

where $n_1(\mathbf{x},0)$ is the initial condition of the $n_1$ field, and $m_2 = r_2 - w_2 r_1/w_1$. The "natural density"[Jensen and Dickman (1993a)], $n_1^{nat}$, then corresponds to the uniform density, $n_1(t=0) = r_1/w_1$, for which the coefficient of the exponential term vanishes, and we get back the Langevin equation for DP (4.45). This derivation provides a simple explanation for the numerical observation of DP exponents in the case of natural initial conditions. However it does not take into account the long-time memory and the fluctuations of passive particles (with power-law time and $p$ dependences [Odor *et al.* (1998)]). Therefore, some of the terms omitted from this derivation (for instance the term proportional to $n_2^2$ in the equation for $n_1$) cannot be safely eliminated [Marques *et al.* (2001)] and this simplified theory does not generate critical fluctuations for its background field. This treatment has still not provided theoretical proof for the initial-density-dependent spreading exponents observed in simulations [Jensen and Dickman (1993a); Jensen (1993a); Mendes *et al.* (1994); Odor *et al.* (1998)] and in the numerical integration of the Langevin equation [López and Muñoz (1997)]. Furthermore the situation is much more complicated when approaching criticality from the **inactive phase**. In particular, the scaling behavior of $n_1$ in this case seems to be unrelated to $n_2$ (this is similar to the diffusive slave field case [Ódor *et al.* (2002)] Sect. 6.7.3). In this case it is more difficult to analyze the field theory and dynamical-percolation-type terms are generated that can be observed in two dimensions by simulations and by mean-field analysis [Muñoz *et al.* (1998, 1996); Wijland (2002)]. Recently it was claimed, based on the field theoretical analysis of the GEP model (see Sect. 4.4) — which exhibits similar long-time memory terms — that the cluster variables should follow stretched exponential decay behavior [J.-Dalmaroni and Hinrichsen (2003)].

In **two dimensions** the critical point of spreading ($p_s$) moves (as the function of initial conditions) and do not necessarily coincide with the bulk critical point ($p_c$). The spreading behavior depends on the coefficient of the exponential, non-Markovian term of (6.30). For a positive coefficient the $p_s$ falls in the inactive phase of the bulk and the spreading follows **dynamical percolation** (see Sect. 4.4). For negative coefficient the $p_s$ falls in the active phase of the bulk and spreading exponents are **nonuniversal** (as in one dimension) but satisfy the hyperscaling law (1.45).

### 6.6.1 *The pair contact process (PCP) model*

Up to now I discussed spreading processes with unary particle production. Now I introduce a family of systems with **binary** particle production (i.e. for a new particle to be produced, two particles are need to collide). The pair contact process (PCP) is defined on the lattice by the following processes:

$$2A \stackrel{1-p}{\longrightarrow} 3A, \quad 2A \stackrel{p}{\to} \emptyset \ , \tag{6.31}$$

such that reactions take place at nearest-neighbor (NN) sites and we allow single-particle occupancy at most. The order parameter is the density of NN pairs $\rho_2$. The PCP exhibits an active phase for $p < p_c$; for $p \geq p_c$ the system eventually falls into an absorbing configuration devoid of NN pairs (but that contains a density $\rho_1$ of isolated particles). The best estimate for the critical parameter **in one dimension** is $p_c = 0.077090(5)$ [Dickman and da Silva (1998)]. Static and dynamic exponents that correspond to initially uncorrelated homogeneous states agree well with those of $1+1$-dimensional DP (4.2). Spreading exponents that involve averaging over all runs, hence involving the survival probability are nonuniversal (see Fig. 6.2). [Odor *et al.* (1998)]. The anomalous critical spreading in this model can be traced to a long memory in the dynamics of the order parameter $\rho_2$, arising from a coupling to an auxiliary field (the local particle density $\rho$), which remains frozen in regions where $\rho_2 = 0$. [Odor *et al.* (1998)] observed a slight variation of the spreading critical point as a function of $\rho_1(0)$ (similarly to the two-dimensional case), but more detailed simulations [Dickman (1999)] suggest that this can be explained by strong corrections to scaling. Simulations provided numerical evidence that $\rho_1$ exhibits anomalous scaling as

$$|\rho_1^{nat} - \rho_1| \propto |p - p_c|^{\beta_1} \ , \tag{6.32}$$

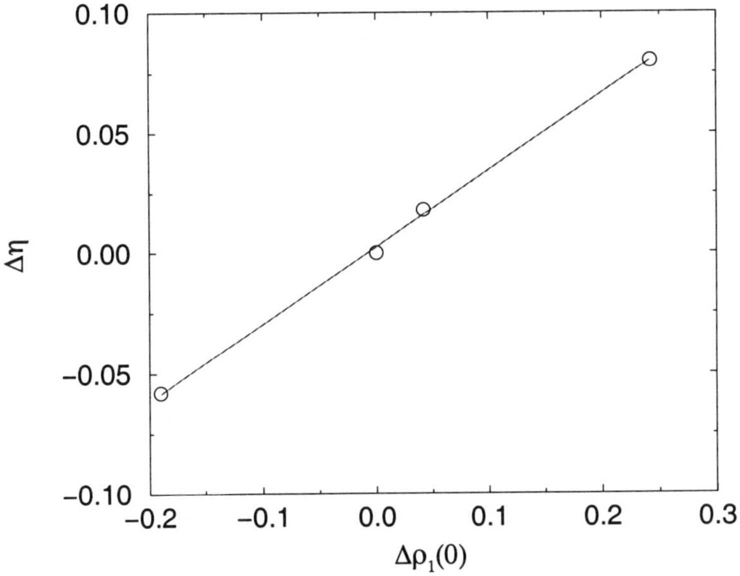


Fig. 6.2 Initial concentration dependence of the exponent $\eta$ in the PCP model [Odor *et al.* (1998)]. Linear regression gives a slope 0.320(7) between $\eta - \eta_{DP}$ and $\rho_1(0) - \rho_1^{nat}$. From [Odor *et al.* (1998)]

to $\rho_1^{nat} = 0.242(1)$ with a DP exponent [Odor *et al.* (1998)] for $p < p_c$ and with $\beta_1 = 0.9(1)$ for $p > p_c$ [Marques *et al.* (2001)]. The **damage-spreading transition** point and the DS exponents of this model coincide with the critical point and the critical exponents of the PCP [Odor *et al.* (1998)].

The effect of an **external particle source** that creates isolated particles, hence does not couple to the order parameter, was investigated by simulations and by GMF+CAM approximations [Dickman *et al.* (2001b)]. While the critical point $p_c$ showed a singular dependence on the source intensity, the critical exponents appear to be unaffected by the presence of the source, except possibly for a small change in $\beta$.

The **crossover from PCP to DP in one dimension** has recently been investigated by numerical simulations [Park and Park (2007)]. By introducing spontaneous particle annihilation $A \to \emptyset$ one can destroy the infinitely degenerated frozen absorbing phase. The crossover exponent estimated to be $\phi = 1.78(6)$ and the scaling function (1.33) was found to describe the crossover well. Park and Park argued that this exponent is not independent, but related to the DP relaxation time exponent: $\phi = \nu_{||} \simeq 1.733$, which is compatible with the numerical estimation within error.

The properties of the **two-dimensional** PCP in homogeneous, uncorrelated initial conditions was investigated by simulations [Kamphorst *et al.* (1999)]. In this case all six NN of a pair were considered for reactions (6.31). The critical point is located at $p_c = 0.2005(2)$. By determining $\alpha$, $\beta/\nu_\perp$ and $Z$ exponents and order-parameter moment ratios by simulations, Kamphorst et al. confirmed the universal behavior of the $2+1$-dimensional DP class (4.2). The spreading exponents are expected to behave as described in Sect. 6.6.

### 6.6.2 *The threshold transfer process (TTP)*

The one-dimensional threshold transfer process (TTP) [Mendes *et al.* (1994)], is a three state model. Each site may be vacant, single or doubly (active) occupied, described by a 3-state variable $\sigma_i = 0, 1, 2$. In each time step, a site is chosen at random. In the absence of active sites, the dynamics is indeed trivial: if $\sigma_i(t) = 0$ (or 1), then $\sigma_i(t+1) = 1(0)$ with probability $r$ (or $1-r$). The system relaxes exponentially to a steady state, where a fraction $r$ of sites have $\sigma_i = 1$ and the others are vacant.

However if $\sigma_i(t) = 2$ (active), then one particle moves to the right, and the other particle hops to the left neighboring site, provided the occupancy there is $< 2$. Therefore $\sigma_i$ is diminished accordingly in this deterministic, particle-conserving transfer. As can easily be seen, the number of active sites either decreases or remains the same in all processes other then $(1,2,1) \to (2,0,2)$; the frequency of these processes depends on the concentration of 1'-s, which is controlled by the parameter $r$.

Any configuration consisting of only 0'-s or 1'-s is absorbing in what concerns the active sites. The absorbing states of the model are fluctuating - in the respective sector of phase space, ergodicity is not broken. [Odor *et al.* (1998)] have shown that the dynamics of 1'-s is strongly affected by the presence of active sites. At the critical point, the concentration of 1'-s relaxes to its steady state value (equal to $r_c$) as a power-law

$$\rho_B(t) - \rho_B^{nat} \propto t^{-\alpha_{DP}} , \tag{6.33}$$

with a 1+1-dimensional DP exponent. The steady state value of $\rho_1$ has also been shown to exhibit anomalous scaling (6.32), with $\rho_1^{nat} = r_c = 0.6894(3)$ [Mendes *et al.* (1994)]. In the supercritical region this exponent is DP like [Odor *et al.* (1998)] (see Fig. 6.3), while in the passive phase $\beta_1 = 1$ [Marques *et al.* (2001)]. The order-parameter ($\rho_2$) exponents belong DP

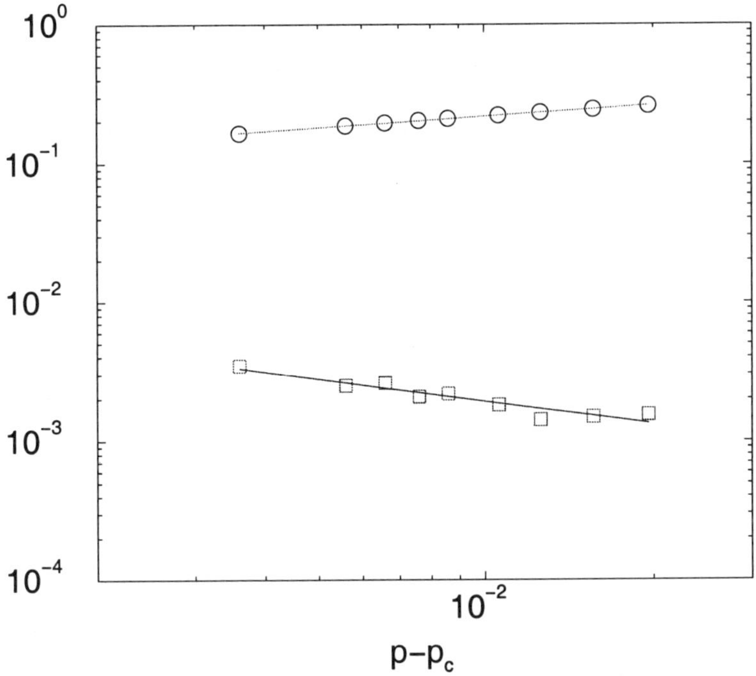


Fig. 6.3 Log-log plot of $\rho_1(r) - \rho_1(r_c)$ (○) and its fluctuation (□) as the function of $p - p_c$ above the critical point of the TTP model. Least-square regression fit resulted in $\beta = 0.27(2)$ and $\gamma = 0.53(6)$. From [Odor *et al.* (1998)].

class (4.2) except for $\delta$ and $\eta$, which are non-universal. They depend on $\rho_1(0)$ linearly and satisfy the hyperscaling relation (1.45).

Note that the TTP bears some resemblance to a sandpile model devised by [Manna (1991)] and to a forest fire model (FFM) proposed by [Bak *et al.* (1987); Grassberger and Kantz (1991)]. Under the correspondence: $2 \to$ burning tree, $1 \to$ live tree, and $0 \to$ ashes, the process $1, 2, 1 \to 2, 0, 2$ describes a burning tree setting its neighbors on fire, and $r$ represents the rate at which new trees emerge from the ashes. But the TTP permits doubly occupied sites to arise only *via* transfer; there is no "lightning" process, as in the FFM. A further contrast is that a rule such as $0, 2, 0 \to 1, 0, 1$ has no analog in the FFM.

[Odor *et al.* (1998)] have found that the **DS transition** and DS exponents of this model coincide with the critical point and critical exponents of the TTP model [Odor *et al.* (1998)].

## 6.7 DP with coupled diffusive field classes

The next question one can pose following Sect. 6.6 is whether a diffusive field coupled to a DP process is relevant. Prominent representatives of such models are binary particle production systems with explicit diffusion of solitary particles. The critical behavior of such systems are still under investigations. The annihilation-fission (AF) process is defined as

$$2A \xrightarrow{\lambda} \emptyset, \qquad 2A \xrightarrow{\sigma} (n+2)A, \qquad A\emptyset \overset{D}{\leftrightarrow} \emptyset A\,. \tag{6.34}$$

The corresponding action for bosonic particles was derived from master equation by [Howard and Täuber (1997)]

$$S = \int d^dx dt[\psi(\partial_t - D\nabla^2)\phi - \lambda(1-\psi^2)\phi^2 \\ +\sigma(1-\psi^n)\psi^2\phi^2]\,. \tag{6.35}$$

Usually bosonic field theories describe well the critical behavior of "fermionic" systems (i.e. those with maximum of one particle per site occupation). This is due to the fact that at absorbing phase transitions the occupation number vanishes. In this case, however the active phases of bosonic and fermionic models differ significantly: in the bosonic model the particle density diverges, while in the fermionic model there is a steady state with finite density. As a consequence the bosonic field theory cannot describe the active phase and the critical behavior of the fermionic AF particle system.

As one can see this theory lacks interaction terms linear in the field variable $\phi$ (massless) in contrast with the DP action (4.40). Although the field theory of **bosonic AF** has turned out to be non-renormalizible, [Howard and Täuber (1997)] concluded that critical behavior cannot be in DP class. In fact the upper critical behavior is $d_c = 2$, which is different from that of DP and PC classes [Ódor *et al.* (2002)]. In the Langevin formulation

$$\frac{\partial \rho(x,t)}{\partial t} = D\nabla^2\rho(x,t) + (n\sigma - 2\lambda)\rho^2(x,t) + \rho(x,t)\eta(x,t) \tag{6.36}$$

the noise is complex:

$$< \eta(x,t) > = 0 \tag{6.37}$$

$$< \eta(x,t)\eta(x',t') > = [n(n+3)\sigma - 2\lambda]\delta^d(x-x')(t-t')\,,$$

a new feature (it is real in case of DP and purely imaginary in case of CDP and PC classes). Eq. (6.36) without noise gives the **MF behavior of the bosonic model**. For $n\sigma > 2\lambda$ the density diverges, while for $n\sigma < 2\lambda$ it decays with a power-law with $\alpha_b^{MF} = 1$. The MF description in the **inactive phase** of the bosonic model was found to be valid for the two-dimensional fermionic AF system too [Ódor *et al.* (2002)]. Here the pair density decays as $\rho_2(t) \propto t^{-2}$ [Ódor *et al.* (2002)] in agreement with the MF approximation. Contrary to this, for $\lambda \leq \lambda_c$ $\rho$ and $\rho_2$ seem to be related by a logarithmic ratio $\rho(T)/\rho_2(t) \propto \ln(t)$. This behavior could not be described by the mean-field approximations and possibly due to the $d = d_c = 2$ spatial dimension.

The field theory suggests that the critical behavior of the **one-dimensional bosonic model** in the inactive phase is dominated by the $3A \to \emptyset$ process, that has an upper critical dimension $d_c = 1$, hence the particle density decays with a power-law exponent: $\alpha = 1/2$ with logarithmic corrections (see Sect. 4.7). This behavior has been confirmed by simulations in case of the one-dimensional annihilation-fission model [Ódor and Menyhárd (2002)].

The field theoretical description of the **fermionic AF** process ran into even more serious difficulties [Täuber (2000)] than that of the bosonic model and predicted an upper an critical dimension $d_c = 1$ that contradicts simulation results [Ódor *et al.* (2002)]. For the fermionic AF system, **mean-field** approximations [Carlon *et al.* (2001); Ódor *et al.* (2002)] gave a continuous transition with exponents

$$\beta = 1, \quad \beta' = 0, \quad Z = 2, \quad \nu_{||} = 2, \quad \alpha = 1/2, \quad \eta = 0. \tag{6.38}$$

These MF exponents are distinct from those of other well known RD classes (DP,PC,VM ...). They were confirmed in a two-dimensional fermionic AF model, with logarithmic corrections, indicating $d_c = 2$ [Ódor *et al.* (2002)]. An explanation for the new type of critical behavior based on symmetry arguments are still lacking, but numerical simulations suggest [Ódor (2000); Hinrichsen (2001b)] that the behavior of this system can be described (at least for strong diffusion) by coupled subsystems: single particles performing annihilating random walk coupled to pairs ($B$) following the DP process: $B \to 2B$, $B \to \emptyset$. The model has two non-symmetric absorbing states: one is completely empty, while in the other a single particle walks randomly. Owing to this (weakly) fluctuating absorbing state, this model does not oppose the conditions of the DP hypothesis. It was conjectured by [Henkel and Hinrichsen (2001)] that this kind of phase transition appears in models where (i) solitary particles diffuse, (ii) particle creation requires two particles and (iii) particle removal requires at least two particles to meet. Other conditions that affect these classes are still under investigation.

### 6.7.1 *The PCPD model*

A model similar to the AF (6.34) was introduced in an early work by [Grassberger (1982b)]. His preliminary simulations in one dimension showed non-DP type transition, but the model has been forgotten for a long time. The diffusive pair contact process (PCPD) introduced by [Carlon *et al.* (2001)] is controlled by two parameters, namely the probability of pair annihilation $p$ and the probability of particle diffusion $D$. The dynamical rules are

$$\begin{aligned} AA\emptyset,\ \emptyset AA &\to AAA && \text{with rate } (1-p)(1-D)/2 \\ AA &\to \emptyset\emptyset && \text{with rate } p(1-D) \\ A\emptyset &\leftrightarrow \emptyset A && \text{with rate } D\,. \end{aligned} \tag{6.39}$$

The *mean-field* approximation gives a continuous transition at $p = 1/3$. For $p \leq p_c(D)$ the particle and pair densities exhibit singular behavior:

$$\rho(\infty) \propto (p_c - p)^{\beta} \qquad \rho_2(\infty) \propto (p_c - p)^{\beta_2}\ , \tag{6.40}$$

while at $p = p_c(D)$ they decay as

$$\rho(t) \propto t^{-\alpha}\ , \quad \rho_2(t) \propto t^{-\alpha_2}\ , \tag{6.41}$$

with the exponents:

$$\alpha = 1/2, \quad \alpha_2 = 1, \quad \beta = 1, \quad \beta_2 = 2\ . \tag{6.42}$$

According to *pair mean-field* approximations the phase diagram can be separated into two regions (see Fig. 6.4). While for $D > 1/7$ the pair approximation gives the same $p_c(D)$ and exponents as the simple MF, for $D < 1/7$-s the transition line breaks and the exponents are different

$$\alpha = 1, \quad \alpha_2 = 1, \quad \beta = 1, \quad \beta_2 = 1\ . \tag{6.43}$$

In the entire inactive phase the decay is characterized by the exponents

$$\alpha = 1, \quad \alpha_2 = 2\ . \tag{6.44}$$

The DMRG method [Carlon *et al.* (2001)] and simulations of the one-dimensional PCPD [Hinrichsen (2001a)] obtained $p_c(D)$ values in agreement, but for the critical exponents no clear picture was found. They could not clarify whether the two distinct universality classes suggested by the pair mean-field approximations were really observable in the $d = 1$ PCPD

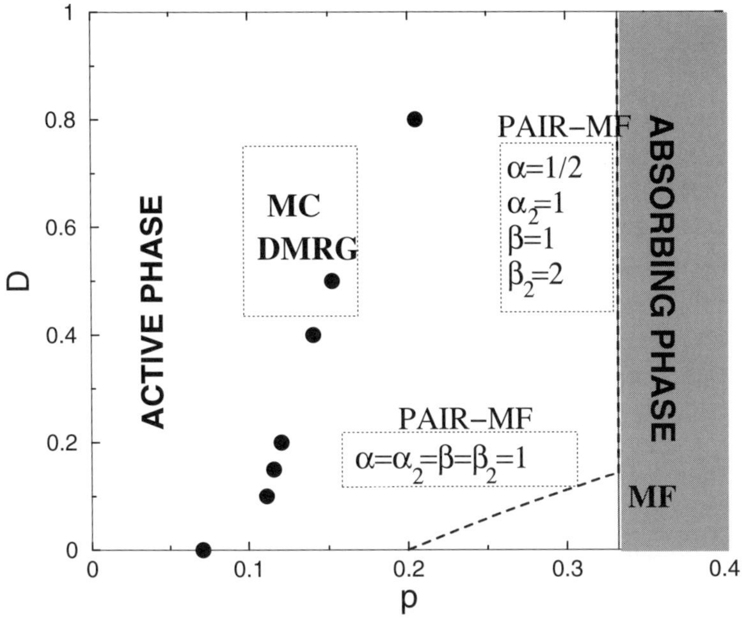


Fig. 6.4 Schematic phase diagram of the one-dimensional PCPD model. Circles correspond to simulation and DMRG results, vertical solid line to mean-field, dashed line to pair-approximation.

model. It is still a debated topic whether one new class, two new classes or continuously changing exponents occur in one dimension. Since the model has two absorbing states (besides the vacuum state there is another one with a single wandering particle) and some exponents were found to be close to those of the PC class ($Z = 1.6 - 1.87$, $\beta/\nu_\perp = 0.47 - 0.51$) [Carlon *et al.* (2001)] suspected that the transition (at least for low-$D$ values) is PC type. However the lack of $Z_2$ symmetry, parity conservation and further numerical data [Hinrichsen (2001a); Ódor (2000)] excluded this possibility. Note, that the MF exponents are also different from those of the PC class. Simulations and CAM calculations for the one dimensional $n = 1$ AF model [Ódor (2000, 2003c)] corroborated the two new universality class candidates (see Fig. 6.5 and Table 6.2). The order parameter exponent ($\beta$) seems to be far from both of the DP and PC class values [Ódor (2000, 2003c)].

The two distinct class behavior may be explained on the basis of competing diffusion strengths of particles and pairs (i.e. for large $D$-s the explicit

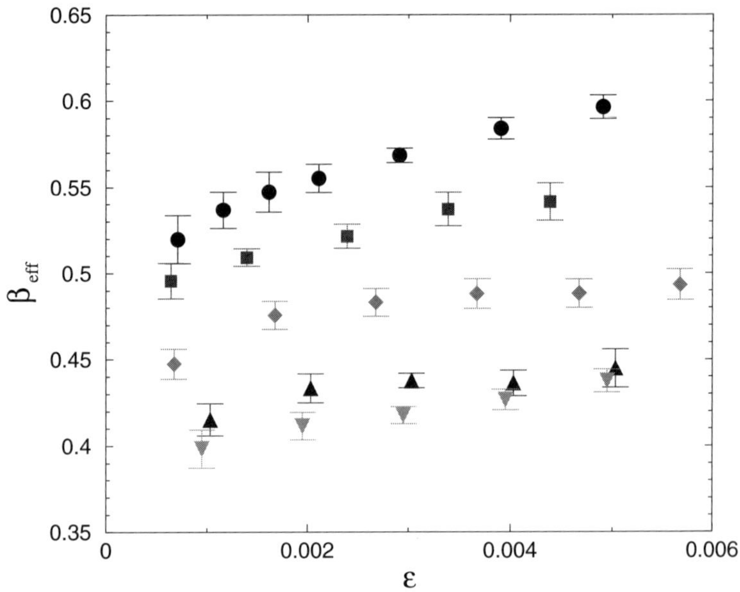


Fig. 6.5 Effective $\beta$ exponents for different diffusion rates. The circles correspond to $D = 0.05$, the squares to $D = 0.1$ the diamonds to $D = 0.2$, the up-triangles to $D = 0.5$ and the down-triangles to $D = 0.7$. From [Ódor (2000)].

Table 6.2 Summary of results for one-dimensional, $n = 1$ AF model. The non-universal critical parameter $p_c$ of the parallel update model is shown here.

| $D$ | 0.05 | 0.1 | 0.2 | 0.5 | 0.9 |
|---|---|---|---|---|---|
| $p_c$ | 0.25078 | 0.24889 | 0.24802 | 0.27955 | 0.4324 |
| $\beta_{CAM}$ | - | 0.58(6) | 0.58(2) | 0.42(4) | - |
| $\beta$ | 0.57(2) | 0.58(1) | 0.58(1) | 0.40(2) | 0.39(2) |
| $\delta$ | 0.273(2) | 0.275(4) | 0.268(2) | 0.21(1) | 0.20(1) |
| $\eta$ | 0.10(2) | - | 0.14(1) | 0.23(2) | 0.48(1) |
| $\delta\prime$ | 0.004(6) | - | 0.004(6) | 0.008(9) | 0.01(1) |

diffusion of lonely particles is stronger). Similar behavior was observed in the case of one-dimensional models with coupled (conserved) diffusive field (see Sect. 6.9). However a full agreement has not been achieved in the literature with respect the precise values of the critical exponents. The low-$D$ $\alpha$ is supported by [Park and Kim (2002)], who considered a case with reaction rates three times of the diffusion rate. On the other hand the high-$D$ $\alpha$ of Table 6.2 coincides with that of [Kockelkoren and Chaté (2003a)], who claim a single value for $0 < D < 1$. By assuming logarithmic corrections it was shown [Ódor (2003c)] that a single universality class can indeed be supported with the exponents

$$\begin{aligned} \alpha = 0.21(1), \quad \beta = 0.40(1), \\ Z = 1.75(15), \quad \beta/\nu_\perp = 0.38(1), \end{aligned} \tag{6.45}$$

but there is no strong evidence for such corrections. Although the upper critical dimension is expected to be at $d_c = 2$ [Ódor *et al.* (2002)] one may not exclude the possibility of a second critical dimension ($d'_c = 1$) or topological effects in one dimension that may cause logarithmic corrections to scaling. The spreading exponent $\eta$ seems to change continuously by varying $D$. Whether this is true asymptotically or the effect of some huge correction to scaling is still not clear.

The simulations by [Ódor (2000)] confirmed that it is irrelevant whether the particles production is spatially symmetric: $A\emptyset A \to AAA$ or spatially asymmetric: $AA\emptyset \to AAA$, $\emptyset AA \to AAA$. Recent simulations and higher level GMF approximations suggest [Ódor *et al.* (2002); Ódor (2003c)] that the peculiarities of the pair approximation are not real; for $N > 2$ cluster approximations the low-$D$ region scaling shrinks or disappears completely. Recently two studies [Dickman and de Menenzes (2002); Hinrichsen (2003)] reported nonuniversality in the dynamical behavior of the PCPD. While the former one explored different sectors (a reactive and a diffusive one) in the time evolution and gave nontrivial exponent estimates, Hinrichsen offered a hypothesis that the ultimate long time behavior should be characterized by DP behavior.

If we replace the annihilation process $2A \to \emptyset$ by coagulation $2A \to A$ in (6.34) we get the annihilation-coagulation model. GMF approximations and simulations of this model resulted in similar phase diagram to those of the PCPD model, albeit without any sign of two distinct regions. In agreement with this CAM approximations and simulations for the $d = 1$ model found the same kind of continuous transition independently from $D$, with exponents in agreement with those of the PCPD in the low-$D$ region [Ódor (2001a); Park and Kim (2002)]. Again the spatial symmetry of particle production was found to be irrelevant. An exact solution was found by [Henkel and Hinrichsen (2001)] for the special case in one dimension, when the diffusion rate is equal to the coagulation rate, corresponding to the inactive phase according to which the particle decay is like of ARW: $\rho \propto t^{-1/2}$.

The **crossover from PCPD to DP in one dimension** has recently been investigated by simulations [Park and Park (2006)]. By introducing a spontaneous particle annihilation term $A \to \emptyset$ one generates a direct channel to the absorbing state, which is an important feature of the DP. The crossover exponent was estimated to be $1/\phi = 0.58(3)$. On the other hand in case of spatial anisotropy (PCPD with a drift see Sect. 6.7.4) this exponent was found to be $1/\phi = 0.49(2)$, which agrees with the mean-field value of the crossover from CDP to DP (see (4.122)).

### 6.7.2 *Cyclically coupled spreading with pair annihilation*

In this section I show an explicit two-component realization of the PCPD class. A cyclically coupled two-component reaction-diffusion system was introduced by [Hinrichsen (2001b)]

$$A \to 2A, \quad A \to \emptyset, \quad A \to B, \quad 2B \to A, \quad B\emptyset \leftrightarrow \emptyset B \tag{6.46}$$

which mimics the PCPD model (Sect. 6.7.1) by mapping pairs to $A$'s and single particles to $B$'s. This model is a coupled DP+ARW system. Its $1+1$-dimensional critical space-time evolution pattern looks very similar to that of the PCPD model. The appearance of the evolution in space-time seems to be a particular feature of this class. It is built up from compact domains with a cloud of lone particles wandering and interacting with them. Furthermore this model also has two nonsymmetric absorbing states: a completely empty one and another with a single wandering $B$. When the annihilation and diffusion rates of $B$-s are fixed ($r = D = 1$), the model exhibits continuous phase transition by varying the production rate of $A$-s and the $A \to B$ transmutation rate. The simulations in one dimension showed that $\rho_A \propto \rho_B$ for large times and resulted in the following critical exponent estimates

$$\alpha = 0.21(2), \quad \beta = 0.38(6), \quad \beta' = 0.27(3), \\ Z = 1.75(5), \quad \nu_{||} = 1.8(1), \tag{6.47}$$

satisfying the generalized hyperscaling relation (1.45). These exponents are similar to those of PCPD model in the high-diffusion region (see Table 6.2), which is reasonable since $D = 1$ is fixed here.

### 6.7.3 *The parity conserving annihilation-fission model*

As we have seen parity conservation plays an important role in unary production systems. In case of BARW processes it changes the universality of the transition from DP (Sect. 4.2.4) to the PC (Sect. 4.6.1) class. The question arises whether one can see similar behavior in the case of binary production systems. Recently [Park *et al.* (2001)] investigated a parity conserving representative ($n = 2$) of the one-dimensional AF model (6.34). By performing simulations for low-$D$-s they found critical exponents in the range of values determined for the corresponding PCPD class. Park et al. claim that the conservation law does not affect the critical behavior and that the binary nature of the offspring production is a sufficient condition for this class (see however Sect. 6.7.2, where there is no such condition).

The two-dimensional version of the parity conserving AF model was investigated by GMF and simulation techniques [Ódor *et al.* (2002)]. While the $N = 1, 2$ GMF approximations showed similar behavior to that of the PCPD model (Sect. 6.7.1), including the two class prediction for $N = 2$, the $N = 3, 4$ approximations do not show $D$ dependence of the critical behavior: $\beta = 1$, $\beta_2 = 2$ were obtained for $D > 0$. Large scale simulations of the particle density confirmed the mean-field scaling behavior with logarithmic corrections. This result can be interpreted as a numerical evidence supporting an upper critical dimension in this model of: $d_c = 2$. The pair density decays in a similar way, but with an additional logarithmic factor to the order parameter. This kind of strongly coupled behavior at and above criticality was observed in case of the PCP model too (see Sect. 6.6.1). At the $D = 0$ end point of the transition line $2 + 1$-dimensional DP criticality (see Sect. 4.2) was found for $\rho_2$ and for $\rho - \rho(p_c)$. In the inactive phase for $\rho(t)$ we can observe the two-dimensional ARW class scaling behavior (see Sect. 4.5.1), while the pair density decays as $\rho_2 \propto t^{-2}$. Again as in $d = 1$ the parity conservation seems to be irrelevant.

### 6.7.4 *The driven PCPD model*

The particle drift has been shown to be irrelevant in major one-component critical RD systems (see Sects. 4.5.5, 4.2.9 and 4.6.9). This is not true for multi-component systems. In the case of the two-component $AB \to \emptyset$ model in one dimension the diffusion drift makes the density decay faster than in isotropy (Sect. 6.1.2). A simulation study in one dimension [Park and Park (2005a)] has shown that the Galilean invariance (4.5.5) is broken in the PCPD model with biased diffusion. In particular the density decay exponent of particles is $\alpha = 0.49(1)$, while for pairs it is $\alpha_2 = 0.56(3)$, which are far from the values of the isotropic PCPD ($\alpha \simeq 0.20$). This strengthens the belief that the PCPD can't be described by a single-species, bosonic field theory and the emergence of non-DP criticality. Furthermore, these exponents are close to the mean-field values of PCPD, suggesting that anisotropy lowers the upper critical dimension to $d_c = 1$. The numerical analysis has also shown that $\rho(t)/\rho_2(t) \sim \ln t$, which is again a signal for upper-critical behavior (the same critical behavior was found in the two-dimensional isotropic PCPD [Ódor *et al.* (2002)]). Note, that the reduction of the upper critical dimension by the biased diffusion is not rare. The most prominent example is the sandpile model related to SOC. It is well known that the directed toppling rules lower the upper critical dimension from 4 to 3 [Dhar and Ramaswamy (1989)]. These results have been confirmed by GMF+CAM analytical calculations [Park and Park (2005b)].

## 6.8 BARWe with coupled non-diffusive field class

Similarly to the PCP (Sect. 6.6.1) the effect of infinitely many frozen absorbing states has been investigated in case of a BARWe model. A parity conserving version of the one-dimensional PCP model was introduced by [Marques and Mendes (1999)], in which pairs follow a BARW2 process, while lone particles are frozen. Simulations showed that while the critical behavior of pairs in case of homogeneous, random initial distribution belongs to the PC class (Sect. 4.6.1), the spreading exponents satisfy hyperscaling (1.45) and change continuously when the initial particle density is varied. These results are similar to those found in the PCP model. Again long-memory effects are responsible for the nonuniversal behavior in case of seedlike (correlated) initial conditions. The slowly decaying memory was confirmed by studying a one-dimensional, interacting monomer-monomer model [Park and Park (2001)] by simulations.

## 6.9 DP with diffusive, conserved slave field classes

In view of the results of **model C** dynamics (Sect. 1.7), the obvious question is, what happens with the phase transition to an absorbing state of a reaction-diffusion system if a conserved, secondary density is coupled to a nonconserved order parameter? One can deduce from the BARW1 spreading process (Sect. 4.2.4) a two-component, reaction-diffusion model (DCF) [Kree *et al.* (1989); Wijland *et al.* (1998)], that exhibits total particle density conservation as follows:

$$A + B \xrightarrow{k} 2B, \quad B \xrightarrow{1/\tau} A\,. \tag{6.48}$$

When the initial particle density ($\rho = \rho_A(0) + \rho_B(0)$) is varied, a continuous phase transition occurs. The general field theoretical treatment was given by [Wijland *et al.* (1998)], while the equal diffusion case, $D_A = D_B$ was investigated by [Kree *et al.* (1989)]). The mean-field exponents that are valid above $d_c = 4$ are shown in Table 6.3. The rescaled action of this model is

$$\begin{aligned} S[\varphi,\overline{\varphi},\psi,\overline{\psi}] = \int d^dxdt \Big[ & \overline{\varphi}(\partial_t - \Delta)\varphi + \overline{\psi}(\partial_t + \lambda(\sigma - \Delta))\psi \\ & + \mu\overline{\varphi}\Delta\psi + g\psi\overline{\psi}(\psi - \overline{\psi}) + u\psi\overline{\psi}(\varphi + \overline{\varphi}) \\ & + v_1\left(\psi\overline{\psi}\right)^2 + v_2\psi\overline{\psi}(\psi\overline{\varphi} - \overline{\psi}\varphi) + v_3\varphi\overline{\varphi}\psi\overline{\psi} \\ & - \rho_B(0)\delta(t)\,\overline{\psi}\Big]\,, \end{aligned} \tag{6.49}$$

where $\psi$ and $\phi$ are auxiliary fields, defined such that their average values coincide with the average density of $B$ particles and the total density of particles respectively. The coupling constants are related to the original parameters of the master equation by

$$\begin{gathered} \mu = 1 - D_B/D_A g = k\sqrt{\rho}/D_A \lambda\sigma = k(\rho_c^{\mathrm{mf}} - \rho)/D_A \\ v_1 = v_2 = -v_3 = k/D_A u = -k\sqrt{\rho}/D_A \lambda = D_B/D_A \\ \rho_B(0) = \rho_B(0)/\sqrt{\rho}\,. \end{gathered} \tag{6.50}$$

If one omits from the action Eq. (6.49) the initial time term proportional to $\rho_B(0)$, then the remainder is, for $\mu = 0$ (i.e. for $D_A = D_B$), invariant under the rapidity reversal symmetry

$$\begin{aligned} \psi(x,t) &\to -\overline{\psi}(x,-t) \\ \overline{\psi}(x,t) &\to -\psi(x,-t) \\ \varphi(x,t) &\to \;\;\overline{\varphi}(x,-t) \\ \overline{\varphi}(x,t) &\to \;\;\varphi(x,-t)\,. \end{aligned} \tag{6.51}$$

The epsilon expansion solution [Kree *et al.* (1989)] and simulation results [de Freitas *et al.* (2000); Fulco *et al.* (2001); Maia and Dickman (2007)] are summarized in Table 6.3. Interestingly the RG predicts $Z = 2$ and $\nu_\perp = 2/d$ in all orders of perturbation theory.

The breaking of this symmetry for $\mu \neq 0$, that is, when the diffusion constants $D_A$ and $D_B$ are different causes different critical behavior for this system. For $D_A < D_B$ the RG predicts new classes with $Z = 2$, $\beta = 1$,

Table 6.3 Summary of results for DCF classes for $D_A = D_B$.

| d | $\beta$ | $Z$ | $\nu_\perp$ |
|---|---|---|---|
| 1 | 0.38(2) | 2 | 2.0(2) |
| $4-\epsilon$ | $1-\epsilon/8$ | 2 | $2/d$ |

$\nu_\perp = 2/d$ [Wijland *et al.* (1998)], but simulations in one-dimension [Fulco *et al.* (2001)] show different behavior (see Table 6.4). For non-Poissonian initial particle density distributions, the critical initial slip exponent $\eta$ varies continuously with the width of the distribution of the conserved density. The $D_A = 0$ extreme case is discussed in Sect. 6.10.

Table 6.4 Summary of results for DCF classes for $D_A < D_B$.

| d | $\beta$ | $Z$ | $\nu_\perp$ |
|---|---|---|---|
| 1 | 0.22(3) | | 2 |
| $4-\epsilon$ | 1 | 2 | 2/d |

For $D_A > D_B$ no stable fixed point solution was found by RG, hence [Wijland *et al.* (1998)] conjectured that there was a first order transition, for which signatures were found in two dimensions by simulations [Oerding *et al.* (2000)]. However $\epsilon$ expansion may break down in case of another critical dimension $d_c' < d_c = 4$, for which simulations in one dimension [Fulco *et al.* (2001)] and [Maia and Dickman (2007)] provided numerical support (see Table 6.5).

Table 6.5 Summary of results for DCF classes for $D_A > D_B$.

| d | $\beta$ | $Z$ | $\nu_\perp$ |
|---|---|---|---|
| 1 | 0.92(3) | | 2.3(3) |
| $4-\epsilon$ | 0 | | |

More recently [Fiore and de Oliveira (2004)] investigated numerically versions of this model family in which singleton, pair or triplet creation is involved. They confirmed that while in case of unary creation the transition remains continuous for any diffusion rate, for binary and triplet creation it becomes first order for strong diffusions. This is in qualitative agreement with the results obtained for diffusive, binary and triplet models without particle conservation [Dickman and Tomé (1991); Cardozo and Fontanari (2006b)].

## 6.10 DP with frozen, conserved slave field classes

If the conserved field coupled to the BARW1 process (Eq. (6.48)) is non-diffusive, non-DP-class (NDCF) behavior is reported [Rossi *et al.* (2000); Pastor-Satorras and Vespagnani (2000)]. These are known as nondiffusive conserved-field (NDCF) classes. The corresponding action can be derived from Eq. (6.49) in the $D_A = 0$ limit:

$$S = \int d^dx dt \Big[\overline{\varphi}(\partial_t + r - D\nabla^2)\varphi + \overline{\psi}(\partial_t - \lambda\nabla^2))\psi \\ + g\psi\overline{\psi}(\psi - \overline{\psi}) + u\psi\overline{\psi}(\varphi + \overline{\varphi}) + v_1\left(\psi\overline{\psi}\right)^2 \\ + v_2\psi\overline{\psi}(\psi\overline{\varphi} - \overline{\psi}\varphi) + v_3\varphi\overline{\varphi}\psi\overline{\psi}\Big] a \; . \qquad (6.52)$$

When we neglect irrelevant terms, Eq. (6.52) is invariant under the shift transformation

$$\psi \to \psi + \Delta, \quad r \to r - v_2\Delta \; , \qquad (6.53)$$

where $\Delta$ is any constant. The field theoretical analysis of this action has run into difficulties [Pastor-Satorras and Vespagnani (2000)]. The upper critical dimension $d_c = 4$ was confirmed by simulations [Lübeck and Hucht (2002)]. In the frozen slave field ($D_A = 0$) case the rapidity reversal symmetry (Eq. (6.51)) is not satisfied, therefore one may expect different hyperscaling relations here as for of the DP class. This may involve more independent exponents ($\beta \neq \beta'$, $\alpha \neq \delta$) than in case of DP, but not necessarily. Other symmetries (like that of the PC class in one dimension or that of the ordinary percolation) can also result in similar exponent relations.

Early simulations [Rossi *et al.* (2000)] reported the breaking of the

$$\alpha = \beta/\nu_{||} \qquad (6.54)$$

relation indeed. In two dimensions for example $\alpha = 0.42(1)$ and $\beta/\nu_{||} = 0.52(1)$ can be read-off from the numerical exponent estimates. A more detailed scaling analysis by [Lübeck and Heger (2003a)] confirmed these deviations below $d_c = 4$. However Lübeck and Heger pointed out, that very strong corrections to scaling (common in coupled systems) can also be blamed for this difference (see Fig. 6.6). Furthermore, spreading simualtions showed that $\delta = \beta/\nu_{||}$ and $\beta = \beta'$ as in case of DP, hence [Lübeck and Heger (2003a)] suggested that the Eq. (4.43 hyperscaling relation is fulfilled in case of NDCF classes as well (see Fig. 6.7).

The critical exponents determined by simulations [Rossi *et al.* (2000); Pastor-Satorras and Vespagnani (2000); Lübeck (2001, 2002); Dickman

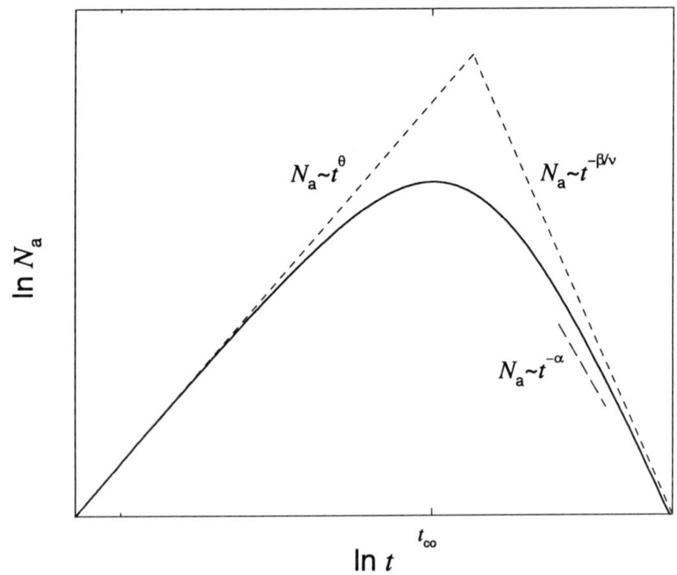


Fig. 6.6 Sketch of the crossover behavior of the number of active sites $N_a$. For $t \ll t_{co}$ $N_a$ scales as $t^\theta$, whereas it is expected to decrease for $t \gg t_{co}$ as $N_a \propto t^{-\beta/\nu_\|}$. It is possible that a too low exponent ($\alpha < \beta/\nu_\|$) is observed in simulations, if one has not reached the asymptotic scaling regime (long-dashed line). From [Lübeck and Heger (2003a)].

*et al.* (2002)] and by the GMF+CAM method in one dimension [Dickman (2002b)] are summarized in Table 6.6. Besides these exponents the universal finite-size scaling functions have also been determined numerically [Lübeck and Heger (2003a)].

Table 6.6 Summary of critical exponent results for NDCF classes.

| d | $\alpha$ | $\beta$ | $Z$ | $\nu_{\|\|}$ | $\sigma$ | $\eta$ |
|---|---|---|---|---|---|---|
| 1 (CTTP) | 0.14(1) | 0.38(1) | 1.4(1) | 2.5(1) | 1.71(4) | 0.350(5) |
| 1 (CLG) | 0.25(1) | 1.01(1) | 2 | 4.0(5) | 2.71(1) | |
| 2 | 0.42(1) | 0.64(1) | 1.54(1) | 1.22(3) | 2.22(3) | 0.30(5) |
| 3 | 0.75(1) | 0.84(1) | 1.82(5) | 1.1(1) | 2.0(4) | 0.15(1) |
| 4 | 1 | 1 | 2 | 1 | 2 | 0 |

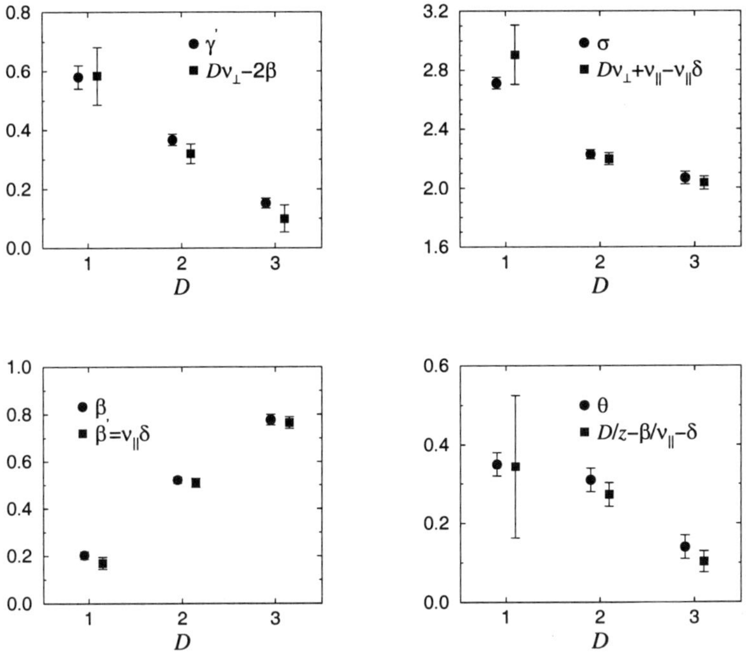


Fig. 6.7 Test of certain scaling laws for the NDCF universality class. The hyperscaling laws $\gamma' = D\nu_\perp - 2\beta$ (upper left), $\sigma = D\nu_\perp + \nu_{||} - \nu_{||}\delta$ (upper right), $\theta = D/z - \beta/\nu_{||} - \delta$ (lower right), as well as the equality $\beta = \beta'$ (lower left) are checked. All scaling laws are fulfilled within the error bars. From. [Lübeck and Heger (2003a)].

### 6.10.1 *Realizations of NDCF classes, SOC models*

The main examples of models in NDCF classes are the conserved threshold transfer process (CTTP) and the conserved lattice gas (CLG) reaction-diffusion model [Rossi *et al.* (2000); Pastor-Satorras and Vespagnani (2000)] (see Fig. 6.8). The CTTP is a modification of the TTP (see Sect. 6.6.2), where lattice sites may be empty, occupied by one particle, or occupied by two particles. Empty and singly occupied sites are considered inactive, whereas double occupied lattice sites are considered active. Active sites transfer both particles to randomly chosen empty or singly occupied nearest neighbor sites. Note, that contrary to the original ($d = 1$) TTP model [Mendes *et al.* (1994)], the particle redistribution is stochastic in CTTP even in one dimension.

In the CLG model lattice sites may be empty or occupied by one particle. In order to mimic a repulsive interaction a given particle is considered as active, if at least one of its neighboring sites on the lattice is occupied by another particle. If all neighboring sites are empty the particle remains inactive. Active particles are moved in the next update step to one of their empty nearest neighbor sites, selected at random.

Furthermore, the models belonging to NDCF classes embrace a large group of stochastic sandpile models [Jensen (1998)], in particular, fixed-energy Manna models [Manna (1991); Dhar (1999a); Dickman *et al.* (1998); Muñoz *et al.* (2001)]. In contrast to the CTTP the Manna model allows unlimited particle occupation of lattice sites (bosonic model). Here lattice sites are considered inactive if the particle occupation is below a certain threshold $n < N_c$. For $n \geq N_c$ the lattice site is active and the particles are moved to nearest neighbors, selected at random for each particle, i.e.,

$$n \longrightarrow 0 \qquad \text{for all sites with} \qquad n \geq N_c\,. \tag{6.55}$$

Now I shall discuss the scaling behavior of the original Manna model with $N_c = 2$.

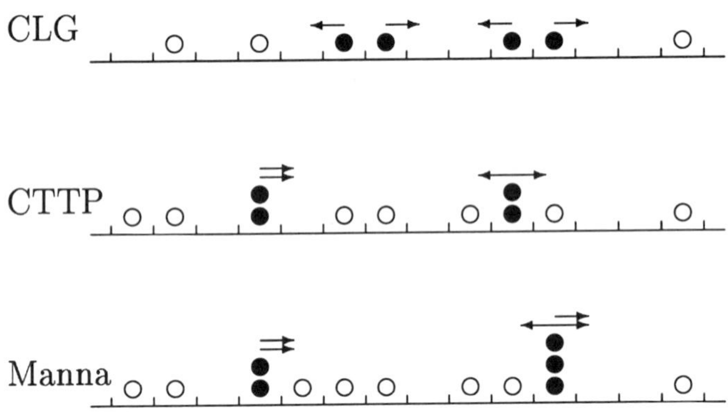


Fig. 6.8 Sketch of the dynamics of three NDCF models in one dimension. Filled circles mark active particles, whereas inactive ones are denoted by open circles. The arrows show how active particles are (probably) moved in the next update step. In the case of the one-dimensional CLG model the particle transfer is deterministic. For the one-dimensional CTTP stochastic (left) as well as deterministic (right) particle movements may occur. Only the one-dimensional Manna model is characterized by a full stochastic dynamics. From Ref. [Lübeck and Heger (2003a)].

As mentioned in the preface one specially interesting family of nonequilibrium critical phenomena are those, which appear in a "self-organized" way, *i.e.* without apparent tuning of parameters. Since the introduction of **self-organized criticality (SOC)** [Bak *et al.* (1987)] understanding the scaling behavior of "self-organized" critical points became a task of high priority. Sandpiles — the archetype example of SOC — are models, in which sand grains are slowly dropped to a lattice and redistributed to neighboring sites, whenever certain instability threshold is exceeded. This may generate cascades of topplings (avalanches), which eventually lead to dissipation of grains at the open boundaries. This mechanism drives the system into a critical state with power-law distributed avalanches (see Eq. (6.56)). It is important to realize that between successive avalanches sandpiles are trapped into one, out of many possible absorbing states. Such state can be characterized by the distribution of heights below the threshold, which is a locally conserved and non-diffusive (grains at sites below threshold do not move) field. After some controversies, it was clearly established that most (isotropic) stochastic sandpiles [Manna (1991)] share the same universal properties [Vespignani and Zapperi (1997); Dickman *et al.* (1998); Dickman (2002a)]. The correspondence is best understood using the fixed-energy ensemble (see [Dickman *et al.* (2000)]). Indeed, (isotropic) stochastic sandpiles belong to a unique universality class, usually called the Manna or NDCF class.

[Alava and Munoz (2001)] have shown that these classes describe the depinning transition of the **quenched Edwards-Wilkinson** (Sect. 7.3) or linear interface models (LIM) [Halpin-Healy and Zhang (1995); Barabási and Stanley (1995)]. This is due to the fact, that quenched disorder can be mapped onto long-range temporal correlations in the activity field [Marsili (1994)]. An analysis of the universal scaling functions, as well as of the critical exponents for $d \geq 2$ [Lübeck and Heger (2003b)] confirmed the original conjecture of [Rossi *et al.* (2000)], that the CLG model, the CTTP, and the Manna model belong to the same universality class.

However this mapping could not be done on the level of Langevin equations of the representatives of NDCF and LIM models, and in one dimension this equivalence breaks down [Alava and Munoz (2001); Dickman *et al.* (2001a); Kockelkoren and Chaté (2003b)]. Considering one-dimensional versions of these systems a "splitting" of the universality class occurs. The reason for this non-universal behavior is that the dimensional reduction changes the stochastic character of the dynamics (see Fig. 6.8.). For instance, in the CLG model the particle movement is deterministic in $d = 1$

and the model exhibits a trivial phase transition with $\beta = 1$ and $\rho_c = 1/2$ (see Ref. [22] in [Rossi *et al.* (2000)]). On the other hand for CTTP, simulations showed that roughly 40% of the relaxation events are deterministic. In case of the Manna model all movements are stochastic. Further difference between these models at criticality is that in one dimension the absorbing state of the CLG exhibits $Z_2$ symmetry ('0101010' and '1010101') [K. Park and mook Kim (2005)], while the CTTP and the Manna model have infinitely many absorbing states similarly to higher dimensions.

This splitting among the the one-dimensional NDCF models is in full agreement with the universality hypothesis advanced for sandpile models [Ben-Hur (1996)]. According to this conjecture the *universality classes of sandpile models are determined by the way the particles are distributed to the next neighbors (deterministic, stochastic, directed, undirected, etc.).* Obviously the Manna universality class is characterized by a stochastic and undirected distribution of particles.

SOC models are usually characterized by the exponents of the probability distribution of size (s), area (a), time (t) and radius (r) of the avalanches, decaying algebraically

$$P_x \propto x^{-\tau_x} , \tag{6.56}$$

where $x \in (s, a, t, r)$, and the $\tau_x$ exponents can be connected to cluster spreading exponents [Muñoz *et al.* (1999)]. Since the hyperscaling relations are fulfilled one can relate the NDCF class steady state exponents to the **SOC avalanche exponents** of the Manna models

$$\tau_r = 1 + Z + d - \frac{\sigma}{\nu_\perp} = 1 + \frac{\beta'}{\nu_\perp}, \tag{6.57}$$

$$\tau_t = 2 + \frac{d}{Z} - \frac{\sigma}{\nu_{||}} = 1 + \frac{\beta'}{\nu_{||}}, \tag{6.58}$$

$$\tau_a = 2 + \frac{Z}{d} - \frac{\sigma}{d\nu_\perp} = 1 + \frac{\beta'}{d\nu_\perp}, \tag{6.59}$$

$$\tau_s = 1 + \frac{\nu_{||} + \nu_\perp d - \sigma}{\nu_{||} + \nu_\perp d - \beta} = 1 + \frac{\beta'}{\nu_{||} + \nu_\perp d - \beta}. \tag{6.60}$$

This has been done and confirmed by [Lübeck and Heger (2003a)] within numerical error bars, which further strengthens our knowledge about NDCF classes.

### 6.10.2 *NDCF with anisotropy*

A directed variant of the sandpile model of [Bak *et al.* (1987)] (see Sect. 6.10.1), which exhibits NDCF (Manna) class scaling behavior was solved exactly by [Dhar and Ramaswamy (1989)]. This model can be mapped onto Takayasu's diffusion and aggregation model [Takayasu (1989)] (Sect. 7.5.3) and Scheidegger's river basin model [Scheidegger (1967)] (for a review see [Dhar (1999b)]), and the redistribution of sand (hence the avalanches) is anisotropic in a given space direction. As a consequence the model exhibits the temporal avalanche exponent $\tau_t = 1.5$ in **two dimensions** and $\tau_t = 2$ in **three and higher dimensions**, which are consistent with the NDCF class values, but the upper critical dimension is reduced from $d = 4$ to $d_c = 3$ and the Galilean invariance (4.5.5) is broken. Such dimensional reduction has been observed recently by simulations of the PCPD model (see Sect. 6.7.4).

### 6.10.3 *NDCF with spatial boundary conditions*

Since the critical exponents, scaling functions and moment ratios of NDCF classes are quite similar to those of the DP, there have been debates if they really represent different universality classes [Mohanty and Dhar (2002)]. A numerical simulation study [Bonachela and Munoz (2007)] supplemented one-dimensional DP and NDCF models with different spatial boundary conditions and determined the cluster spreading exponents. As the Table 6.7 shows the differences in the exponents, in case of absorbing and reflecting boundary conditions, are relevant and indicate non-DP behavior. Here $\tau$ is

Table 6.7 Critical exponents of DP and NDCF, without a wall and in the presence of absorbing and reflecting walls. From Ref.[Bonachela and Munoz (2007)].

| | $\eta$ | $\delta$ | $\tau$ | $\tau_t$ |
|---|---|---|---|---|
| $DP$ | 0.313(1) | 0.159(1) | 1.108(1) | 1.159(1) |
| $NDCF$ | 0.350(5) | 0.170(5) | 1.11(2) | 1.17(2) |
| $DP_{abs}$ | 0.045(2) | 0.426(2) | 1.28(3) | 1.426(2) |
| $DP_{ref}$ | 0.046(2) | 0.425(2) | 1.25(3) | 1.425(2) |
| $NDCF_{abs}$ | −0.33(2) | 0.85(2) | 1.56(2) | 1.81(2) |
| $NDCF_{ref}$ | 0.35(3) | 0.16(3) | 1.11(3) | 1.15(3) |

the avalanche size (1.9) and $\tau_t$ is the avalanche time distribution exponent, defined as

$$n_t \propto t^{-\tau_t} . \tag{6.61}$$

Values in rows 3 ($DP_{abs}$) and 4 ($DP_{ref}$) coincide, also those in 2 (NDCF) and 6 ($NDCF_{ref}$) are also equal within error-bars.

## 6.11 Coupled $N$-component DP classes

From the basic reaction-diffusion systems one can generate $N$-component ones coupled by interactions symmetrically or asymmetrically. [Janssen (1997b, 2001)] introduced and analyzed by field theoretical RG method (up to two-loop order) a quadratically coupled, $N$-species generalization of the DP process of the form:

$$\begin{aligned} A_\alpha &\leftrightarrow 2A_\alpha \\ A_\alpha &\rightarrow \emptyset \\ A_\alpha + A_\beta &\rightarrow kA_\alpha + lA_\beta \ , \end{aligned} \tag{6.62}$$

where $k, l$ may take the values $(0, 1)$. He has shown that the multicritical behavior is always described by the Reggeon field theory (DP class), but this is unstable and leads to unidirectionally coupled DP systems (see Fig. 6.9). Janssen has also shown that by this model the linearly, unidirectionally coupled DP (UCDP) (see Sect. 7.6) behavior can be described. The universality class behavior of UCDP is discussed in Sect. 7.6. However, in **one dimension**, if BARWo-type processes are coupled (which alone exhibit DP class transition (see Sect. 4.2.4) **hard-core interactions** can modify the phase transition universality (see Sect. 6.13.1).

## 6.12 Coupled $N$-component BARW2 classes

Bosonic, $N$-component BARW2 systems with two offsprings (N-BARW2), of the form

$$A_\alpha \rightarrow 3A_\alpha \tag{6.63}$$

$$A_\alpha \rightarrow A_\alpha + 2A_\beta \tag{6.64}$$

$$2A_\alpha \rightarrow \emptyset \tag{6.65}$$

were introduced and investigated by [Cardy and Täuber (1998)] via field theoretical RG method. These models exhibit parity conservation of each species and permutation symmetry on $N$ types (generalization for $O(N)$ symmetry is violated by the annihilation term). The $A \rightarrow 3A$ processes turns out to be irrelevant, because like pairs annihilate immediately. Models

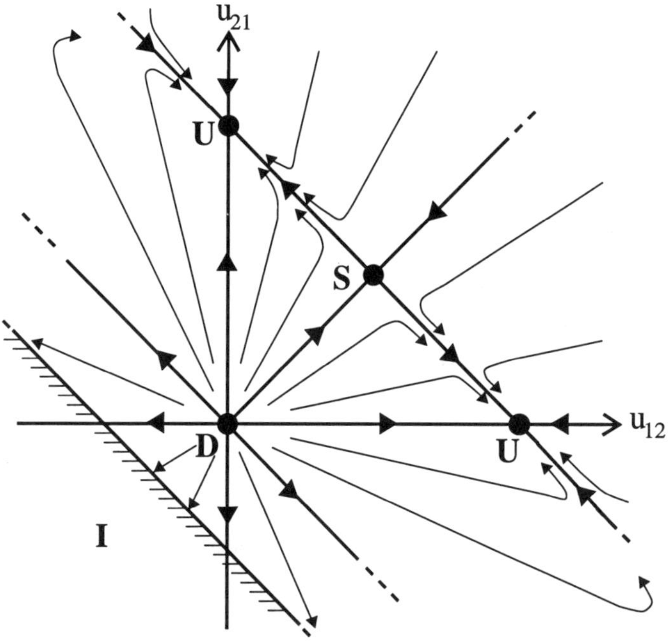


Fig. 6.9 Flow of the interspecies couplings in the two-component, DP model under renormalization. D, decoupled; S, symmetric; U, unidirectional fixed points. From [Janssen (2001)].

with Eq. (6.64) branching terms exhibit continuous phase transitions at **zero branching rate**. The universality class is expected to be independent from $N$ and coincides with that of the $N \to \infty$ model, which can be solved exactly. The critical dimension is $d_c = 2$, and for $d \leq 2$ the exponents are

$$\beta = 1, \quad Z = 2, \quad \alpha = d/2, \quad \nu_{||} = 2/d, \quad \nu_{\perp} = 1/d \ , \tag{6.66}$$

while at $d = d_c = 2$ logarithmic correction to the density decay are expected. Simulations on a $d = 2$ (fermionic) lattice model confirmed these results [Ódor (2001c)].

In one dimension it turns out that **hard-core interactions** between particles can be relevant and different universal behavior can emerge in fermionic variants (see Sect. 6.13). The phase transition behavior of bosonic and fermionic models in one dimension is equivalent (at least for static exponents) in case of *pairwise initial conditions* (see Sect. 6.2), when different types of particles do not make up blockades for each other. Such a situation happens when these particles are generated as domain walls of $N+1$ component systems exhibiting $S_{N+1}$ symmetric absorbing states (see Sects. 6.12.1 and 4.6.2).

### 6.12.1 *Generalized contact processes with $n > 2$ absorbing states in one dimension*

The generalized contact process has already been introduced in Sect. 4.6.3 with the main purpose of showing an example of a PC class universality class transition in case of $Z_2$ symmetric absorbing states. The more general case with $n > 2$ permutation symmetric absorbing states was investigated using DMRG method by [Hooyberghs *et al.* (2001)] and turned out to exhibit an N-BARW2 transition. In the one-dimensional model, where each lattice site can be occupied by at most one particle ($A$) or can be in any of $n$ inactive states ($\emptyset_1$, $\emptyset_2 \ldots \emptyset_n$) the reactions are:

$$AA \to A\emptyset_k, \emptyset_k A \qquad \text{with rate } \lambda/\text{n} \tag{6.67}$$

$$A\emptyset_k, \emptyset_k A \to \emptyset_k\emptyset_k \qquad \text{with rate } \mu_\text{k} \tag{6.68}$$

$$A\emptyset_k, \emptyset_k A \to AA \qquad \text{with rate 1} \tag{6.69}$$

$$\emptyset_k\emptyset_l \to A\emptyset_l,\ \emptyset_k A \qquad (k \neq l) \text{ with rate 1.} \tag{6.70}$$

The original contact process (Sect. 4.2.1), corresponds to the $n = 1$ case, from which the reaction (6.70) is obviously absent. The reaction (6.70) in the case $n \geq 2$ ensures that configurations like $(\emptyset_i\emptyset_i \ldots \emptyset_i\emptyset_i\emptyset_j\emptyset_j \ldots \emptyset_j\emptyset_j)$, with $i \neq j$ are not absorbing. Such configurations do evolve in time until the different domains coarsen and one of the $n$ absorbing states $(\emptyset_1\emptyset_1 \ldots \emptyset_1)$, $(\emptyset_2\emptyset_2 \ldots \emptyset_2)$, $\ldots (\emptyset_n\emptyset_n \ldots \emptyset_n)$ is reached. For generalized contact processes with $n = 2$, simulations [Hinrichsen (1997)] and a DMRG study [Hooyberghs *et al.* (2001)] proved that the transition falls in the PC class if $\mu_1 = \mu_2$, or in the DP class if the symmetry between the two absorbing states is broken ($\mu_1 \neq \mu_2$).

The DMRG study for $n = 3$ and $n = 4$ showed that, the model is in the active phase in the whole parameter space, and the critical point is shifted to the limit of infinite reaction rates. In this limit the dynamics of the model can be mapped onto the zero temperature $n$-state Potts model (see also the simulation results of [Lipowski and Droz (2002a)]). It was conjectured by [Hooyberghs *et al.* (2001)] that the model is in the same N-BARW2 universality class for all $n \geq 3$. If we consider a region between $\emptyset_i$ and $\emptyset_j$ as a domain wall $X_{ij}$, we can follow the dynamics of such variables. In the limit $\lambda \to \infty$, $X_{ij}$ coincides with the particle $A$ and in the limit $\mu \to \infty$ $X_{ij}$ coincides with the bond variable $\emptyset_i\emptyset_j$. For finite values of these parameters one still can apply this reasoning at a coarse-grained level. In this case $X_{ij}$ is not a sharp domain wall, but an object with a fluctuating thickness. For $n = 2$ it was shown in Sect. 4.6.3 that such domain walls follow BARW2 dynamics (4.174). For $n > 2$ one can show that besides the N-BARW2 reactions (6.63),(6.64) involving a maximum of two types of particles, reaction types occur involving three different domains ($i \neq j$, $i \neq k$ and $j \neq k$):

$$X_{ij} \to X_{ik}X_{kj} \qquad X_{ik}X_{kj} \to X_{ij} \tag{6.71}$$

with increasing importance as $n \to \infty$. These reactions break the parity conservation of the N-BARW2 process. Therefore the numerical findings in [Hooyberghs *et al.* (2001)] for $n = 3, 4$ indicate that they are probably **irrelevant** or the conditions for N-BARW2 universal behavior could be relaxed. Owing to the fact that the $X_{ij}$ variables are domain walls they appear in pairwise manner, hence **hard-core exclusion** effects are ineffective for the critical behavior in one dimension. For the effect of pairwise initial conditions on dynamical exponents see Sect. 6.2.

For $n = 3$, upon breaking the global $S_3$ symmetry to a lower one, one gets a transition either in the directed percolation (Sect. 4.2), or in the parity conserving class (Sect. 4.6), depending on the choice of parameters [Hooyberghs *et al.* (2001)]. Simulations indicate [Lipowski and Droz (2002b)] that for this model local symmetry breaking may also generate PC class transition.

## 6.13 Hard-core 2-BARW2 classes in one dimension

Besides the effects of coupling interactions in low dimensions, blockades generated by hard-core particles may also play an important role. The effect of particle exclusion (i.e. $AB \not\leftrightarrow BA$) in 2-BARW2 models (Sect. 6.12) was investigated by [Ódor (2001c); Kwon *et al.* (2000)]. In $d = 2$ the bosonic field theoretical predictions [Cardy and Täuber (1998)] were confirmed [Ódor (2001c)]; mean-field class transition with logarithmic corrections. In one dimension, however two types of phase transitions were identified at zero branching rate ($\sigma = 0$) depending on the arrangement of offspring relative to the parent in process (6.64). That is, if the parent separates the two offspring (2-BARW2s):

$$A \stackrel{\sigma}{\to} BAB\ , \tag{6.72}$$

the steady state density is higher than in the case when they are created on the same site (2-BARW2a)

$$A \xrightarrow{\sigma} ABB \tag{6.73}$$

at a given branching rate, because in the former case they are unable to annihilate each other. This results in different order parameter exponents for the symmetric (2-BARW2s) and for the asymmetric (2-BARW2a) cases

$$\beta_s = 1/2, \quad \beta_a = 2\ . \tag{6.74}$$

This result is in contrast with a widespread belief that the bosonic field theory (where $AB \leftrightarrow BA$ is allowed) can describe these systems (because in that case the critical behavior is different (see (6.66)). This observation led [Kwon *et al.* (2000)] to the conjecture that in one-dimensional, reaction-diffusion systems a series of new universality classes should appear if particle exclusion is present. Note, however, that since the transition is at $\sigma = 0$ in both cases, the on-critical exponents do not depend on how particles are created and they can be identified with those described in Sect. 6.2. [Ódor (2001c)] determined a set of critical exponents satisfying scaling relations for this two new classes shown in Table 6.8.

Table 6.8 Summary of critical exponents in one dimension for N-BARW2 like models. The N-BARW2 data are quoted from [Cardy and Täuber (1996)]. Data divided by "|" correspond to random vs pairwise initial conditions [Hooyberghs *et al.* (2001); Ódor and Menyhárd (2000); Menyhárd and Ódor (2002)]. Exponents denoted by * exhibit slight initial density dependence.

| exponent | N-BARW2 | N-BARW2s | N-BARW2a |
|---|---|---|---|
| $\nu_{\|\|}$ | 2 | 2.0(1)\|0.915(2) | 8.0(4)\|3.66(2) |
| $Z$ | 2 | 4.0(2)\|1.82(2)* | 4.0(2)\|1.82(2)* |
| $\alpha$ | 1/2 | 0.25(1)\|0.55(1)* | 0.25(1)\|0.55(1)* |
| $\beta$ | 1 | 0.50(1)\|1.00(1) | 2.0(1)\|1.00(1) |

### 6.13.1 *Hard-core 2-BARWo models in one dimension*

Hard-core interactions in the two-component, one-offspring production model (2-BARW1) were investigated by [Ódor (2001d)]. Without interaction between different species one would expect a DP class transition. By introducing the $AB \not\leftrightarrow BA$ blocking in the two-component model,

$$A \xrightarrow{\sigma} AA \quad B \xrightarrow{\sigma} BB \tag{6.75}$$

$$AA \xrightarrow{1-\sigma} \emptyset \quad BB \xrightarrow{1-\sigma} \emptyset \tag{6.76}$$

[Ódor (2001d)] located a DP class transition at $\sigma = 0.81107$. Note, that the effect exerted by different species on each other are irrelevant now, unlike for the case of coupled ARW (see Sect. 6.2). On the other hand if we couple the two subsystems by offspring production:

$$A \xrightarrow{\sigma/2} AB \quad A \xrightarrow{\sigma/2} BA \tag{6.77}$$

$$B \xrightarrow{\sigma/2} AB \quad B \xrightarrow{\sigma/2} BA \tag{6.78}$$

$$AA \xrightarrow{1-\sigma} \emptyset \quad BB \xrightarrow{1-\sigma} \emptyset \,, \tag{6.79}$$

a continuous phase transition emerges at $\sigma = 0$ rate. Therefore the on-critical exponents are the same as those described in Sect. 6.2 and the order parameter exponent is found to be $\beta = 1/2$. This transition belongs to the same class as the 2-BARW2s model (see Sect. 6.13).

The **parity conservation** law, which is relevant in the case of one component BARW systems turns out to be irrelevant here. This finding reduces the expectations suggested by [Kwon *et al.* (2000)]. In fact the blockades introduced by exclusions generate robust classes. [Ódor (2001d)] suggested that **in one-dimensional, coupled branching and annihilating random walk systems of $N$-types of excluding particles at $\sigma = 0$ two universality classes exist, those of 2-BARW2s and 2-BARW2a models**, depending on whether the reactants can immediately annihilate (i.e. when similar particles are not separated by other type(s) of particle(s)) or not. Recent investigations in similar models [Park and Park (2001); Lipowski and Droz (2001)] are in agreement with this hypothesis.

### 6.13.2 *Coupled binary spreading processes*

Two-component versions of the PCPD model (Sect. 6.7.1) with particle exclusion in one-dimension were introduced and investigated by simulations in [Ódor (2002)] with the aim of extending the hypothesis of [Ódor (2001d)] for $N$-component BARW systems (Sect. 6.13.1). The following models with the same diffusion and annihilation terms ($AA \to \emptyset$, $BB \to \emptyset$) as in (Sect. 6.2) but different production processes were investigated.

1) Production and annihilation random walk model (2-PARW):

$$AA \xrightarrow{\sigma/2} AAB, \quad AA \xrightarrow{\sigma/2} BAA, \tag{6.80}$$

$$BB \xrightarrow{\sigma/2} BBA, \quad BB \xrightarrow{\sigma/2} ABB \,. \tag{6.81}$$

2) Symmetric production and annihilation random walk model

(2-PARWs):

$$AA \xrightarrow{\sigma} AAA, \tag{6.82}$$

$$BB \xrightarrow{\sigma} BBB \,. \tag{6.83}$$

These two models exhibit active steady states for $\sigma > 0$ with a continuous phase transition at $\sigma_c = 0$. Therefore the exponents defined on the critical point are those of the two-component ARW model with exclusion (Sect. 6.2). Together with the exponent $\beta = 2$ result for both cases this indicates that they belong to the N-BARW2a class. This also means that the hypothesis set up for N-BARW2 systems [Ódor (2001d)] (Sect. 6.13.1) can be extended.

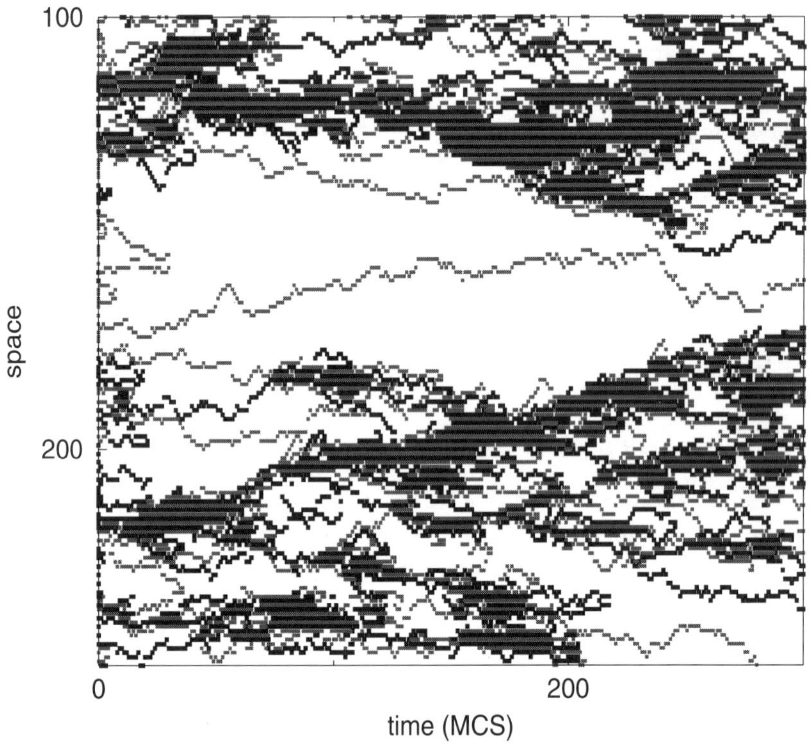


Fig. 6.10 Space-time evolution of the 2-PARWas model at the critical point. Black pixels correspond to $A$ particles, others to $B$'s. From [Ódor (2002)].

3) Asymmetric production and annihilation random walk model (2-PARWa):

$$AB \xrightarrow{\sigma/2} ABB, \quad AB \xrightarrow{\sigma/2} AAB, \tag{6.84}$$

$$BA \xrightarrow{\sigma/2} BAA, \quad BA \xrightarrow{\sigma/2} BBA\,. \tag{6.85}$$

This model does not have an active steady state. The $AA$ and $BB$ pairs annihilate themselves on contact, while if an $A$ and $B$ particle meet an $AB \to ABB \to A$ process eliminates blockades and the densities decay with the $\rho \propto t^{-1/2}$ law for $\sigma > 0$. For $\sigma = 0$ the blockades persist, and in case of random initial state $\rho \propto t^{-1/4}$ decay (see Sect. 6.2) can be observed here.

4) Asymmetric production and annihilation random walk model with spatially symmetric creation (2-PARWas):

$$AB \xrightarrow{\sigma/2} ABA, \quad AB \xrightarrow{\sigma/2} BAB, \tag{6.86}$$

$$BA \xrightarrow{\sigma/2} BAB, \quad BA \xrightarrow{\sigma/2} ABA\,. \tag{6.87}$$

In this case $AB$ blockades proliferate from production events. As a consequence, an active steady state appears for $\sigma_c > 0.3253(1)$ with a continuous phase transition. The space-time evolution from a random initial state shows (Fig. 6.10) that compact domains of alternating $\ldots ABAB\ldots$ sequences are formed, separated by lone wandering particles. This pattern is very similar to what was seen in the case of one-component binary spreading processes [Hinrichsen (2001b)]: compact domains within a cloud of lone random walkers, except that now domains are built up from alternating sequences only. This means that the ..$AAAA$... and ...$BBBB$... domains decay by this annihilation rate and the particle blocking is responsible for the formation of compact clusters. In the language of coupled DP + ARW model [Hinrichsen (2001b)] the pairs following the DP process are now the $AB$ pairs, which cannot decay spontaneously but only through an annihilation process: $AB + BA \to \emptyset$. They interact with two types of particles executing annihilating random walk with exclusions. The simulations resulted in the critical exponent estimates: $\beta = 0.37(2)$, $\alpha = 0.19(1)$ and $Z = 1.81(2)$, which agree fairly well with those of the PCPD model in the high-diffusion-rate region [Ódor (2000)].

# Chapter 7

# Surface-Interface Growth Classes

During the last years there has been an interest in the description of the self-affine kinetic roughening of surfaces, the micro-structure and the scaling behavior of films (see for instance [Yanguas-Gil *et al.* (2006)]). Related topics are the depinning transition of elastic systems in disordered media [Narayan and Fisher (1993); Nattermann *et al.* (1992); Fisher (1985); Kolton *et al.* (2006)] and the wetting transition [Majumdar *et al.* (2000a); Ginelli *et al.* (2006)]. In application to parallel and distributed computations, the important consequence of the derived scaling is the existence of the upper bound for the desynchronization in conservative update algorithm for parallel discrete-event simulations [Kolakowska *et al.* (2004)].

Interface growth classes are strongly related to the basic universality classes discussed so far and can be observed in experiments more easily. For example one of the few experimental realizations of the robust DP class (4.2) is related to a depinning transition in inhomogeneous porous media [Buldyrev *et al.* (1992)] (see Sect. 7.5). The interface models can be defined either by continuum equations or by lattice models of solid-on-solid (SOS) or restricted solid-on-solid (RSOS) types. In the latter case the height variables $h_i$ of adjacent sites are restricted

$$|h_i - h_{i+1}| \leq 1\,. \tag{7.1}$$

The morphology of a growing interface is usually characterized by its width

$$W(L,t) = \Big[\frac{1}{L}\sum_i h_i^2(t) - \Big(\frac{1}{L}\sum_i h_i(t)\Big)^2\Big]^{1/2}. \tag{7.2}$$

In the absence of any characteristic length, growth processes are expected to show power-law behavior of the correlation functions in space and height and the surface is described by the *Family-Vicsek* scaling [Family and Vicsek (1985)] form:

$$W(L,t) = t^{\tilde{\alpha}/Z} f(L/\xi_{||}(t)), \tag{7.3}$$

with the scaling function $f(u)$

$$f(u) \sim \begin{cases} u^{\tilde{\alpha}} & \text{if } u \ll 1 \\ \text{const.} & \text{if } u \gg 1 \end{cases} . \tag{7.4}$$

Here $\tilde{\alpha}$ is the **roughness exponent** and characterizes the deviation from a flat surface in the stationary regime in which correlation length $\xi_{||}(t)$ has reached a value larger than the system size $L$. The exponent, defined by the ratio

$$\tilde{\beta} = \tilde{\alpha}/Z \tag{7.5}$$

is called the **growth exponent** and characterizes the short time behavior of the surface. Similarly to equilibrium critical phenomena, these exponents do not depend on the microscopic details of the system under investigation. Using these exponents it is possible to divide growth processes into universality classes [Barabási and Stanley (1995); Krug (1997)]. The scaling form (7.3) of $W^2$ is invariant under $\Lambda$ rescaling

$$x \to \Lambda x, \qquad t \to \Lambda^{Z} t, \qquad h(x,t) \to \Lambda^{-\tilde{\alpha}} h(x,t) \ . \tag{7.6}$$

Recently **anomalous roughening** has been observed in many growth models and experiments. In these cases the measurable $\tilde{\alpha}_{loc}$ roughness exponent is different from $\tilde{\alpha}$ and may satisfy a different scaling law.[1]

Surfaces in $d+1$ dimensional systems can be **mapped** onto a time step of a $d$ dimensional particle reaction-diffusion or spin models. For example the one-dimensional Kawasaki spin model corresponding to the $K \leftrightarrow 3K$ process with random walk of kinks is mapped onto the $1+1$-dimensional surface as shown in Fig. 7.1. This means that a spatial profile $\{h_i(t)\}$ corresponds to a unique $\{s_j\}$ spin configuration at time $t$ by accumulating the spin values

$$h_i(t) = \sum_{j=1}^{i} s_j \ . \tag{7.7}$$

In one dimension the surface can also be considered as a random walker with fluctuation

$$\Delta x \propto t^{1/Z_w} \ . \tag{7.8}$$

Hence the roughness exponent is related to the dynamical exponent $Z_w$ as

$$\tilde{\alpha} = 1/Z_w \ . \tag{7.9}$$

[1]See, for example, [Krug (1997); Schroeder *et al.* (1993); López and Rodríguez (1996); Sarma *et al.* (1996); Dasgupta *et al.* (1996); López and Rodríguez (1997); Oliveira (1992); H. Yang and Lu (1994); Jeffries *et al.* (1996); López and Schmittbuhl (1998); Morel *et al.* (1998); Bru *et al.* (1998)].

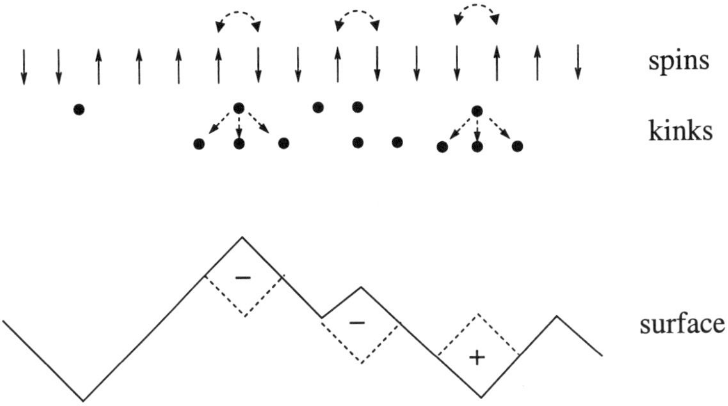


Fig. 7.1 Mapping between spins, kinks and surfaces in one dimension.

The $\tilde{\alpha} = 1/2$ corresponds to uncorrelated (or finite correlation length) random walks. If $\tilde{\alpha} > 1/2$ the surface exhibits correlations, while if $\tilde{\alpha} < 1/2$ the displacements in the profile are anti-correlated. Since the surfaces may exhibit drifts the fluctuations around the mean is measured defining the *local roughness (Hurst) exponent.* Using this surface mapping [Sales *et al.* (1997)] have characterized the different classes of Wolfram's one-dimensional cellular automata [Wolfram (1983)].

One can show that by coarse graining the one-dimensional Kawasaki dynamics

$$w_i = \frac{1}{4\tau}\left[1 - \sigma_i\sigma_{i+1} \ , +\lambda(\sigma_{i+1} - \sigma_i)\right] \tag{7.10}$$

a mapping can be made onto the KPZ equation (7.24), and the surface dynamics for $\lambda \neq 0$ (corresponding to anisotropic case) are in the KPZ class (Sect. 7.4) while for $\lambda = 0$ they are in the Edward Wilkinson class (Sect. 7.2). While these classes are related to simple random walk with $Z_w = 1/\tilde{\alpha} = 2$ the question arises what surfaces are related to other kind of random walks (example Levy-flights or correlated random walks etc.). Recently it was shown that globally constrained random walks (i.e. when a walker needs to visit each site an even number of times) can be mapped onto surfaces with dimer-type dynamics [Noh *et al.* (2001)] with $Z_w = 3 = 1/\tilde{\alpha}$.

By studying the correspondence between lattice models with absorbing states and models of pinned interfaces in random media [Dickman and Munoz (2000)] established the scaling relation

$$\tilde{\beta} = 1 - \beta/\nu_{||} \ , \tag{7.11}$$

which was confirmed numerically for $d = 1, 2, 3, 4$ contact processes (Sect. 4.2.1). The local roughness exponent was found to be smaller than the global value indicating anomalous surface growth in DP class models. In interface models different types of transitions may take place.

*(1) Roughening transitions* may occur between smooth phase characterized by finite width $W$ (in an infinite system) and rough phase when the width and diverges in an infinite system (but saturates in finite ones) as a result of varying some control parameters ($\epsilon$). Near the transition point the spatial ($\xi_{||}$) and growth direction correlations ($\xi_{\perp}$) diverge as

$$\xi_{||} \propto \epsilon^{\nu_{||}} \tag{7.12}$$

$$\xi_{\perp} \propto \epsilon^{\nu_{\perp}} \; . \tag{7.13}$$

Note, that in RD systems $\xi_{||}$ denotes the temporal correlation length. In the smooth phase the heights $h_i(t)$ are correlated below $\xi_{\perp}$. While in equilibrium models roughening transitions exist in $d > 1$ dimensions only, in nonequilibrium system this may occur in $d = 1$ as well.

*(2) Depinning transitions* occur when, as the consequence of changing some control parameter (usually an external force $F$), the surface starts to propagate with a speed $v$ and evolves into a rough state. Close to the transition $v$ is expected to scale as

$$v \propto (F - F_c)^{\tilde{\theta}} \; , \tag{7.14}$$

with the velocity exponent $\tilde{\theta}$ and the correlation lengths diverge. Known depinning transitions (in random media) are related to absorbing phase transitions with conserved quantities (see Sects. 7.3 and 7.4.3).

*(3) A so-called faceting phase transition* may also take place in the rough phase when up-down symmetrical facets appear. In this case the surface scaling behavior changes (see Sect. 7.7).

## 7.1 The random deposition class

Random deposition is the simplest surface growth process that involves uncorrelated adsorption of particles on top of each other. Therefore columns grow independently, linearly without bounds. The roughness exponent $\tilde{\alpha}$ (and correspondingly $Z$) is not defined here. The width of the surface grows as $W \propto t^{1/2}$ hence $\tilde{\beta} = 1/2$ in all dimensions. An example for such behavior is shown in a dimer growth model in Sect. 7.7.1.

## 7.2 Edwards-Wilkinson (EW) classes

As was mentioned in Sect. 6.13.2, growth models of this class in one dimension can easily mapped onto spins with symmetric Kawasaki dynamics or to particles with ARW (Sect. 4.5.1). If we postulate the translation and reflection symmetries for a noisy surface growth process

$$\mathbf{x} \to \mathbf{x} + \mathbf{\Delta x} \quad t \to t + \Delta t \quad h \to h + \Delta h \quad \mathbf{x} \to -\mathbf{x} \quad h \to -h \tag{7.15}$$

we are led to the Edwards-Wilkinson (EW) equation [Edwards and Wilkinson (1982)]

$$\partial_t h(\mathbf{x}, t) = v + \sigma \nabla^2 h(\mathbf{x}, t) + \eta(\mathbf{x}, t), \tag{7.16}$$

which is the simplest stochastic differential equation that describes a surface growth with these symmetries. The coupling $v$ corresponds to the mean growth velocity, $\sigma$ to the surface tension and $\eta$ is a zero-average Gaussian noise field with variance

$$< \eta(\mathbf{x}, t)\eta(\mathbf{x}', t') >= 2D\delta^{d-1}(\mathbf{x} - \mathbf{x}')(t - t') \ . \tag{7.17}$$

This equation is linear and exactly solvable. The critical exponents of EW classes are

$$\tilde{\beta} = (\frac{1}{2} - \frac{d}{4}), \quad Z = 2 \ , \tag{7.18}$$

and the upper critical dimension is $d_c = 2$. For $d > d_c = 2$ the roughness exponent $\tilde{\beta} = \tilde{\alpha} = 0$ and the interface becomes flat.

## 7.3 Quench disordered EW classes (QEW)

In random media linear interface growth is described by the so called quenched Edwards-Wilkinson (QEW) equation

$$\partial_t h(\mathbf{x}, t) = \sigma \nabla^2 h(\mathbf{x}, t) + F + \eta(\mathbf{x}, h(\mathbf{x}, t)), \tag{7.19}$$

where $F$ is a constant, external driving term and $\eta(\mathbf{x}, h(\mathbf{x}, t))$ is the *quenched noise.* The corresponding linear interface models (LIM) exhibit a *depinning transition* at $F_c$. The universal behavior of these models were investigated by [Nattermann *et al.* (1992); Narayan and Fisher (1993); Kim *et al.* (2001)] and it was shown to be **equivalent with NDCF** classes (Sect. 6.10). [Narayan and Fisher (1993)] analyzed analytically by PRG method the critical behavior of the $d$ dimensional QEW model near the depinning threshold. The roughness exponent in the critical region was shown to be independent from the type of disorder, in contrast with the equilibrium static behavior, where there are two different universality classes corresponding to random-bond and random-field disorder. This critical dynamic roughening exponent is argued to be equal to its equilibrium static random-field value:

$$\tilde{\alpha} = \epsilon/3 \tag{7.20}$$

to all orders in $\epsilon = 3 - d$ dimensions. All other critical exponents can be obtained using the scaling relations and $Z$, which is calculated up to order $O(\epsilon)$:

$$Z = 2 - (2/9)\epsilon \ . \tag{7.21}$$

These results agree fairly well with those of numerical simulations. As in the NDCF (SOC) models **avalanches** can also occur, such that the probability of their diameter exceeding $l$ decays as $l^{-\kappa}$, just below the critical threshold with $\kappa = d - 2 + \alpha$.

### 7.3.1 *EW classes with boundaries*

In practice surface growth has boundaries in all directions. One of the most important boundary is a hard-core substrate at zero height. This corresponds to **wetting** phenomena. When a phase ($A$) is in contact with a substrate, wetting occurs if a macroscopic layer of a coexisting phase ($B$) is adsorbed at the substrate. The wetting transition is characterized by the divergence of the $B$-layer thickness; the $AB$ interface goes arbitrarily far from the substrate. While wetting of surfaces at or near thermal equilibrium is well understood [Dietrich (1986)], the study of wetting phenomena far from equilibrium is a challenging new field. In the past decade most of these studies are based on particular lattice models [Hinrichsen *et al.* (1997a, 2000)] or phenomenological Langevin equations [Tu *et al.* (1997)]. If we increase the growth rate a **wetting transition** from a bound to a moving phase may occur.

The substrate — as a boundary — may change the scaling properties, and can introduce additional order parameter and associated critical exponent(s).[2] The new, extended universality classes are called (following [Barato *et al.* (2007)]) **bounded Edwards-Wilkinson (bEW)**, and **bounded Kardar-Parisi-Zhang (bKPZ)** classes. In case of KPZ

[2]Some authors say that it does not change the universality class of the growth process, but in this book we define a class by the whole set of critical exponents, therefore new independent exponents define new class behaviors.

(Sect. 7.4), where the reflection symmetry is broken the new exponents depend on the sign of the non-linear term, hence these are denoted by **bKPZ+** and **bKPZ-**. The numerically obtained scaling behavior and crossover phenomena are discussed here and in Sect. 7.4.4.

A model for nonequilibrium wetting proposed by [Hinrichsen *et al.* (1997a)] is defined on a one-dimensional lattice, with $L$ sites and periodic boundary conditions. Each site $i$ is associated with a height variable $h_i = 0, 1, 2, 3....$ The interface obeys the restricted solid-on-solid (RSOS) condition

$$|h_i - h_{i\pm 1}| \leq 1, \tag{7.22}$$

and evolves by random-sequential updates. At each randomly selected site one of the following reactions take place:

- deposition of a particle ($h_i \to h_i + 1$) with rate $q$,
- evaporation of a particle ($h_i \to h_i - 1$) at the edges of plateaus with rate $r$,
- evaporation of a particle ($h_i \to h_i - 1$) from the middle of a plateau with rate $p$.

The simulations were started from the flat substrate level and the corresponding phase diagram is shown on Fig. 7.2. Note, that small values of $p$ are more realistic since evaporation at the edges of a plateau is usually more likely than in the middle. The order parameter for the wetting transition is the density of sites at zero height $\rho_0$. In the bound phase near the transition point $q_c(p)$ the zero level density vanishes as

$$\rho_0 \sim (q_c - q)^{\beta_0} . \tag{7.23}$$

Depending on the parameters $p$, $q$ the model exhibits different wetting transitions from a bound to a moving phase. In particular for $p = 0$ and $q_c = 0.3993(1)$ it exhibits the scaling of unidirectionally coupled DP (see Sects. 7.6 and 7.6.1),[3] for $p \neq q$ the transition belongs to **bKPZ** classes (see Sect. 7.4.4), while at the $p = q$ symmetry point the transition is of **bEW** type. In the last case, in the bounded stationary state an exact solution [Hinrichsen *et al.* (1997a)] provides the exponents shown in Table 7.1.

[3] In the lack of hole digging process, the model does not feel the substrate once a layer is filled.

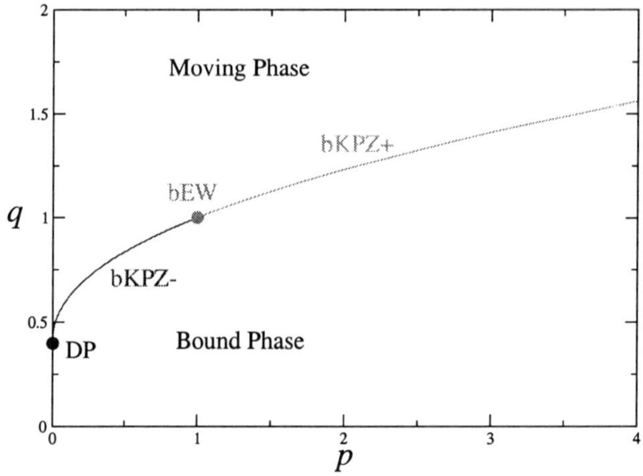


Fig. 7.2 Phase diagram of the wetting model with $r = 1$. From Ref. [Barato *et al.* (2007)].

Table 7.1 Scaling exponents of the bEW class in $d = 1$. From [Hinrichsen *et al.* (1997a)].

| $\alpha_0$ | $\beta_0$ | $\tilde{\alpha}$ | $\tilde{\beta}$ | $Z$ | $\nu_\perp$ |
|---|---|---|---|---|---|
| 3/4 | 1 | 1/2 | 1/4 | 2 | 2/3 |

## 7.4 Kardar-Parisi-Zhang (KPZ) classes

In experiments the reflection symmetry in the height direction is usually broken. If we drop the $h \to -h$ symmetry from Eq. (7.15) we can add a term to the Langevin equation (7.16), which is the most relevant one in renormalization group sense, breaking the up-down symmetry:

$$\partial_t h(\mathbf{x},t) = v + \sigma \nabla^2 h(\mathbf{x},t) + \lambda(\nabla h(\mathbf{x},t))^2 + \eta(\mathbf{x},t) \,. \tag{7.24}$$

That is the Kardar-Parisi-Zhang (KPZ) equation [Kardar *et al.* (1986)]. Here again $v$ describes the mean growth velocity, $\sigma$ the surface tension and $\eta$ the zero-average Gaussian noise field with variance

$$< \eta(\mathbf{x},t)\eta(\mathbf{x}',t') >= 2D\delta^{d-1}(\mathbf{x}-\mathbf{x}')(t-t') \,. \tag{7.25}$$

The KPZ equation is nonlinear, but exhibits a tilting symmetry as the result of the Galilean invariance:

$$h \to h' + \epsilon \mathbf{x}, \qquad \mathbf{x} \to \mathbf{x}' - \lambda \epsilon t, \qquad t \to t' \tag{7.26}$$

where $\epsilon$ is an infinitesimal angle. As a consequence the scaling relation

$$\tilde{\alpha} + Z = 2 \tag{7.27}$$

**holds in any dimensions**. This equation was invented to describe the kinetic roughening of a growing interface, but turned out to model other important physical phenomena, such as the randomly stirred fluid (Burgers equation) [Forster *et al.* (1977)], directed polymers in random media [Kardar (1987)], dissipative transport [van Beijeren *et al.* (1985); Janssen and Schmittmann (1986)] and the magnetic flux lines in superconductors [Hwa (1992)].

In **one dimension** the model is exactly solvable [Kardar (1987)], the critical exponents are known exactly, whereas for $d > 1$ dimensions numerical estimates exist [Barabási and Stanley (1995)] (see Table 7.2). In **higher**

Table 7.2 Scaling exponents of KPZ classes.

| $d$ | $\tilde{\alpha}$ | $\tilde{\beta}$ | $Z$ |
|---|---|---|---|
| 1 | 1/2 | 1/3 | 3/2 |
| 2 | 0.38 | 0.24 | 1.58 |
| 3 | 0.30 | 0.18 | 1.66 |

**dimensions** the theoretical understanding of the KPZ equation is still not complete. It is known, that for $d > 2$, there exist two different regimes, separated by a critical value $\lambda_c$ of the non-linear coefficient [Kardar *et al.* (1986)]. While in the weak-coupling regime ($\lambda < \lambda_c$), the behavior is governed by the $\lambda = 0$ fixed point corresponding to the EW equation (7.16), in the strong-coupling (rough) regime ($\lambda > \lambda_c$), the non-linearity becomes relevant and despite considerable efforts, the statistical properties for $d \geq 2$ remain controversial.

The **upper critical dimension** of this model is debated. Mode coupling theories and various phenomenological field theoretical schemes [Halpin-Healy (1990); Lässig (1995); Lässig and Kinzelbach (1997)] settle to $d_c = 4$, whereas functional renormalization group to two-loop order suggests $d_c \simeq 2.5$ [Le Doussal and Wiese (2003, 2005)]. A set of related theories, such as replica symmetry breaking [Mézard and Parisi (1991)], variational studies [Garel and Orland (1997)] and Flory-Imry-Ma arguments [Monthus and Garel (2004)] predicted $d_c = 2$.

Contrary to analytical approaches numerical solution of the KPZ equation [Moser *et al.* (1991)], simulations [Kim and Kosterlitz (1989); Wolf and Kertész (1987); Tang *et al.* (1992); Ala-Nissila *et al.* (1993); Marinari *et al.* (2000)], and the results of real-space renormalization group calculations [Castellano *et al.* (1998b,a, 1999)] provide **no evidence for a finite** $d_c$. Furthermore, a numerical study [Tu (1994)] of the mode-coupling equations gives no indication for the existence of a finite $d_c$ either. Recently simulations of the restricted solid-on-solid growth models were used to build the width distributions of $d = 2 - 5$ dimensional KPZ interfaces. The universal scaling function associated with the steady-state width distribution was found to change smoothly as $d$ is increased, thus strongly suggesting that $d = 4$ is not an upper critical dimension for the KPZ equation. The dimensional trends observed in the scaling functions indicate that the upper critical dimension is at infinity [Marinari *et al.* (2002)].

For at least a partial resolution of these findings a very recent mode coupling theory by [Canet and Moore (2007)] was suggested within the framework of $N$**-component KPZ** [Doherty *et al.* (1994)]. Here systematic $1/N$ expansion found a new branch of solutions (universality class 'S') for $d < 2$, which seems to be similar to the results of studies claiming $d_c = 2$. The other solution ('F') exists for $0 < d < 4$, with $d_c = 4$ (see Fig. 7.3). Canet and Moore assumed that the S branch is realized if the noise of KPZ exhibits some long-range correlations. However this S behavior can only occur in $d = 1$ physical dimension, and there it coincides with the other one (F).

### 7.4.1 *KPZ with anisotropy*

The presence of an anisotropy along the surface may drastically change the scaling properties of the KPZ equation. In two dimensions this leads to

$$\frac{\partial h}{\partial t} = \nu_x \partial_x^2 h + \nu_y \partial_y^2 h + \frac{\lambda_x}{2}(\partial_x h)^2 + \frac{\lambda_y}{2}(\partial_y h)^2 + \eta(x,y,t), \tag{7.28}$$

which means that the surface tension ($\nu$) and the coupling of the nonlinear term $\lambda$ are different in the two directions. This equation was introduced by [Villain (1991)] and its properties were studied by [Wolf (1991)]. The scaling properties depend on the signs of the coefficients $\lambda_x$ and $\lambda_y$. When $\lambda_x \lambda_y < 0$, a surface described by Eq. (7.28) exhibits the universality class behavior of EW (see Sect. 7.2). However, when $\lambda_x \lambda_y > 0$ the scaling

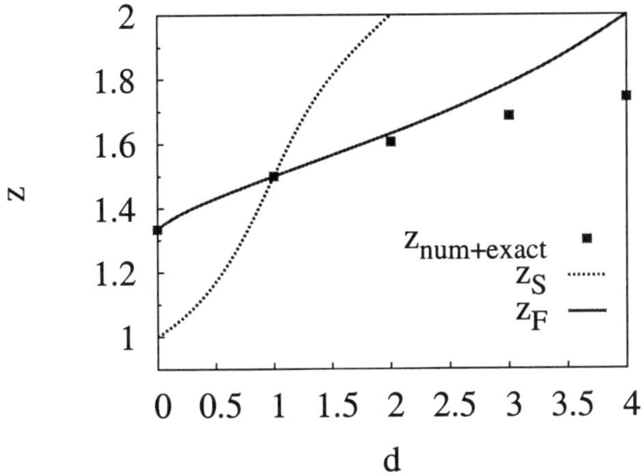


Fig. 7.3 The two solutions for the dynamical exponent $Z_S(d)$ and $Z_F(d)$, compared with numerical [Marinari *et al.* (2000)] results for $d = 2, 3, 4$ and the exact results in $d = 0, 1$, all represented by squares. From [Canet and Moore (2007)].

properties are described by the isotropic KPZ (see Sect. 7.4) equation.

### 7.4.2 *The Kuramoto-Sivashinsky (KS) Equation*

We can extend the KPZ equation with a higher order derivative term

$$\partial_t h = \sigma \nabla^2 h + \lambda (\nabla h)^2 - K(\nabla^2)^2 h + \eta(x,t). \tag{7.29}$$

and arrive to the so called (noisy) Kuramoto-Sivashinsky (KS) Equation [Kuramoto and Tsuzuki (1977); Sivashinsky (1979)]. Here the surface tension coefficient $\sigma$ is negative (in contrast with the KPZ), whereas $K$ is a positive surface diffusion coefficient [Herring (1950); Mullins (1957)]. As usual $\eta(x,t)$ is a Gaussian white noise, with zero mean and short-range correlations (see Eq. (7.17)). The combination of $\nabla^2 h$ and $|\nabla h|^2$ terms model the effect of particles being knocked out by the incoming ions. These terms can be derived in a simpler model for ion bombardment, the linear, Bradley-Harper (BH) model (see Sect. 7.5.2) [Bradley and Harper (1988)].

In KS even the deterministic variant ($\eta = 0$) exhibits spatiotemporal chaos and is useful to describe pattern formation, such as chemical turbulence and flame-front propagation [Kuramoto and Tsuzuki (1977)]. There are other physical systems, including ion sputtering, where the noisy version of the KS equation is used [Cuerno and Lauritsen (1995)].

An important, unanswered question is whether the KPZ and KS equations belong to the same or to different universality classes. In **one dimension** numerical [Zaleski (1989); Sneppen *et al.* (1992)] and analytical [L'vov *et al.* (1993); Yakhot (1981); Cuerno and Lauritsen (1995)] results show that the KS and KPZ equations indeed exhibit the same scaling behavior. The short time scale solution of KS equation reveals an unstable pattern-forming behavior, with a morphology reminiscent of **ripples** [L'vov *et al.* (1993)].

However, it is still an open question whether the KS and the KPZ equations fall into the same universality class in **two dimensions**. PRG calculations [Cuerno and Lauritsen (1995)] ran in the same problem as in case of KPZ: the strong coupling fixed point was unaccessible to the dynamic RG perturbative approach. Computer simulations are somewhat contradictory. In particular the simulations of [Drotar *et al.* (1999)] in $2+1$ dimensions found the initial time exponents: $\tilde{\beta} = 0.22-0.25$, $\tilde{\alpha} = 0.75-0.8$ and $Z = 3.0-4.0$. The long time exponents, $\tilde{\beta} = 0.16-0.21$, $\tilde{\alpha} = 0.25-0.28$ are close but significantly less than those of the KPZ (see Table 7.2).

### 7.4.3 *Quench disordered KPZ (QKPZ) classes*

In random media nonlinear interface growth is described by the so-called **quenched KPZ equation** (QKPZ) [Barabási and Stanley (1995)],

$$\partial_t h(\mathbf{x},t) = \sigma \nabla^2 h(\mathbf{x},t) + \lambda(\nabla h(\mathbf{x},t))^2 + F + \eta(\mathbf{x}, h(\mathbf{x},t)) \ , \tag{7.30}$$

where $F$ is a constant, external driving term and $\eta(\mathbf{x}, h(\mathbf{x},t))$ is the *quenched noise* (which does not fluctuate in time). For $\lambda = 0$ the problem reduces to the QEW case (see Sect. 7.3). If we increase the force above a critical value $F_c$ the interface keeps moving for arbitrary large times, however if $F < F_c$ it becomes pinned after some time due to the randomly distributed sites with $\eta < 0$.

In **one dimension** close to the pinning transition [Csahók *et al.* (1993)] performed direct simulation of the continuum equation (7.30) and came to the conclusion: The numerical values for the exponents $\tilde{\alpha}$ and $\tilde{\beta}$ in $d = 1$ are in agreement with the scaling theory of Eq. (7.30), which yields

$$\tilde{\alpha} = (4-d)/4 \ , \quad \tilde{\beta} = (4-d)/(4+d) \ . \tag{7.31}$$

This suggests an engineering upper critical dimension $d_c = 4$, above which the surface becomes flat. The universal behavior was investigated by [Buldrev *et al.* (1993); Leschhorn (1996); Amaral *et al.* (1995)] extensively and

Table 7.3 Critical exponents of the QKPZ for dimensions measured directly from the simulations.

| | $(1+1)$ | $(2+1)$ | $(3+1)$ | $(4+1)$ | $(5+1)$ | $(6+1)$ |
|---|---|---|---|---|---|---|
| $\nu_{\|\|}$ | $1.73 \pm 0.02$ | $1.16 \pm 0.05$ | $0.95 \pm 0.1$ | $0.66 \pm 0.10$ | $0.6 \pm 0.1$ | $0.5 \pm 0.1$ |
| $\tilde{\alpha}$ | $0.63 \pm 0.01$ | $0.48 \pm 0.03$ | $0.38 \pm 0.04$ | $0.27 \pm 0.05$ | $0.25 \pm 0.05$ | $0.2 \pm 0.2$ |
| $z$ | $1.01 \pm 0.02$ | $1.15 \pm 0.05$ | $1.36 \pm 0.05$ | $1.58 \pm 0.05$ | $1.7 \pm 0.1$ | $1.8 \pm 0.2$ |
| $\nu_{\perp}$ | $1.10 \pm 0.02$ | $0.57 \pm 0.05$ | $0.34 \pm 0.05$ | $0.2 \pm 0.1$ | $0.15 \pm 0.05$ | $0.1 \pm 0.1$ |
| $\beta$ | $0.63 \pm 0.01$ | $0.41 \pm 0.05$ | $0.29 \pm 0.07$ | $0.2 \pm 0.1$ | $0.1 \pm 0.1$ | $0.1 \pm 0.1$ |
| $\theta$ | $0.58 \pm 0.07$ | $0.8 \pm 0.2$ | $1.0 \pm 0.2$ | $1.0 \pm 0.2$ | | |

predicted simulation values are summarized in Table 7.3. As we can see the the roughness exponent does not become zero for $d \geq 4+1$ and the naive $d_c = 4$ scaling prediction is not fulfilled.

[Tang and Leschhorn (1992)] have shown numerically that in **one dimension** this class is described by 1+1-dimensional directed percolation depinning. This has the following heuristic explanation. If the interface is pinned by a path of impurities, it percolates from one side to the other, blocking the whole interface front. But since this path is directed, DP is expected to control its scaling. However this argument applies in one dimension only. In **higher dimensions** quenched KPZ is related to percolating directed surfaces [Barabási *et al.* (1996)]. More rigorous, field theoretical connection between these classes has not been established.

In the case of a negative KPZ non-linearity either a first order or a DP type of depinning transition was reported by simulations [Jeong *et al.* (1999)]. This is similar to the scenario, which occurs if one replaces the potential in KPZ with an attractive wall and a complex wetting transition occurs with tricritical behavior (see [Muñoz (2004)]).

Due to mapping of surface growth (see for instance Fig. 7.1) onto RD systems the effect of impurities in KPZ can also be studied by disordered ASEP models (see the discussion in Sect. 3.1.3).

### 7.4.4 *KPZ classes with boundaries*

The introduction of a limiting or bounding wall into the KPZ model leads to quite different phenomenologies, depending on whether the wall is an upper or a lower one (for a recent review see [Muñoz (2004)]), due to the absence of height-inversion symmetry. In such systems two different phases may exist: a depinned one, in which the KPZ-interface moves freely away from the wall, and a pinned one, with a finite stationary average distance from it. These phases are separated by a **wetting transition**, but its universal behavior is manifested in other systems (like transitions occurring in DNA alignment problems, Burgers' turbulence, bounded directed-polymers in random-media etc. for references see [Muñoz (2004)]). The Langevin equation, describing such system with an additional bounding term: $-be^{(p-1)h}$ is a generalization of the KPZ Eq. (7.24):

$$\partial_t h(r,t) = v - be^{(p-1)h} + \lambda(\nabla h(r,t))^2 + D\nabla^2 h(r,t) + \sigma\eta(r,t), \qquad (7.32)$$

where $\lambda$, $D$, and $\sigma$ are constants, and $\eta$ is a Gaussian white noise.

When we fix a given type of limiting wall – for instance, a lower, rigid substrate on top of which the interface grows – we can observe two different classes of depinning transitions, depending on the sign of $\lambda$.[4] By performing a Cole-Hopf transformation (4.218) this can be mapped onto the multiplicative noise equation and depending on the sign of the nonlinearity two different universality classes (MN1 or MN2) emerge (discussed in Sect. 4.10).[5] [Muñoz (2004)] discusses the behavior of MN1 and MN2 classes in detail. The limit: $p \to \infty$ describes a *hard* (impenetrable) wall: the interface cannot cross the $h = 0$ limit in either the upper ($h(x,t) < 0$) nor the lower wall case ($h(x,t) \geq 0$). But this hardness is not a relevant feature and the **MN1 and MN2 are the same universality classes as bKPZ+ and bKPZ− respectively**.

Now I present the latest numerical results obtained for the **one-dimensional** lattice version of wetting transition corresponding to Eq. (7.32). As discussed in Sect. 7.3.1 in case of a solid lower boundary (a substrate) the KPZ model exhibits a wetting transition. In **one dimension** a model has been introduced [Hinrichsen *et al.* (1997a)], which exhibits various nonequilibrium wetting transitions by changing the evaporation $p$ and deposition $q$ parameters of that model (see Fig. 7.2). If $0 < p \neq q$ this transition is of a bounded KPZ class type. The critical exponents have been determined by simulations [Barato *et al.* (2007)] and are shown in Table 7.4. [Barato *et al.* (2007)] argue that the wetting process for small $p$ values (small hole digging probability at the middle of plateaus) can be interpreted as a DP process in an external field. Therefore in the $p \to 0$ limit one can see a **crossover from bKPZ− to the DP class** with a crossover

[4]One can show that the KPZ equation with a positive non-linearity and a lower-wall is equivalent to the KPZ with a negative non-linearity coefficient and an upper-wall [Muñoz (2004)].

[5]Note that the direct numerical integration of KPZ-like Langevin equations is uncontrollable due to well-documented numerical instabilities.

Table 7.4 Scaling exponents of bKPZ classes in $d = 1$. From [Barato *et al.* (2007)].

| class | $\alpha_0$ | $\beta_0$ | $\tilde{\alpha}$ | $\tilde{\beta}$ | $Z$ | $\nu_\perp$ |
|---|---|---|---|---|---|---|
| bKPZ- | 1.184(10) | 1.776(15) | 1/2 | 1/3 | 3/2 | 1 |
| bKPZ+ | 0.228(5) | 0.342(8) | 1/2 | 1/3 | 3/2 | 1 |

exponent $1/\phi = 0.44(2)$ and for $p = 0$ one observes the so-called unidirectionally coupled DP class behavior (see Sect. 7.6), with level dependent exponents.

## 7.5 Other continuum growth classes

For continuum growth models exhibiting the symmetries

$$\mathbf{x} \to \mathbf{x} + \mathbf{\Delta x} \quad t \to t + \Delta t \quad h \to h + \Delta h \quad \mathbf{x} \to -\mathbf{x} \tag{7.33}$$

there are several possible general Langevin equations, with relevant terms, classified as follows [Barabási and Stanley (1995)].

- The deterministic part describes conservative or nonconservative process (i.e. the integral over the entire system may be zero or not). The conservative terms are $\nabla^2 h$, $\nabla^4 h$ and $\nabla^2 (\nabla h)^2$. The only relevant nonconservative terms is the $(\nabla h)^2$.
- The system may be linear or not.
- The noise term may be conservative (i.e., the result of some surface diffusion) with the correlator

$$\langle \eta_d(\mathbf{x},t)\eta_d(\mathbf{x}',t')\rangle = (-2D_d\nabla^2 + D'_d\nabla^4)\delta(\mathbf{x}-\mathbf{x}')\delta(t-t') \tag{7.34}$$

  or it may be nonconservative as in Eq. (7.17), as the result of adsorption, desorption mechanisms.

When the surface growth properties of such systems other than the EW and KPZ classes were analyzed five other universality classes were identified (see Table 7.5).

In the next subsection I discuss one of them in more detail.

### 7.5.1 *Molecular beam epitaxy classes (MBE)*

By technological applications molecular beam epitaxy (MBE) is an important tool for modifying surfaces. The relevant surface mechanisms at high temperatures are surface diffusion and fluctuations in the ion beam flux

Table 7.5 Summary of continuum growth classes discussed in this section. Here $\eta$ denotes uncorrelated Gaussian noise, while $\eta_d$ a conservative noise. Following [Barabási and Stanley (1995)].

| Langevin equation | $\tilde{\alpha}$ | $\beta$ | Z |
|---|---|---|---|
| $\partial_t h = -K\nabla^4 h + \eta$ | $\frac{4-d}{2}$ | $\frac{4-d}{8}$ | 4 |
| $\partial_t h = \nu\nabla^2 h + \eta_d$ | $\frac{-d}{2}$ | $\frac{-d}{4}$ | 2 |
| $\partial_t h = -K\nabla^4 h + \eta_d$ | $\frac{2-d}{2}$ | $\frac{2-d}{8}$ | 4 |
| $\partial_t h = -K\nabla^4 h + \lambda_1\nabla^2(\nabla h)^2 + \eta$ | $\frac{4-d}{3}$ | $\frac{4-d}{8+d}$ | $\frac{8+d}{3}$ |
| $d \leq 1$ | | | |
| $\partial_t h = -K\nabla^4 h + \lambda_1\nabla^2(\nabla h)^2 + \eta_d$ | $\frac{2-d}{3}$ | $\frac{2-d}{10+d}$ | $\frac{10+d}{3}$ |
| $d > 1$ | | | |
| | $\frac{2-d}{2}$ | $\frac{2-d}{8}$ | 4 |

(for a review see [Makeev *et al.* (2002)]). The standard approach to describe such system is to set up a macroscopic Langevin equation starting from phenomenology. If we assume that surface diffusion is the only relaxation mechanism present, the height $h(\boldsymbol{r}, t)$ of a surface obeys a continuity equation of the form

$$\frac{\partial h}{\partial t} + \tilde{\nabla} \cdot \boldsymbol{j} = 0 \ , \tag{7.35}$$

where $\boldsymbol{j}$ is a surface current density tangent to the surface, and $\tilde{\nabla}$ is calculated in a frame with axes parallel to the surface ($x$,$y$ direction). The current can be expressed as the gradient of a chemical potential $\mu(\boldsymbol{r}, t)$

$$\boldsymbol{j} \propto -\tilde{\nabla}\mu(\boldsymbol{r}, t) = -\tilde{\nabla}^2 \frac{\delta F[h]}{\delta h} , \tag{7.36}$$

where $\mu$ minimizes the free energy $F[h]$, which is proportional to the total surface area

$$F[h] = \int d\boldsymbol{r}\sqrt{1 + (\partial_x h)^2 + (\partial_y h)^2} \ . \tag{7.37}$$

Neglecting higher than second order derivatives of $h$ one arrives at

$$\partial t h = -K\nabla^2(\nabla^2 h) = -K\nabla^4 h, \tag{7.38}$$

with a positive constant $K$, which called the MBE equation [Mullins (1957)]. Taking into account random noise this is just the first Langevin equation in Table 7.5, which is also called Mullins diffusion. Such universal growth behavior was also reported in tumor growth processes [Bru *et al.* (1998); Martins *et al.* (2007)]. In atomic deposition the randomness is mainly due to the fluctuations in the intensity $J$ of the incoming ion beam

$$\eta(\mathbf{r}, t) \equiv \delta J(\mathbf{r}, t) \ , \tag{7.39}$$

which is an uncorrelated Gaussian noise (7.17). Hence the stochastic growth equation describing the surface diffusion and fluctuations in an erosion process has the form

$$\frac{\partial h}{\partial t} = -K\nabla^4 h - J + \eta(\mathbf{x}, t). \tag{7.40}$$

In tumor growth the source term represents the cell division rate. However experimental studies indicate that ion bombardment leads to an enhancement of the surface adatom mobility and thus may drastically change the relaxation mechanism, as compared to regular surface diffusion.

### 7.5.2 *The Bradley-Harper (BH) model*

By combining the linear gradient terms of the EW model (Sect. 7.2) and the MBE (Sect. 7.5.1) one arrives to the so-called combined linear, or Bradley-Harper (BH) model [Bradley and Harper (1988)]

$$\frac{\partial h}{\partial t} = A\nabla^2 h + K\nabla^4 h + \eta(\mathbf{x}, t). \tag{7.41}$$

where $\eta$ is a Gaussian white noise again. This is a simple model for ion-implantation, where the first term describes the particle desorption, the second one the surface diffusion. The stability requires $A \geq 0$ and $K \leq 0$ Simple power-counting shows that the asymptotic behavior in an infinite system is dominated by first term, therefore EW class scaling is expected for $t \to \infty$. However for short times the Mullins diffusion behavior, corresponding to the second term, can also be observed. In a finite system for suitable choice of the crossover length scale the MBE growth can be made dominant for all times [Majaniemi *et al.* (1996)].

### 7.5.3 *Classes of mass adsorption-desorption aggregation and chipping models (SOC)*

Based on the interest in self-organized critical systems (SOC), in which different physical quantities exhibit power-law distributions in the steady state over a wide region of the parameter space [Bak *et al.* (1987)], a family of lattice models, in which masses diffuse, aggregate on contact, or chip off a single unit mass was introduced by [Majumdar *et al.* (1998, 2000b)]. Self-organized criticality has been studied in a variety of model systems ranging from sandpiles to earthquakes. Note, that simple SOC models can be mapped onto NDCF particle models (see Sect. 6.10) and the models presented in this section are also similar to those discussed in Sect. 7.3.1.

However in the lattice-gas representation they correspond to long-range particle hopping processes, therefore they exhibit different fixed points and scaling behavior.

A particularly simple lattice model due to [Takayasu and Takayasu (1997)], with mass diffusion, aggregation upon contact and adsorption of unit masses from outside at a constant rate, exhibits self-organized criticality. In this model the steady-state mass distribution exhibits a nontrivial power-law decay for large masses in all dimensions [Takayasu *et al.* (1988)]. These mass adsorption-desorption models in one dimension are defined as follows. A site $i$ is chosen randomly and then one of the following events can occur:

(1) Adsorption: With rate $q$, a single particle is adsorbed at site $i$; thus $m_i \to m_i + 1$.
(2) Desorption: With rate $p$, a single particle is desorbed from site $i$; thus $m_i \to m_i - 1$ provided $m_i \geq 1$.
(3) Diffusion and Aggregation: With rate 1, the mass $m_i$ at site $i$ moves to a nearest neighbor site [either $(i-1)$ or $(i+1)$] chosen at random. If it moves to a site that already has some particles, then the total mass just adds up; thus $m_i \to 0$ and $m_{i\pm1} \to m_{i\pm1} + m_i$.
(4) Chipping (single-particle dissociation): With rate $w$ a bit of mass at the site "chips" off: provided $m_i \geq 1$ a single particle leaves site $i$ and moves with equal probability to one of the neighboring sites.

While the original **Takayasu model** ($p = 0$, $w = 0$) does not have a phase transition in the steady state, by introducing a nonzero desorption rate $p$ induces a critical line $p_c(q)$ in the $p - q$ plane. For fixed $q$, if one increases $p$ from 0, one finds that for all $p < p_c(q)$, the steady state mass distribution has the same large $m$ behavior as in the Takayasu case, i.e.,

$$P(m) \propto m^{-\tau_t} , \tag{7.42}$$

where the exponent $\tau_t$ is the 'Takayasu exponent' and is independent of $q$. For $p = p_c(q)$, we find the steady state mass distribution still decays algebraically for large $m$, but with a new critical exponent $\tau_c$ which is bigger than the Takayasu exponent $\tau_t$. For $p > p_c(q)$, we find that

$$P(m) \sim \exp(-m/m^*) \tag{7.43}$$

for large $m$ where $m^*$ is a characteristic mass that diverges if one approaches $p_c(q)$ from the $p > p_c(q)$ side. The critical exponent $\tau_c$ is the same everywhere on the critical line $p_c(q)$. This phase transition occurs in all spatial dimensions including $d = 1$. The $\tau$ exponents were determined for the mean-field and one dimensional cases ([Takayasu *et al.* (1988); Majumdar *et al.* (2000a)])

$$\tau_t^{MF} = 3/2, \quad \tau_c^{MF} = 5/2, \quad \tau_t^{1d} = 4/3, \quad \tau_c^{1d} = 1.833 \, , \tag{7.44}$$

although the location of $d_c$ is not known. In **one dimension** this model can also be mapped onto an interface dynamics, (see Fig. 7.1 for a similar construction), if we interpret the configuration of masses as an interface profile regarding $m_i$ as a local height variable. The phase transition of that model can be qualitatively interpreted as a nonequilibrium **wetting transition** of the interface. The corresponding surface growth exponents are [Majumdar *et al.* (2000a)]

$$\tilde{\beta}^{MF} = 1/6, \quad Z = 2, \quad \tilde{\beta}^{1d} = 0.358, \quad Z = 2 \, . \tag{7.45}$$

In the $p = q = 0$ conserved-mass aggregation model (CMAM), a nonequilibrium phase transition occurs when the chipping rate or the average mass per site $\rho$ is varied. There is a critical line $\rho_c(w)$ in the $\rho$-$w$ plane that separates two types of asymptotic behaviors of $P(m)$. For fixed $w$, as $\rho$ is varied across the critical value $\rho_c(w)$, the large $m$ behavior of $P(m)$ was found to be

$$P(m) \sim \begin{cases} e^{-m/m^*} & \rho < \rho_c(w), \\ m^{-\tau} & \rho = \rho_c(w), \\ m^{-\tau} + \text{infinite aggregate} & \rho > \rho_c(w). \end{cases} \tag{7.46}$$

As one increases $\rho$ beyond $\rho_c$, this asymptotic algebraic part of the critical distribution remains unchanged but in addition an infinite aggregate forms. This means that all the additional mass $(\rho - \rho_c)V$ (where $V$ is the volume of the system) condenses onto a single site and does not disturb the background critical distribution. This is analogous, in spirit, to the **condensation** of a macroscopic number of bosons onto the single $k = 0$ mode **in an ideal Bose gas** as the temperature goes below a certain critical value. [Rajesh and Majumdar (2001)] proved analytically that the mean field phase boundary has the shape $\rho_c(w) = \sqrt{w+1} - 1$, *exactly* and independently of the spatial dimension $d$. They also provided unambiguous numerical evidence that the exponent $\tau = 5/2$ is independent of $d$ too. The corresponding interface growth exponents are:

$$Z = 2, \; \tilde{\alpha} = 2/3 \tag{7.47}$$

[Majumdar *et al.* (2000b)]. Even though the single site distribution $P(m)$ may be given exactly by the mean field solution, that does not prove that mean field theory or product measure is the exact stationary state in all dimensions. The **lattice gas equivalent** of this model in the $w \to \infty$ limit (only chipping) is the symmetric exclusion process [Ligget (1985); Halpin-Healy and Zhang (1995)]. This means that in the corresponding interface model nonlocal moves are absent. The continuum version on this model is well described by the EW equation (7.16), which exhibits the universal behavior discussed in Sect. 7.2.

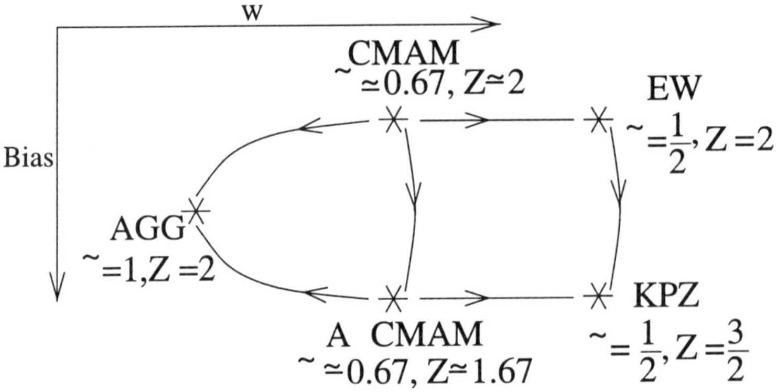


Fig. 7.4 Schematic depiction of fixed point in the conserved aggregation models. Following [Majumdar *et al.* (2000a)].

The **left-right asymmetric version** of the CMAM model (A-CMAM) was also studied by [Majumdar *et al.* (2000b)]. This has a qualitatively similar phase transition in the steady state as the CM model but exhibits a different class phase transition owing to the mass current in this system. Simulations in one dimension predict: $Z = 1.67$, $\tilde{\alpha} = 0.67$. In the $w \to \infty$ limit one can make a mapping onto a lattice gas again: the ASEP model (see Sect. 3.1.2), which exhibits the KPZ class interface universality (Sect. 7.4).

Finally the $w \to 0$, **no chipping limit** of the CMAM and A-CMAM models corresponds to the so-called aggregation model (AGG) both, where masses only diffuse as a whole and aggregate with each other. In the lattice gas language this corresponds to long-range hopping to the farthest available hole. In the spin model representation this is a coarsening with domains growing as $l(t) \propto t^{1/2}$ (thus $Z = 2$) and a finite system of size $L$ eventually breaks up into two domains of opposite signs. In the interface model this means mound formation (see [Krug (1997)]) and the scaling behavior is characterized by the exponents

$$Z = 2, \ \tilde{\alpha} = 1 \tag{7.48}$$

[Majumdar *et al.* (2000b)]. The latter (roughness) exponent can be understood from the fact that the maximum height of the single mound is $O(L)$, hence the width of the fluctuations is also of the order $O(L)$. The fixed points and flows corresponding to different CMAM versions (limits) are shown on Fig. 7.4.

## 7.6 Unidirectionally coupled DP classes

As was mentioned in Sect. 6.11 in the case of coupled, multi-species DP processes field theoretical RG analysis [Janssen (1997b)] predicts DP criticality with an unstable, symmetrical fixed point, such that subsystems with unidirectionally coupled DP (UCDP) behavior emerge. This was shown to be valid for linearly coupled $N$-component, contact processes too. Unidirectionally coupled DP systems of the form

$$\begin{array}{ll} A \leftrightarrow 2A & A \rightarrow A + B \\ B \leftrightarrow 2B & B \rightarrow B + C \\ C \leftrightarrow 2C & C \rightarrow C + D \\ \ldots & \end{array} \tag{7.49}$$

were investigated by [Täuber *et al.* (1998); Goldschmidt *et al.* (1999)] with the motivation that such models can describe interface growth models, where adsorption-desorption are allowed at terraces and edges (see Sect. 7.6.1). The simplest set of Langevin equations for such systems was set up by [Alon *et al.* (1998)]:

$$\begin{aligned} \partial_t \phi_k(\mathbf{x}, t) = \sigma \phi_k(\mathbf{x}, t) - \lambda \phi_k^2(\mathbf{x}, t) + D \nabla^2 \phi_k(\mathbf{x}, t) \\ + \mu \phi_{k-1}(\mathbf{x}, t) + \eta_k(\mathbf{x}, t) \ , \end{aligned} \tag{7.50}$$

where $\eta_k$ are independent, multiplicative noise fields for level $k$ with correlations

$$\begin{aligned} \langle \eta_k(\mathbf{x}, t) \rangle &= 0 \, , \qquad (7.51) \\ \langle \eta_k(\mathbf{x}, t) \eta_l(\mathbf{x}', t') \rangle &= 2\Gamma \, \phi_k(\mathbf{x}, t) \, \delta_{k,l} \, \delta^d(\mathbf{x} - \mathbf{x}') \, \delta(t - t') \ , \end{aligned}$$

for $k > 0$, while for the lowest level ($k = 0$) $\phi_{-1} \equiv 0$ is fixed. The parameter $\sigma$ controls the offspring production, $\mu$ is the coupling and $\lambda$ is the coagulation rate. As one can see the $k = 0$ equation is just the Langevin equation of DP (4.45). The **mean-field** solution of these equations that is valid above $d_c = 4$ results in critical exponents for the level $k$:

$$\beta_{MF}^{(k)} = 2^{-k} \,. \tag{7.52}$$

and $\nu_\perp^{MF} = 1/2$ and $\nu_{||}^{MF} = 1$ independently of $k$. For $d < d_c$ field theoretical RG analysis of the action for $k < K$ levels

$$S = \sum_{k=0}^{K-1} \int d^d x \, dt \left\{ \psi_k \left( \tau\partial_t - D\nabla^2 - \sigma \right) \phi_k \right.$$
$$\left. - \mu\psi_k\phi_{k-1} + \frac{\Gamma}{2} \psi_k \Big( \phi_k - \psi_k \Big) \phi_k \right\} , \tag{7.53}$$

was performed by [Täuber *et al.* (1998); Goldschmidt *et al.* (1999)]. The RG treatment ran into several difficulties. Infrared-divergent diagrams were encountered [Goldschmidt (1998)] and the coupling constant $\mu$ was shown to be a relevant quantity (which means it diverges under RG transformations). [Goldschmidt *et al.* (1999)] argued that this is the reason why scaling seems to break down in simulations for large times (in lattice realization $\mu$ is limited). The exponents of the one-loop calculations for the first few levels, corresponding to the interactive fixed line as well as results of lattice simulations are shown in Table 7.6. These scaling exponents can be observed for intermediate times, but it is not clear if in the asymptotically long time they drift to the decoupled values or not.

The main representatives of these classes are certain monomer adsorption-desorption models (Sect. 7.6.1) and polynuclear growth models (PNG) with depinning transitions [Kertész and Wolf (1989); Lehner *et al.* (1990); Toom (1994a,b)]. The latter types of systems are defined by **parallel** update dynamic rules and coupled DP processes emerge in a *co-moving* frame.

### 7.6.1 *Monomer adsorption-desorption at terraces*

Here I show an example, how a coupled particle system is related to an interface growth model. [Alon *et al.* (1998, 1996)] defined SOS and RSOS models that can be mapped onto UCDP (Sect. 7.6). In these models adsorption or desorption processes may take place at terraces and edges (these models correspond to a special case discussed in Sect. 7.3.1). For each update a site $i$ is chosen at random and an atom is adsorbed

$$h_i \to h_i + 1 \text{ with probability } q \tag{7.54}$$

Table 7.6 Critical exponents of UCDP. From [Goldschmidt *et al.* (1999)].

| | $d=1$ | $d=2$ | $d=3$ | $d=4-\epsilon$ |
|---|---|---|---|---|
| $\beta_1$ | 0.280(5) | 0.57(2) | 0.80(4) | $1-\epsilon/6+O(\epsilon^2)$ |
| $\beta_2$ | 0.132(15) | 0.32(3) | 0.40(3) | $1/2-\epsilon/8+O(\epsilon^2)$ |
| $\beta_3$ | 0.045(10) | 0.15(3) | 0.17(2) | $1/4-O(\epsilon)$ |
| $\delta_1$ | 0.157(4) | 0.46(2) | 0.73(5) | $1-\epsilon/4+O(\epsilon^2)$ |
| $\delta_2$ | 0.075(10) | 0.26(3) | 0.35(5) | $1/2-\epsilon/6+O(\epsilon^2)$ |
| $\delta_3$ | 0.03(1) | 0.13(3) | 0.15(3) | $1/4-O(\epsilon)$ |
| $\eta_1$ | 0.312(6) | 0.20(2) | 0.10(3) | $\epsilon/12+O(\epsilon^2)$ |
| $\eta_2$ | 0.39(2) | 0.39(3) | 0.43(5) | $1/2+O(\epsilon^2)$ |
| $\eta_3$ | 0.47(2) | 0.56(4) | 0.75(10) | $3/4-O(\epsilon)$ |
| $2/Z_1$ | 1.26(1) | 1.10(2) | 1.03(2) | |
| $2/Z_2$ | 1.25(3) | 1.12(3) | 1.04(2) | $1+\epsilon/24+O(\epsilon^2)$ |
| $2/Z_3$ | 1.23(3) | 1.10(3) | 1.03(2) | |
| $\nu_{\perp,1}$ | 1.12(4) | 0.70(4) | 0.57(4) | |
| $\nu_{\perp,2}$ | 1.11(15) | 0.69(15) | 0.59(8) | $1/2+\epsilon/16+O(\epsilon^2)$ |
| $\nu_{\perp,3}$ | 0.95(25) | 0.65(15) | 0.62(9) | |
| $\nu_{\parallel,1}$ | 1.78(6) | 1.24(6) | 1.10(8) | |
| $\nu_{\parallel,2}$ | 1.76(25) | 1.23(17) | 1.14(15) | $1+\epsilon/12+O(\epsilon^2)$ |
| $\nu_{\parallel,3}$ | 1.50(40) | 1.15(30) | 1.21(15) | |

or desorbed at the edge of a plateau

$$\begin{aligned} h_i &\to \min(h_i, h_{i+1}) \text{ with probability } (1-q)/2 \, , \\ h_i &\to \min(h_i, h_{i-1}) \text{ with probability } (1-q)/2 \, . \end{aligned} \tag{7.55}$$

Identifying empty sites at a given layer as $A$ particles, the adsorption process can be interpreted as the decay of $A$ particles ($A \to \emptyset$), while the desorption process corresponds to $A$ particle production ($A \to 2A$). In the language of particles there is a DP process (see Sect. 4.2) on each layer. These processes generate reactions on subsequent layers, hence they are coupled. The simulations in one dimension have shown that this coupling is relevant in the upward direction only, hence the model is equivalent to the UCDP process. Defining the order parameters on the $k$-th layer as

$$n_k = \frac{1}{N} \sum_i \sum_{j=0}^{k} \delta_{h_i,j} \, , \tag{7.56}$$

where $h_i$ is the height at site $i$, they are expected to scale as

$$n_k \sim (q_c - q)^{\beta^{(k)}} \, . \qquad k = 1, 2, 3, \ldots \tag{7.57}$$

When the growth rate ($q$) is increased, these models exhibit a roughening transition at $q_c = 0.189$ (for RSOS) and at $q_c = 0.233(1)$ (for SOS) from a non-moving, smooth phase to a moving, rough phase in one spatial dimension. The $\beta^k$ (as well as other exponents) take those of the one-dimensional UCDP class values (see Table 7.6).

The scaling behavior of the interface width is characterized by many different length scales. At criticality it increases as

$$W^2(t) \propto \tau \ln t \tag{7.58}$$

with $\tau \simeq 0.102(3)$ for RSOS [Hinrichsen and Mukamel (2003)].

## 7.7 Unidirectionally coupled PC classes

Having seen the connection of UCDP and monomer surface growth processes, now we turn towards reactions of dimers. The interesting results presented in (Sects. 7.6 and 7.6.1) motivated the study of similar processes with parity conservation (Sect. 4.6.1). The following unidirectionally coupled BARW2 (Sect. 4.6.1) models have been introduced [Hinrichsen and Ódor (1999b)]:

$$\begin{array}{lll} A \to 3A & B \to 3B & C \to 3C \\ 2A \to \emptyset & 2B \to \emptyset & 2C \to \emptyset \\ A \to A+B & B \to B+C & C \to C+D, \dots \end{array} \tag{7.59}$$

generalizing the concept of UCDP. The mean-field approximation of the reaction scheme (7.59) looks like

$$\begin{aligned} \partial_t n_A &= \sigma n_A - \lambda n_A^2 \,, \\ \partial_t n_B &= \sigma n_B - \lambda n_B^2 + \mu n_A \,, \\ \partial_t n_C &= \sigma n_C - \lambda n_C^2 + \mu n_B \,, \dots \end{aligned} \tag{7.60}$$

where $n_A, n_B, n_C$ correspond to the densities $n_0, n_1, n_2$ in the growth models. Here $\sigma$ and $\lambda$ are the rates for offspring production and pair annihilation respectively. The coefficient $\mu$ is an effective coupling constant between different particle species. Since these equations are coupled in only one direction, they can be solved by iteration. Obviously, the mean-field critical point is $\sigma_c = 0$. For small values of $\sigma$ the stationary particle densities in the active state are given by

$$n_A = \frac{\sigma}{\lambda}\,, \quad n_B \simeq \frac{\mu}{\lambda}\left(\frac{\sigma}{\mu}\right)^{1/2}\,, \quad n_C \simeq \frac{\mu}{\lambda}\left(\frac{\sigma}{\mu}\right)^{1/4}\,, \tag{7.61}$$

corresponding to the mean field critical exponents

$$\beta_A^{MF} = 1, \quad \beta_B^{MF} = 1/2, \quad \beta_C^{MF} = 1/4, \dots . \tag{7.62}$$

These exponents should be valid for $d > d_c = 2$. Solving the mean-field equations, for the asymptotic temporal behavior one finds $\nu_{||} = 1$, implying that $\delta_k^{MF} = 2^{-k}$.

The effective action of unidirectionally coupled BARW2's should be given by

$$
\begin{aligned}
S[\psi_0, \psi_1, \psi_2, \ldots, \bar{\psi}_0, \bar{\psi}_1, \bar{\psi}_2, \ldots] \\
= \int d^d x \, dt \sum_{k=0}^{\infty} \Big\{ \bar{\psi}_k (\partial_t - D\nabla^2) \psi_k - \lambda (1 - \bar{\psi}_k^2) \psi_k^2 \\
+ \sigma (1 - \bar{\psi}_k^2) \bar{\psi}_k \psi_k + \mu (1 - \bar{\psi}_k) \bar{\psi}_{k-1} \psi_{k-1} \Big\} ,
\end{aligned}
\tag{7.63}
$$

where $\psi_{-1} = \bar{\psi}_{-1} \equiv 0$. Here the fields $\psi_k$ and $\bar{\psi}_k$ represent the configurations of the system at level $k$. Since even the RG analysis of the one component BARW2 model suffered serious problems [Cardy and Täuber (1996)] the solution of the theory of (7.63) seems to be hopeless. Furthermore one expects similar infrared diagram problems and diverging coupling strengths as in case of UCDP. Simulations of a three-component model in one-dimension, coupled by instantaneous particle production of the form (7.59), resulted in decay exponents for the order parameter defined as (7.56):

$$
\delta_A = 0.280(5) \ , \ \delta_B = 0.190(7) \ , \ \delta_C = 0.120(10) \ , \tag{7.64}
$$

For further critical exponents see Sect. 7.7.1.

It would be interesting to investigate parity-conserving growth processes in higher dimensions. Since the upper critical dimension $d_c'$ is less than 2, one expects the roughening transition – if still exists – to be described by mean-field exponents. In higher dimensions, $n$-mers might appear in different shapes and orientations. The main representatives of these classes are certain dimer adsorption-desorption models (Sect. 7.7.1) and polynuclear growth models (PNG) with depinning transitions [Hinrichsen and Ódor (1999a)]. The latter type of systems are defined by **parallel** update dynamic rules and coupled PC processes emerge in a *co-moving* frame.

### 7.7.1 *Dimer adsorption-desorption at terraces*

Similarly to the monomer case (Sect. 7.6.1) dimer adsorption-desorption models were invented by [Hinrichsen and Ódor (1999b,a); Noh *et al.* (2000)]. With the restriction that desorption may only take place at the edges of a plateau, they can be mapped onto the unidirectionally coupled BARW2

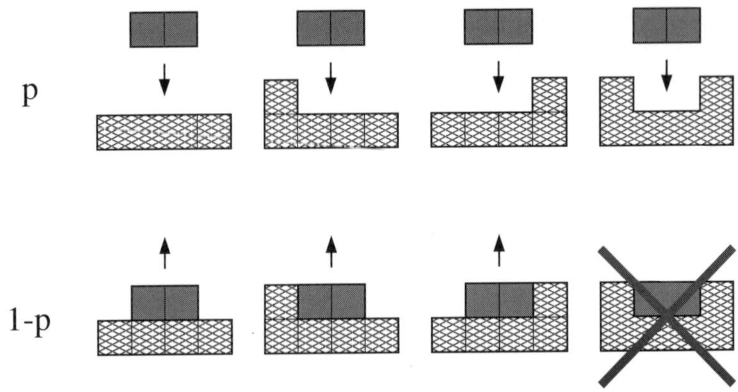


Fig. 7.5 Definition of the dimer adsorption-desorption model. Dimers can be adsorbed with probability $p$ and desorbed at the edges of terraces with probability $1-p$. Evaporation from the middle of plateaus is not allowed. From [Hinrichsen and Ódor (1999a)].

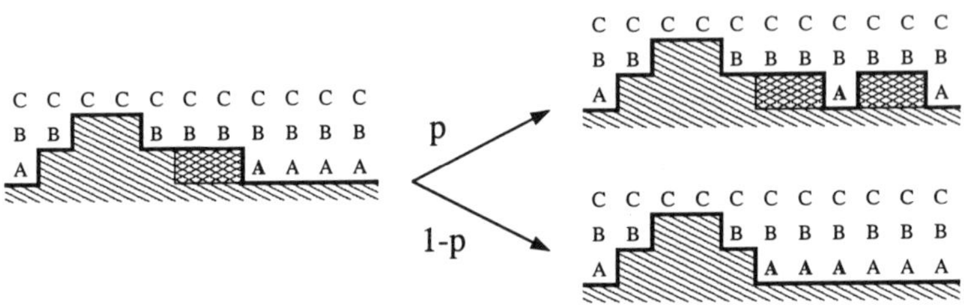


Fig. 7.6 Particle interpretation of the dimer adsorption-desorption model. Dimers adsorption corresponds to the particle annihilation $2A \to \emptyset$; the desorption describes a $A \to 3A$ branching process at the bottom layer. Similar processes take place at higher levels. From [Hinrichsen and Ódor (1999a)].

(Sect. 7.7). These systems exhibit very rich phase diagrams and scaling behavior, therefore I shall discuss them in more detail in this section. The dynamical rules in $d = 1$ are defined in Fig. 7.5. The mapping onto unidirectionally coupled BARW2 can be seen on Fig. 7.6. Such dimer models can be defined in arbitrary spatial dimensions. [Hinrichsen and Ódor (1999a)] investigated four variants in one dimension:

(1) Variant A is a restricted solid-on-solid (RSOS) model evolving by random sequential updates.
(2) Variant B is a solid-on-solid (SOS) model evolving by random sequential updates.

(3) Variant C is a restricted solid-on-solid (RSOS) model evolving by parallel updates.
(4) Variant D is a solid-on-solid (SOS) model evolving by parallel updates.

Variants A and B exhibit transitions in contrast with PNG models, in which only parallel update rules permit roughening transitions. When the adsorption rate $p$ is varied, the phase diagram shown in Fig. 7.7 emerges for the RSOS and SOS cases. If $p$ is very small, only a few dimers are adsorbed at the surface, staying there for a short time before they evaporate back into the gas phase. Thus, the interface is anchored to the actual bottom layer and does not propagate. In this smooth phase the interface width growths logarithmically, until it saturates to a finite value (even for $L \to \infty$). As $p$ increases, a growing number of dimers covers the surface, and large islands of several layers stacked on top of each other are formed. Approaching a certain critical threshold $p_c$ the mean size of the islands diverges and the interface evolves into a rough state, with the finite-size scaling form

$$W^2(L,t) \simeq a \ln\Big[t\, G(t/L^Z)\Big]\,. \tag{7.65}$$

The order parameter defined on the $k$-th layer as Eq. (7.56) exhibits unidirectionally coupled BARWe critical behavior. The transition rates and exponents are summarized in Table 7.7.

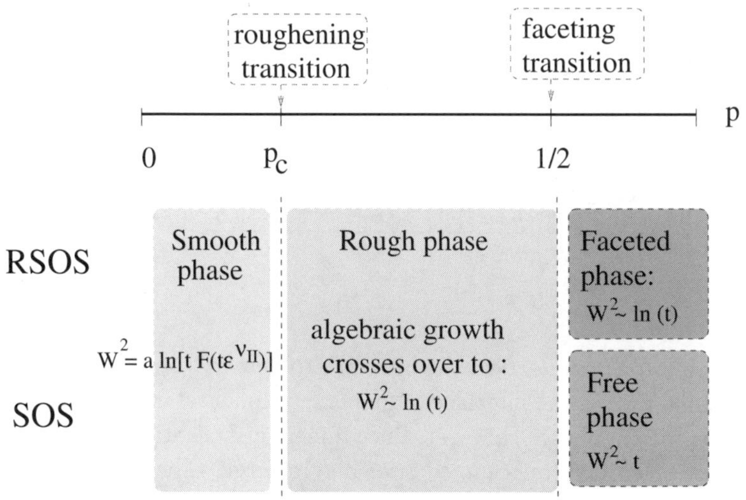


Fig. 7.7 Phase diagram of one-dimensional dimer models. From [Hinrichsen and Ódor (1999a)].

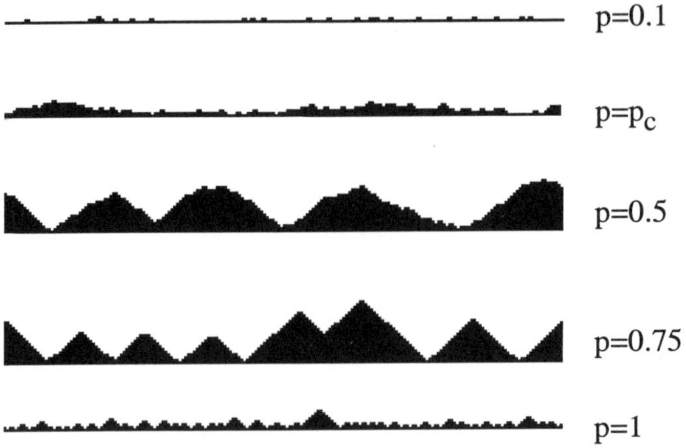


Fig. 7.8 Typical interface configurations of the restricted dimer model (variant A) for various values of $p$ (see text). From [Hinrichsen and Ódor (1999a)].

Table 7.7 Numerical estimates for the four variants of the dimer model at the roughening transition $p = p_c$ (upper part) and at the transition $p = 0.5$ (lower part). From [Hinrichsen and Ódor (1999a)].

| variant | A | B | C | D |
|---|---|---|---|---|
| restriction | yes | no | yes | no |
| updates | random | random | parallel | parallel |
| $p_c$ | 0.3167(2) | 0.292(1) | 0.3407(1) | 0.302(1) |
| $a$ | 0.172(5) | 0.23(1) | 0.162(4) | 0.19(1) |
| $Z$ | 1.75(5) | 1.75(5) | 1.74(3) | 1.77(5) |
| $\delta_0$ | 0.28(2) | 0.29(2) | 0.275(10) | 0.29(2) |
| $\delta_1$ | 0.22(2) | 0.21(2) | 0.205(15) | 0.21(2) |
| $\delta_2$ | 0.14(2) | 0.14(3) | 0.13(2) | 0.14(2) |
| $\tilde{\alpha}$ | 1.2(1) | undefined | 1.25(5) | undefined |
| $\tilde{\beta}$ | 0.34(1) | 0.50(1) | 0.330(5) | 0.49(1) |

Above $p_c$ one may expect the interface to detach from the bottom layer in the same way as the interface of monomer models starts to propagate in the supercritical phase. However, since dimers are adsorbed at neighboring lattice sites, solitary unoccupied sites may emerge. These pinning centers prevent the interface from moving and lead to the formation of 'droplets'. Due to interface fluctuations, the pinning centers can slowly diffuse to the left and to the right. When two of them meet at the same place, they annihilate and a larger droplet is formed. Thus, although the interface remains pinned, its roughness increases continuously. The width initially increases algebraically until it slowly crosses over to a logarithmic increase $W(t) \sim \sqrt{a \ln t}$.

The restricted as well as the unrestricted variants undergo a second phase transition at $p = 0.5$ [Noh *et al.* (2000)], where the width increases *algebraically* with time as $W \sim t^{\tilde{\beta}}$. In the RSOS case ordinary *Family-Vicsek* scaling Eq. (7.3) occurs, with the exponents given in Table 7.7. The dynamic exponent is $Z = \tilde{\alpha}/\tilde{\beta} \simeq 3$.

This value stays the same if one allows **dimer digging at the faceting transition (dimer-F)** at $p = 0.5$, but other surface exponents

$$\tilde{\alpha} \simeq 0.29(4) \quad , \tilde{\beta} \simeq 0.111(2) \tag{7.66}$$

will be different [Noh *et al.* (2000)]. An explanation of the latter exponents is given in [Noh *et al.* (2001)] based on mapping to *globally constrained random walks.*

In the SOS cases (variants B, D) large spikes are formed (see Fig. 7.8), and the surface roughens much faster with a growth exponent of $\tilde{\beta} \simeq 0.5$. The interface evolves into configurations with large columns of dimers separated by pinning centers. These spikes can grow or shrink almost independently. As the columns are spatially decoupled, the width does not saturate in finite systems, i.e., the dynamic exponents $\tilde{\alpha}$ and $Z$ have no physical meaning.

For $p > 0.5$ the restricted models A and C evolve into faceted configurations (see Fig. 7.9). The width first increases algebraically until the pinning centers become relevant and the system crosses over to a logarithmic increase in the width. Therefore, the faceted phase may be considered as a rough phase. The unrestricted models B and D, however, evolve into spiky interface configurations. The spikes are separated and grow independently by deposition of dimers. Therefore, $W^2$ increases *linearly* with time, defining the *free* phase of the unrestricted models.

In the simulations mentioned up until now the interface was grown from flat initial conditions. It turns out that when one starts with **random initial conditions** $h_i = 0, 1$, the densities $n_k$ turn out to decay much more slowly. For restricted variants an algebraic decay of $n_0$ with an exponent

$$\delta_0 \simeq 0.13 \tag{7.67}$$

was observed. Similarly, the critical properties of the faceting transition at $p = 0.5$ are affected by random initial conditions. The nonuniversal

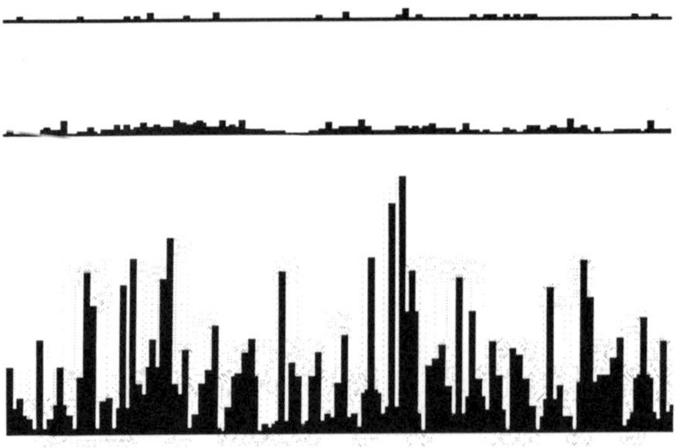


Fig. 7.9 Typical interface configurations of the unrestricted dimer model (variant B) for various values of $p$. At the transition $p = 0.5$ large spikes of stacked dimers are formed. For $p > 0.5$ these spikes are biased to grow with constant velocity. From [Hinrichsen and Ódor (1999a)].

behavior for random initial conditions is related to an *additional* parity conservation law. The dynamic rules not only conserve parity of the particle number but also conserve the parity of droplet sizes. Starting with a flat interface the lateral size of droplets is always even, allowing them to evaporate entirely. However, for a random initial configuration, droplets of odd size may be formed, which have to recombine in pairs before they can evaporate, slowing down the dynamics of the system. In the language of BARW2 processes, the additional parity conservation law is due to the absence of nearest-neighbor diffusion. Particles can only move by a combination of offspring production and annihilation, i.e., by steps of *two* lattice sites. Therefore, particles at even and odd lattice sites have to be distinguished. Only particles of different parity can annihilate. Starting with a fully occupied lattice all particles have alternating parity throughout the whole temporal evolution, leading to the usual critical behavior at the PC transition. For random initial conditions, however, particles of equal parity cannot annihilate, slowing down the decay of the particle density. Similar **sector decomposition** has been observed in diffusion of $k$-mer models [Barma *et al.* (1993); Barma and Dhar (1994)].

# Chapter 8

# Summary and Outlook

In summary dynamical extensions of critical universality classes of classical, equilibrium models were introduced in the first part of this book. New exponents, concepts, subclasses, mixing dynamics and some unresolved problems have been discussed. If the time-reversal symmetry is not broken, systems evolve towards some Gibbs state from an out of equilibrium initial condition, and dynamical scaling behavior occurs, which arises similarly as the surface scaling due to some spatial boundary conditions in equilibrium models.

If the detailed balance condition is not satisfied, phase transitions, or other fixed points of scaling in the RG sense, can occur with scaling behavior characterized by the exponents (and scaling functions) of the so-called "genuine nonequilibrium dynamical classes". The discussion of these classes is divided into several chapters. For classes, with fluctuating ordered states — mainly driven diffusive systems — the universal behavior has not been documented very well yet. The fewer number of known classes here can also be attributed to the fact that particle current destroys the fluctuations, resulting in first order of mean-field class phase transitions.

In case of classes, which occur at phase transitions to absorbing states, possessing weak fluctuations at most, of RD or spin systems, the exploration of universality is more advanced. Here the lower critical dimension is not restricted to two as for equilibrium systems, hence phase transitions in one dimension may also occur. Although such models may seem to be artificial to those who are interested in two or three dimensional applications, which are more common in real life, they appear in effectively one-dimensional models, like traffic or RNA. It is important, that the knowledge of these classes provides a base for understanding of the higher dimensional ones. Furthermore, one-dimensional RD and spin models can be mapped onto

1+1-dimensional surface/interfaces, which are discussed in a separate chapter due to their utmost importance.

Table 8.1 Summary of families of known universality classes of absorbing phase transitions in homogeneous, isotropic systems without long range interactions.

| CLASS ID | Features | Section |
|---|---|---|
| DP | time reversal symmetry | 4.2 |
| DIP | long memory | 4.4 |
| VM | $Z_2$ symmetry | 4.5 |
| 1R | non-accessible vacuum | 4.8 |
| PCP | coupled frozen field | 6.6 |
| NDCF | coupled, conserved frozen field | 6.10 |
| PC | $Z_2$ symmetry, BARW2 conservation | 4.6.1 |
| PCPD | DP coupled to ARW | 6.7 |
| DCF | coupled diffusive, conserved field | 6.9 |
| N-BARW2 | N-component BARW2 conservation | 6.12 |
| N-BARW2s | symmetric NBARW2 + exclusion | 6.13 |
| N-BARW2a | asymmetric NBARW2 + exclusion | 6.13 |

The class behavior is usually determined by the spatial dimension, symmetries, boundary conditions and inhomogeneities like in case of the equilibrium models, but in low dimensions hard-core exclusion, causing segregation of frozen states, was found to be a relevant factor too, resulting in different scaling behavior in fermionic and bosonic models. Moreover such topological constraints can introduce different scaling behavior in case of special initial conditions (sector dependence). Existing, bosonic field theories can't capture their behavior. A conjecture is advanced (see Sect. 6.13.1) that *in coupled branching and annihilating random walk systems of $N$-types of excluding particles for continuous transitions at $\sigma = 0$ branching rate two universality classes exist, in one dimension those of 2-BARW2s and 2-BARW2a models*, depending on whether the reactants can immediately annihilate (i.e. when similar particles are not separated by other type(s) of particle(s)) or not. This has been confirmed by simulations for different two-component models.

The role of symmetries and conservation laws is not so evident as in case of equilibrium models. They are expressed in terms of the relations of fields and response fields most precisely. For example the parity conservation in unary, branching and annihilating random walks is relevant, while in hard-core and in binary spreading models it is irrelevant. Furthermore in recently discovered coupled system with binary, triplet or quadruplet particle production models no special symmetry seems to be responsible for novel type of critical behavior and there are still debates, whether these models belong to the robust DP class or not. Recently it turned out that the existence of a direct channel to the absorbing state (i.e. $A \to \emptyset$ spontaneous decay) is a relevant factor for the emergence of non-DP class behavior. Furthermore it is not enough to see in the bare reactions (or bare action) the lack of such term, because they can be generated via fluctuations, hence a full (non-perturbative) RG study is necessary in most cases. The web of topological phase space structure provides a useful guidance to understand such class behavior in many cases.

Mean-field classes of one-component RD models depend on the number of reacting particles and it is also important if the transition occurs at zero or non-zero branching rate. In case of competing reactions diffusion can push the class behavior in one or the other fixed points. The mean-field classes can also give a guide to distinguishing classes below $d_c$. Table 8.2 collects the mean-field exponents and upper critical dimensions of the known absorbing-state model classes. However in some cases a proper field theoretical analysis seems to be feasible only by considering models (like the PCPD) as multi-component ones. But even in that case the topological effects, relevant in low dimension, may hamper field theoretical analysis. The results for coupled, multi-component RD models are discussed in a separate chapter.

Table 8.2 Mean-field classes of known, homogeneous absorbing-state transitions.

| CLASS | $\beta$ | $\beta'$ | $Z$ | $\nu_{\|\|}$ | $\alpha$ | $\delta$ | $\eta$ | $d_c$ |
|---|---|---|---|---|---|---|---|---|
| DP | 1 | 1 | 2 | 1 | 1 | 1 | 0 | 4 |
| DIP | 1 | 1 | 2 | 1 | 1 | 1 | 0 | 6 |
| VM | 0 | 1 | 2 | 1 | 0 | 1 | 0 | 2 |
| PC | 1 | 0 | 2 | 1 | 1 | 0 | -1/2 | 2 |
| 1R | 1 | 1 | 2 | 1 | 1 | 1 | 0 | 2 |
| BmARW | $\frac{1}{m-1}$ | 0 | 2 | 1 | $\frac{1}{m-1}$ | 0 | 0 | $\frac{2}{m-1}$ |
| PARWs | 1 | 0 | 2 | n | 1/n | 0 | 0 | |
| PARWa | $\frac{1}{m-n}$ | | 2 | $\frac{m-1}{m-n}$ | $\frac{1}{m-1}$ | | | |
| NDCF | 1 | 1 | 2 | 1 | 1 | 1 | 0 | 4 |
| NBARW2 | 1 | 0 | 2 | 1 | 1 | 0 | 0 | 2 |

Table 8.1 summarizes the best-known families of absorbing phase transition classes of homogeneous, spatially isotropic systems. Those which are below the horizontal line exhibit *weakly fluctuating absorbing states.* However these fluctuations are much weaker than those of Chapter 3, usually they are generated by remaining diffusive particles, which cannot react.

This can be due to to topological constraints, for example, which prohibits their contact. The necessary and sufficient conditions for these classes are usually unknown. In $d > 1$ dimension the mapping of spin-systems onto RD systems of particles is not straightforward; instead one should consider the theory of branching interfaces [Cardy (1998)]. Preliminary simulations of generalized Potts models, exhibiting absorbing states [Lipowski and Droz (2002a)] have found interesting critical phenomena.

The disorder, present always in nature is an other interesting aspect of scaling behavior, therefore its effect is also discussed in this book. In many cases disorder — if relevant — can cause non-universal scaling or ultra-slow (logarithmic) activated scaling. Spatial anisotropy is an other factor, which can destroy Galilean invariance (GI) of multi-component models and change the upper critical dimension. The anisotropy of reactions may also introduce novel kind of fixed points. The generalization of GI, the local scale invariance, has been found to be satisfied (at least approximately) in many nonequilibrium system, exhibiting scaling at the critical point or at $T = 0$. It is an open question whether it be regarded as an exact symmetry in models of $Z \neq 2$ space-time anisotropy.

Table 8.3 Asymptotic surface growth class exponents of 2 + 1-dimensional models.

| CLASS ID | $\partial h_t$ | $\tilde{\alpha}$ | $\tilde{\beta}$ | $Z$ | Section |
|---|---|---|---|---|---|
| Random dep. | $\eta$ | - | 1/2 | - | 7.1 |
| EW | $\nabla^2 h + \eta$ | 0 | 0 | 2 | 7.2 |
| | $\nabla^2 h + \eta_d$ | -1 | -1/4 | 2 | 7.5 |
| MBE | $-\nabla^4 h + \eta$ | 1 | 1/4 | 4 | 7.5.1 |
| | $\nabla^4 h + \eta_d$ | 0 | 0 | 4 | 7.5 |
| QEW | $\nabla^2 h + F + \eta$ | 2/3 | 6/19 | 19/9 | 7.3 |
| KPZ | $\nabla^2 h + \|\nabla h\|^2 + \eta$ | 0.38 | 0.24 | 1.58 | 7.4 |
| QKPZ | $\nabla^2 h + \|\nabla h\|^2 + F + \eta$ | 0.48(3) | 0.41(5) | 1.15(5) | 7.4.3 |
| BH | $\nabla^2 h - \nabla h^4 + \eta$ | 0 | 0 | 2 | 7.5.2 |
| KS | $-\nabla^2 h - \nabla h^4 + \|\nabla h\|^2 + \eta$ | 0.27(2) | 0.19(3) | | 7.4.2 |
| | $-\nabla h^4 + \nabla^2 \|\nabla h\|^2 + \eta$ | 2/3 | 1/5 | 10/3 | 7.5 |
| | $-\nabla h^4 + \nabla^2 \|\nabla h\|^2 + \eta_d$ | 0 | 0 | 4 | 7.5 |

The understanding of dynamical scaling at nonequilibrium first order phase transitions and tricritical points is still quite limited. Despite this fact I have discussed it in a short chapter. Contradicting conjectures about the existence of discontinuous transitions in nonequilibrium models are presented. It is an open question if tricritical DP behavior can occur in one dimension. The recently discovered tricritical classes of some RD models are also presented here.

Perhaps surface growth classes are the closest to application ranging from solid state technology to medical research. Therefore the surface growth exponents of them are summarized in Tables 8.3 and 8.4. Interface models can exhibit various (roughening, wetting, depinning, faceting ...) phase transitions, which are discussed in detail in the last chapter of this book. The transitions can be continuous of discontinuous and tricritical phenomena may also appear.

Table 8.4 Asymptotic surface growth class exponents of 1+1-dimensional models.

| CLASS ID | $\partial h_t$ | $\tilde{\alpha}$ | $\tilde{\beta}$ | $Z$ | Section |
|---|---|---|---|---|---|
| Random dep. | $\eta$ | - | 1/2 | - | 7.1 |
| EW | $\nabla^2 h + \eta$ | 1/2 | 1/4 | 2 | 7.2 |
| | $\nabla^2 h + \eta_d$ | -1/2 | -1/4 | 2 | 7.5 |
| MBE | $-\nabla^4 h + \eta$ | 3/4 | 3/8 | 4 | 7.5.1 |
| | $\nabla^4 h + \eta_d$ | 1/2 | 1/8 | 4 | 7.5 |
| QEW | $\nabla^2 h + F + \eta$ | 1 | 3/4 | 4/3 | 7.3 |
| bEW | | 1/2 | 1/4 | 2 | 7.3.1 |
| KPZ | $\nabla^2 h + \|\nabla h\|^2 + \eta$ | 1/2 | 1/3 | 3/2 | 7.4 |
| QKPZ | $\nabla^2 h + \|\nabla h\|^2 + F + \eta$ | 0.63(1) | 0.63(1) | 1.01(2) | 7.4.3 |
| BH | $\nabla^2 h - \nabla h^4 + \eta$ | 1/2 | 1/4 | 2 | 7.5.2 |
| Takayasu | | 0.72 | 0.358 | 2 | 7.5.3 |
| CMAM | | 2/3 | 1/3 | 2 | 7.5.3 |
| A-CMAM | | 0.67 | 0.4 | 1.67 | 7.5.3 |
| AGG | | 1 | 1/2 | 2 | 7.5.3 |
| dimer-A-C | | 1.2(1) | 0.33(1) | 3.6(2) | 7.7.1 |
| dimer-B-D | | - | 0.50(1) | - | 7.7.1 |
| dimer-F | | 0.29(4) | 0.111(2) | 3 | 7.7.1 |
| | $-\nabla h^4 + \nabla^2\|\nabla h\|^2 + \eta$ | 1 | 1/3 | 3 | 7.5 |
| | $-\nabla h^4 + \nabla^2\|\nabla h\|^2 + \eta_d$ | 1/3 | 1/11 | 11/3 | 7.5 |

Further research is needed to explore the universality classes of nonequilibrium phase transitions that occur in external current driven systems and of other models exhibiting fluctuating ordered states [Evans (2000); Evans *et al.* (1998)] in more detail. Nonequilibrium phase transitions in quantum systems [Rácz (2002)] or by irregular graph or network based systems are also of current interest of research. Finally, having settled the problems raised by fundamental nonequilibrium models, one should turn towards the study of more applied systems.

# Appendix

Table A.1 Commonly used abbreviations with the section number of definition.

| | | |
|---|---|---|
| ABC | active boundary condition | (4.2.5) |
| ADS | anisotropic diffusive system | (3.1.1) |
| A-CMAM | asymmetric conserved-mass aggregation model | (7.5.3) |
| AGG | aggregation model | (7.5.3) |
| ASEP | asymmetric exclusion process | (3.1.2) |
| ARW | annihilating random walk ($AA \to \emptyset$) | (4.5.1) |
| AF | annihilation-fission process | (6.7) |
| BmARW | branching and $m$-particle annihilation | (4.1.5) |
| BARW | branching and annihilation random walk | (4.2) |
| BARWe | even-offspring branching and annihilation random walk | (4.6.1) |
| BARWo | odd-offspring branching and annihilation random walk | (4.2.4) |
| BARW2 | two-offspring branching and annihilation random walk | (4.6.1) |
| bEW | bounded Edwards-Wilkinson | (7.3.1) |
| bKPZ+ | KPZ with lower boundary condition | (4.10) |
| bKPZ- | KPZ with upper boundary condition | (4.10) |
| BH | Bradley-Harper model (class) | (7.5.2) |
| BP | binary production | (6.7) |
| CA | cellular automaton | (4.2.3) |
| CI | conformal invariance | (1.6.2) |
| CLG | conserved lattice gas | (6.10.1) |
| CAM | coherent anomaly method | (4.2) |
| CDP | compact directed percolation | (4.2.3) |
| CMAM | conserved-mass aggregation model | (7.5.3) |
| CTTP | conserved threshold transfer process | (6.10.1) |
| DCF | diffusive conserved field | (6.9) |
| DDS | driven diffusive system | (3.1.1) |
| DI | directed Ising | (4.6) |
| DK | Domany-Kinzel cellular automaton | (4.2.3) |
| DP | directed percolation | (4.2) |
| DLG | driven lattice gas | (3.1) |
| DS | damage spreading | (1.5.1) |

Table A.1 (*Continued*)

| | | |
|---|---|---|
| DIP | dynamical percolation | (4.4) |
| EW | Edwards-Wilkinson | (7.2) |
| FA | Fredrickson-Anderson model | (4.8) |
| FFM | forest fire model | (6.6.2) |
| FOT | first order transition | (4.10) |
| DMRG | density matrix renormalization group | (3.1.4) |
| GEP | generalized epidemic process | (4.4) |
| GDK | generalized Domany-Kinzel cellular automaton | (4.6.3) |
| GI | Galilean invariance | (4.5.5) |
| GMF | generalized mean-field approximation | (4.10) |
| IBC | inactive boundary condition | (4.2.5) |
| IRFP | infinite randomness fixed point | (4.2.10) |
| KPZ | Kardar-Parisi-Zhang | (7.4) |
| KS | Kuramoto-Sivashinsky equation | (7.4.2) |
| KT | Kosterlitz-Thouless | (2.5) |
| LIM | linear interface model | (7.3) |
| LSI | local scale invariance | (1.6.2) |
| MBE | molecular beam epitaxy | (7.5.1) |
| MCP | multi-critical point | (4.2.10) |
| MF | mean-field approximation | (1.6) |
| MN | multiplicative noise | (4.10) |
| N-BARW2 | even-offspring, $N$-component branching and annihilation random walk | (6.12) |
| NDCF | non-diffusive conserved field | (6.10) |
| NEKIM | nonequilibrium Ising model | (4.6.2) |
| NEKIMA | anisotropic NEKIM model | (4.5.5) |
| NEKIMCA | stochastic CA version of NEKIM | (4.6.3) |
| n-PC | $n$-particle production PC model | (4.9) |
| NPRG | non-perturbative renormalization group | (1.6) |
| PC | parity conserving class | (4.6) |
| PCP | pair contact process | (6.6) |
| PCPD | pair contact process with particle diffusion | (6.7.1) |
| 2-PARW | two-component, production and annihilating random walk | (6.13.2) |
| 2-PARWa | asymmetric production 2-PARW model | (6.13.2) |
| 2-PARWs | symmetric production 2-PARW model | (6.13.2) |
| 2-PARWas | 2-PARWa model with spatially symmetric creation | (6.13.2) |
| PARWa | asymmetric production and m-particle annihilating random walk | (4.1.5) |
| PARWs | symmetric production and m-particle annihilating random walk | (4.1.3) |
| PNG | polynuclear growth model | (7.6) |
| PRG | perturbative renormalization group | (1.6) |
| RBC | reflecting boundary condition | (4.2.5) |
| q-MAM | symmetrical, multi-species $A_i + A_j \to \emptyset$ class | (6.3) |
| QEW | quench disordered EW | (7.3) |
| QKPZ | quench disordered KPZ | (7.4.3) |
| RIM | random field Ising model | (2.3.7) |
| RD | reaction-diffusion | (1.6) |

Table A.1 (*Continued*)

| | | |
|---|---|---|
| RG | renormalization group | (1.6) |
| RSOS | restricted solid on solid model | (6.13.2) |
| SCA | stochastic cellular automaton | (4.2.3) |
| SOC | self organized criticality | (4.2.3) |
| SOS | solid on solid model | (6.10.1) |
| TASEP | totally asymmetric exclusion process | (3.1.2) |
| TDP | tricritical directed percolation | (5.1) |
| TDGL | time dependent Ginzburg Landau equation | (2.1) |
| TTP | threshold transfer process | (6.6) |
| UCDP | unidirectionally coupled directed percolation | (7.6) |
| VM | voter model | (4.5) |
| ZRP | zero range process | (3.1.2) |

# Bibliography


Abarbanel, H. D. I., Bronzan, J. B., Sugar, R. L. and White, A. R. (1975). *Phys. Rep.* **21**, p. 119.

Abriet, S. and Karevski, D. (2004). *Eur. Phys. J. B* **41**, p. 79.

Abriet, S. and Karevskia, D. (2004). *Eur. Phys. J. B* **37**, p. 47.

Achahbar, A., Alonoso, J. J. and Munoz, M. (1996). *Phys. Rev. E* **54**, p. 4838.

Achahbar, A., Garrido, P. L., Marro, J. and Muñoz, M. A. (2001). *Phys. Rev. Lett.* **87**, p. 195702.

Aharony, A. and Harris, A. B. (1996). *Phys. Rev. Lett.* **77**, p. 3700.

Aharony, A., Harris, A. B. and Wiseman, S. (1998). *Phys. Rev. Lett.* **81**, p. 252.

Ala-Nissila, T., Hjelt, T., Kosterlitz, J. M. and Venäläinen, O. (1993). *J. Stat. Phys.* **72**, p. 207.

Alava, M. and Munoz, M. A. (2001). *Phys. Rev. E* **65**, p. 026145.

Albano, E. V. (1994). *J. Phys. A* **27**, p. L881.

Albano, E. V. (1997). *Phys. Rev. E* **55**, p. 7144.

Albano, E. V. and Saracco, G. (2002). *Phys. Rev. Lett.* **88**, p. 145701.

Albert, R. and Barabási, A.-L. (2002). *Rev. Mod. Phys.* **74**, p. 47.

Alcaraz, F., Droz, M., Henkel, M. and Rittenberg, V. (1994). *Ann. Phys. (NY)* **230**, p. 250.

Aldana, M., Dosetti, V., Huepe, C. and Larralde, H. (2007). *Phys. Rev. Lett.* **98**, p. 095702.

Aldana, M. and Larralde, H. (2004). *Phys. Rev. E* **70**, p. 066130.

Alon, U., Evans, M., Hinrichsen, H. and Mukamel, D. (1996). *Phys. Rev. Lett.* **76**, p. 2746.

Alon, U., Evans, M., Hinrichsen, H. and Mukamel, D. (1998). *Phys. Rev. E* **57**, p. 4997.

Alonso, I. J. and Munoz, M. A. (2001). *Eur. Phys. Lett.* **56**, p. 485.

Amaral, L. A. N., Barabási, A.-L., Buldyrev, S. V., Harrington, S. T., Havlin, S., Sadr-Lahijany, R. and Stanley, H. E. (1995). *Phys. Rev. E* **51**, p. 4655.

Amit, D. J. (1984). *Field theory, the renormalization group and critical phenomena (2nd edn.)* (World Scientific, Singapore).

Antal, T., Droz, M., Lipowski, A. and Ódor, G. (2001). *Phys. Rev. E* **64**, p. 036118.

Arndt, P. F., Heinzel, T. and Rittenberg, V. (1998). *J. Phys. A* **31**, p. 833.

Arndt, P. F., Heinzel, T. and Rittenberg, V. (1999). *J. Stat. Phys.* **97**, p. 1.

Bagnuls, C. and Bervillier, C. (2001). *Phys. Rep.* **348**, p. 91.

Bak, P. and Sneppen, K. (1993). *Phys. Rev. Lett.* **71**, p. 4083.

Bak, P. and Tang, C. (1989). *J. Geophys. Res.* **94**, p. 15635.

Bak, P., Tang, C. and Wiesenfeld, K. (1987). *Phys. Rev. Lett.* **59**, p. 381.

Barabási, A. L. (2002). *Linked: The New Science of Networks* (Perseus, Cambridge).

Barabási, A. L., Grinstein, G. and Munoz, M. A. (1996). *Phys. Rev. Lett.* **76**, p. 1481.

Barabási, A. L. and Stanley, H. E. (1995). *Fractal Concepts in Surface Growth* (Cambridge University Press, Cambridge).

Barato, A. C., Hinrichsen, H. and de Oliveira, M. J. (2007). *Phys. Rev. E* **77**, p. 011101.

Barma, M. and Dhar, D. (1994). *Phys. Rev. Lett.* **73**, p. 2135.

Barma, M., Grynberg, M. D. and Stinchcomb, R. B. (1993). *Phys. Rev. Lett.* **70**, p. 1033.

Barma, M. and Ramaswamy, R. (1993). *Non-linearity and breakdown in boft condensed matter* (Springer, Berlin).

Baronchelli, A., Dall'Asta, L., Barrat, A. and Loreto, V. (2007). *Phys. Rev. E* **76**, p. 051102.

Bassler, K. E. and Browne, D. A. (1996). *Phys. Rev. Lett.* **77**, p. 4094.

Bassler, K. E. and Browne, D. A. (1997). *Phys. Rev. E* **55**, p. 5225.

Bassler, K. E. and Browne, D. A. (1998). *J. Phys. A* **31**, p. 6309.

Bassler, K. E. and Rácz, Z. (1994). *Phys. Rev. Lett.* **73**, p. 1320.

Bassler, K. E. and Rácz, Z. (1995). *Phys. Rev. E* **52**, pp. R9–12.

Bassler, K. E. and Schmittman, B. (1994). *Phys. Rev. Lett.* **73**, p. 3343.

Baumann, F. and Gambassi, A. (2007). *J. Stat. Mech.* p. P01002.

Baumann, F. and Henkel, M. (2007). *J. Stat. Mech.* p. P01012.

Baumann, F., Henkel, M. and Pleimling, M. (2007). Phase-ordering kinetics of two-dimensional disordered ising models, Eprint: arXiv:0709.3228.

Baxter, R. J. (1982). *Exactly Solved Models in Statistical Mechanics* (Academic, London).

Belanger, D. P. (2000). *Braz. J. Phys.* **30**, p. 682.

Belavin, A. A., Polyakov, A. M. and Zamolodchikov, A. B. (1984). *Nucl. Phys. B* **241**, p. 333.

ben Avraham, D., Burschka, M. A. and Doring, C. (1990). *J. Stat. Phys.* **60**, p. 695.

ben Avraham, D., Leyvraz, F. and Redner, S. (1994). *Phys. Rev. E* **50**, p. 1843.

ben Avraham, D. and Redner, S. (1986). *Phys. Rev. A* **34**, p. 501.

Ben-Hur, A. (1996). *Phys. Rev. E* **53**, p. R1317.

Ben-Naim, E. and Krapivsky, P. L. (1994). *J. Phys. A* **27**, p. L481.

Benzoni, J. and Cardy, J. L. (1984). *J. Phys. A* **17**, p. 179.

Bergersen, B. and Rácz, Z. (1991). *Phys. Rev. Lett.* **67**, p. 3047.

Berges, J., Tetradis, N. and Wetterich, C. (2002a). *Phys. Rep.* **363**, p. 223.

Berges, J., Tetradis, N. and Wetterich, C. (2002b). *Phys. Rep.* **363**, pp. 223–386.

Berlin, T. H. and Kac, M. (1952). *Phys. Rev.* **86**, p. 821.
Berry, H. (2003). *Phys. Rev. E* **67**, p. 031907.
Berthier, L., Barrat, J.-L. and Kurchan, J. (1999). *Eur. Phys. J. B* **11**, p. 635.
Bervillier, C., Juttner, A. and Litim, D. F. (2007). *Nucl. Phys. B* **783**, p. 213.
Bialas, P. *et al.* (2000). *Nucl. Phys. B* **583**, p. 368.
Bidaux, R., Boccara, N. and Chaté, H. (1989). *Phys. Rev. A* **39**, p. 3094.
Binder, K. (1981). *Z. Phys. B* **43**, p. 119.
Binder, K. and Stauffer, D. (1974). *Phys. Rev. Lett.* **33**, p. 1006.
Binder, K. and Wang, J. S. (1989). *J. Stat. Phys.* **55**, p. 87.
Birner, T., Lippert, K., Müller, R., Künel, A. and Behn, U. (2002). *Phys. Rev. E* **65**, p. 046110.
Blanchard, P. *et al.* (2000). *J. Phys. A* **33**, p. 8603.
Blote, H. W. J., Heringa, J. R., Hoogland, A. and Zia, R. K. P. (1990). *J. Phys. A* **23**, p. 3799.
Boccara, N. and Roger, M. (1993). in E. Tirapegui and W. Zeller (eds.), *Instabilities and Nonequilibrium Structures*, Vol. 4 (Kluver Academic, Dordrecht), p. 109.
Bonachela, J. A. and Munoz, M. A. (2007). *Physica A* **384**, pp. 89–93.
Bouchaud, J. P. and Bray, A. J. (2004). in M. E. Cates and M. R. Evans (eds.), *Soft and Fragile Matter* (Bristol: IOP).
Bouchaud, J. P. and Georges, A. (1990). *Phys. Rep.* **195**, p. 127.
Bradley, R. M. and Harper, J. (1988). *J. Vac. Sci. Tech. A* **6**, p. 2390.
Bramson, M. and Lebowitz, J. (1988). *Phys. Rev. Lett.* **61**, p. 2397.
Bray, A. J. (1994). *Adv. Phys.* **43**, p. 357.
Bray, A. J., Briant, A. J. and Jervis, D. K. (2000). *Phys. Rev. Lett.* **84**, p. 1503.
Bray, A. J., Derrida, B. and Godreche, C. (1994). *Eur. Phys. Lett.* **27**, p. 175.
Bray, A. J., Humayun, K. and Newman, T. J. (1991). *Phys. Rev. B* **43**, p. 3699.
Broadbent, S. R. and Hammersley, J. M. (1957). *Proc. Camb. Phil. Soc.* **53**, p. 629.
Bronzan, J. B. and Dash, J. W. (1974). *Phys. Lett. B* **51**, p. 496.
Brower, R., Furman, M. A. and Moshe, M. (1978). *Phys. Lett. B* **76**, p. 213.
Brown, K. S., Bassler, K. E. and Browne, D. A. (1997). *Phys. Rev. E* **56**, p. 3953.
Bru, A., Pastor, J. M., Fernaud, I., Bru, I. and Melle, S. (1998). *Phys. Rev. Lett.* **81**, p. 4008.
Brunel, V., Oerding, K. and van Wijland, F. (2000). *J. Phys. A* **33**, p. 1085.
Brunstein, A. and Tomé, T. (1998). *Physica A* **257**, p. 334.
Brutovsky, B., Horvath, D. and Lisy, V. (2006). Simulation of multicellular tumor spheroids growth dynamics, Eprint: physics/0610134.
Buldrev, S. V., Havlin, S. and Stanley, H. E. (1993). *Physica A* **200**, p. 200.
Buldyrev, S. V., Barabási, A. L., Caserta, F., Havlin, S., Stanley, H. E. and Vicsek, T. (1992). *Phys. Rev. A* **45**, p. R8313.
Bunde, A. and Havlin, S. (1991). *Fractals and disordered systems* (Springer Verlag, Heidelberg).
Burgers, J. M. (1974). *The Nonlinear Diffusion Equation* (Riedel, Boston).
Burlatskii, S. F. and Ovchinnikov, A. A. (1978). *Russ. J. Phys. Chem.* **52**, p. 1635.

Burschka, M., Doering, C. R. and ben Avraham, D. (1989). *Phys. Rev. Lett.* **63**, p. 700.

Cafiero, R., Gabrielli, A. and Muñoz, M. A. (1998). *Phys. Rev. E* **57**, p. 5060.

Calabrese, P. and Gambassi, A. (2002). *Phys. Rev. E* **66**, p. 066101.

Calabrese, P. and Gambassi, A. (2005). *J. Phys. A: Math. Gen.* **38**, p. R133.

Calabrese, P. and Gambassi, A. (2007). *J.Stat.Mech* , p. P01001.

Canet, L., Delamotte, B., Deloubrière, O. and Wschebor, N. (2004). *Phys. Rev. Lett.* **92**, p. 195703.

Canet, L. and Delamotte, H. C. B. (2004). *Phys. Rev. Lett.* **92**, p. 255703.

Canet, L., Delamotte, H. C. B., Dornic, I. and Muñoz, M. A. (2005). *Phys. Rev. Lett.* **95**, p. 100601.

Canet, L. and Moore, M. A. (2007). *Phys. Rev. Lett.* **98**, p. 200602.

Cannas, S., Stariolo, D. and F.A.Tamarit (2001). *Physica A* **294**, p. 362.

Cardozo, G. and Fontanari, J. (2006a). *Physica A* **359**, p. 478.

Cardozo, G. O. and Fontanari, J. (2006b). *Eur. Phys. J. B* **51**, p. 555.

Cardy, J. L. (1983a). *J. Phys. A* **16**, p. 3617.

Cardy, J. L. (1983b). *J. Phys. A* **16**, p. L709.

Cardy, J. L. (1987). Conformal invariance, in C. Domb and J. L. Lebowitz (eds.), *Phase Transitions and Critical Phenomena*, Vol. 11 (Academic Press).

Cardy, J. L. (1995). *J. Phys. A* **28**, p. L19.

Cardy, J. L. (1996). Scaling and renormalization in statistical physiscs, in P. Goddard and J. Yeomans (eds.), *Cambridge Lecture Notes in Physics*, Vol. 5 (Cambridge University Press).

Cardy, J. L. (1997). Renormalisation group approach to reaction-diffusion problems, Eprint: cond-mat/9607163.

Cardy, J. L. (1998). Renormalisation group theory of branching potts interfaces, Eprint: cond-mat/9806098.

Cardy, J. L. and Grassberger, P. (1985). *J. Phys. A* **18**, p. L267.

Cardy, J. L. and Sugar, R. L. (1980). *J. Phys. A* **13**, p. L423.

Cardy, J. L. and Täuber, U. C. (1996). *Phys. Rev. Lett.* **77**, p. 4780.

Cardy, J. L. and Täuber, U. C. (1998). *J. Stat. Phys.* **90**, p. 1.

Carlon, E., Henkel, M. and Schollwöck, U. (1999). *Eur. Phys. J. B* **12**, p. 99.

Carlon, E., Henkel, M. and Schollwöck, U. (2001). *Phys. Rev. E* **63**, pp. 036101–1.

Castellano, C., Gabrielli, A., Marsili, M., Munoz, M. and Pietronero, L. (1998a). *Phys. Rev. E* **58**, p. R5209.

Castellano, C., Marsili, M., Munoz, M. and Pietronero, L. (1999). *Phys. Rev. E* **59**, p. 6460.

Castellano, C., Marsili, M. and Pietronero, L. (1998b). *Phys. Rev. Lett.* **80**, p. 3527.

Castro, M. *et al.* (1998). *Phys. Rev. E* **57**, p. R2491.

Chamon, C. (2005). *Phys. Rev. Lett.* **94**, p. 040402.

Chayes, J. T., Chayes, L., Fisher, D. S. and Spencer, T. (1986). *Phys. Rev. Lett.* **57**, p. 2999.

Chopard, B. and Droz, M. (1998). *Cellular Automaton Modelling of Physical Systems* (Cambridge Univ. Press, Cambridge).

Chow, T. S. (2000). *Mesoscopic Physics of Complex Materials Texts in Contemporary Physiscs* (Springer Verlag, Berlin).

Chowdhury, D., Santen, L. and Schadschneider, A. (2000). *Phys. Reports* **329**, p. 199.

Chowdhury, D., Schadschneider, A. and Nishinari, K. (2007). Traffic phenomena in biology: from molecular motors to organisms, Eprint: arXiv:physics/0702194.

Cieplak, M. and Robbins, M. (1988). *Phys. Rev. Lett.* **60**, p. 2042.

Coniglio, A. and Klein, W. (1980). *J. Phys. A* **13**, p. 2775.

Corberi, F., Gambassi, A., Lippiello, E. and Zannetti, M. (2007) Eprint: arXiv:0711.1527.

Corberi, F., Lippiello, E. and Zannetti, M. (2006). *Phys. Rev. E* **74**, p. 041106.

Cornell, S. J., Kaski, K. and Stinchcomb, R. B. (1991). *Phys. Rev. B* **44**, p. 12263.

Creutz, M. (1986). *Ann. Phys.* **167**, p. 62.

Crisanti, A. and Grassberger, P. (1994). *J. Phys. A* **27**, p. 6955.

Csahók, Z., Honda, K. and Vicsek, T. (1993). *J. Phys. A* **L171**.

Cuerno, R. and Lauritsen, K. B. (1995). *Phys. Rev. E* **52**, p. 4853.

Cugliandolo, L. F. (2003). Slow relaxation and nonequilibrium dynamics in condensed matter, in J. L. Barrat, J. Dalibard, J. Kurchan and M. V. Feigelman (eds.), *Les Houches Session 77* (Springer).

Czirók, A., Barabási, A. and Vicsek, T. (1999). *Phys. Rev. Lett.* **82**, p. 209.

Czirók, A., Stanley, H. E. and Vicsek, T. (1997). *J. Phys. A* **30**, p. 1375.

da Silva, L., Tamarit, F. A. and Magalhães, A. C. N. (1997). *J. Phys. A* **30**, p. 2329.

Dahmen, S. R., Sittler, L. and Hinrichsen, H. (2007). *J. Stat. Mech.* p. P01011.

Dasgupta, C., Sarma, S. D. and Kim, J. M. (1996). *Phys. Rev. E* **54**, p. R4552.

de Dominics, C. and Peliti, L. (1978). *Phys. Rev. B* **18**, p. 353.

de Freitas, J. E., Lucena, L. S., da Silva, L. R. and Hilhorst, H. J. (2000). *Phys. Rev. E* **61**, p. 6330.

de Gennes, P. G. (1969). *Phys. Lett. A* **30**, p. 454.

de Silva, R., Alves, N. A. and de Felicio, J. R. D. (2002). *Physics Lett. A* **298**, p. 325.

De'Bell, K. and Essam, J. W. (1983). *J. Phys. A* **16**, p. 385.

Deem, M. W. and Park, J.-M. (1998). *Phys. Rev. E* **57**, p. 2681.

Delamotte, B. (2007). An introduction to the nonperturbative renormalization group, Eprint: cond-mat/0702365.

Delamotte, B. and Canet, L. (2005). *Cond. Mat. Phys.* **8**, p. 163.

Deloubriére, O., Hilhorst, H. J. and Täuber, U. C. (2002). *Phys. Rev. Lett.* **89**, p. 250601.

Deloubriére, O. and van Wijland, F. (2002). *Phys. Rev. E* **65**, p. 046104.

DeMasi, A., Ferrari, P. A. and Lebowitz, J. L. (1985). *Phys. Rev. Lett.* **55**, p. 1947.

DeMasi, A., Ferrari, P. A. and Lebowitz, J. L. (1986). *J. Stat. Phys.* **44**, p. 589.

Derrida, B. (1995). *J. Phys. A* **28**, p. 1481.

Derrida, B. (1998). *Phys. Rep.* **301**, p. 65.

Derrida, B., Bray, A. J. and Godréche, C. (1994). *J. Phys. A* **27**, p. L357.

Derrida, B., Evans, M. R., Hakim, V. and Pasquier, V. (1992). *J. Phys. A* **26**, p. 1493.

Derrida, B., Hakim, V. and Pasquier, V. (1995). *Phys. Rev. Lett.* **75**, p. 751.

Derrida, B. and Weisbuch, G. (1987). *Europhys. Lett.* **4**, p. 657.

Dhar, D. (1999a). *Physica A* **263**, p. 4.

Dhar, D. (1999b). Studying self-organized criticality with exactly solved models, Eprint: arXiv:cond-mat/9909009.

Dhar, D. and Ramaswamy, R. (1989). *Phys. Rev. Lett.* **63**, p. 1659.

Dickman, R. (1990). *Phys. Rev. A* **42**, p. 6985.

Dickman, R. (1994). *Phys. Rev. E* **50**, p. 4404.

Dickman, R. (1999). Nonuniversality and critical point shift in systems with infinitely many absorbing configurations, Eprint: cond-mat/9909347.

Dickman, R. (2002a). *Physica A* **306**, p. 90.

Dickman, R. (2002b). N-site approximations and cam analysis for a stochastic sandpile, Eprint: cond-mat/0204608.

Dickman, R. (2007). Absorbing-state phase transitions: exact solutions of small systems, Eprint: arXiv:0709.3774v1.

Dickman, R., Alava, M., Munoz, M. A., Peltola, J., Vespignani, A. and Zapperi, S. (2001a). *Phys. Rev. E* **64**, p. 056104.

Dickman, R. and ben Avraham, D. (2001). *Phys. Rev. E* **64**, p. 020102.

Dickman, R. and da Silva, J. K. L. (1998). *Phys. Rev. E* **58**, p. 4266.

Dickman, R. and de Menenzes, M. A. F. (2002). *Phys. Rev. E* **66**, p. 045101.

Dickman, R. and de Silva, J. K. (1998). *Phys. Rev. E* **58**, p. 4266.

Dickman, R. and Jensen, I. (1991). *Phys. Rev. Lett.* **67**, p. 2391.

Dickman, R. and Munoz, M. A. (2000). *Phys. Rev. E* **62**, p. 7631.

Dickman, R., Munoz, M. A., Vespignani, A. and Zapperi, S. (2000). *Braz. J. of Physics* **30**, p. 27.

Dickman, R., Rabelo, W. R. M. and Ódor, G. (2001b). *Phys. Rev. E* **65**, p. 016118.

Dickman, R. and Tomé, T. (1991). *Phys. Rev. A* **44**, p. 4833.

Dickman, R., Tomé, T. and de Oliveira, M. J. (2002). Sandpiles with height restrictions, Eprint: cond-mat/0203565.

Dickman, R. and Tretyakov, A. Y. (1995). *Phys. Rev. E* **52**, p. 3218.

Dickman, R., Vespignani, A. and Zapperi, S. (1998). *Phys. Rev. E* **57**, p. 5095.

Dietrich, S. (1986). in C. Domb and J. Lebowitz (eds.), *Phase Transitions and Critical Phenomena*, Vol. 12 (Academic Press, London), p. 1.

Díez-Minguito, M., Garrido, P. and Marro, J. (2005). *Phys. Rev. E* **72**, p. 026103.

Doherty, J. P., Moore, M. A., Kim, J. M. and Bray, A. J. (1994). *Phys. Rev. Lett.* **72**, p. 2041.

Doi, M. (1976). *J. Phys. A* **9**, p. 1465.

Domany, E. and Kinzel, W. (1984). *Phys. Rev. Lett.* **53**, p. 311.

Dornic, I., Chaté, H., Chave, J. and Hinrichsen, H. (2001). *Phys. Rev. Lett.* **87**, p. 045701.

Dorogovtsev, S. N. and Mendes, J. (2003). *Evolution of Networks: From Biological Nets to the Internet and WWW* (Oxford University Press).

Doussal, P. L. and Monthus, C. (1999). *Phys. Rev. E* **60**, p. 1212.

Drossel, H. and Schwabl, F. (1993). *Physica A* **199**, p. 183.

Drotar, J. T., Zho, Y.-P., Lu, T.-M. and Wang, G.-C. (1999). *Phys. Rev. E* **59**, p. 177.

Drouffe, J. M. and Godrèche, C. (1999). *J. Phys. A* **32**, p. 249.

Drouffe, J. M. and Godrèche, C. (2001). *Eur. Phys. J. B* **20**, p. 281.

Droz, M., Rácz, Z. and Schmidt, J. (1989). *Phys. Rev. A* **39**, p. 2141.

Droz, M., Rácz, Z. and Tartaglia, T. (1990). *Phys. Rev. A* **41**, p. 6621.

Droz, M. and Sasvári, L. (1993). *Phys. Rev. E* **48**, p. R2343.

Durret, R. (1988). Lecture notes on particle systems and percolation, (Wadsworth, Pacific Grove, CA).

Edwards, S. F. and Anderson, P. W. (1975). *J. Phys. F* **5**, p. 965.

Edwards, S. F. and Wilkinson, D. R. (1982). *Proc. R. Soc.* **381**, p. 17.

Elgart, V. and Kamenev, A. (2006). *Phys. Rev. E* **74**, p. 041101.

Eloranta, K. and Nummelin, E. (1992). *J. Stat. Phys.* **69**, p. 1131.

Enns, T., Henkel, M., Picone, A. and Schollwöck, U. (2004). *J. Phys. A* **37**, p. 10479.

Essam, J. W. (1989). *Phys. A* **22**, p. 4927.

Essam, J. W., Guttmann, A. and De'Bell, K. (1988). *J. Phys. A* **21**, p. 3815.

Essam, J. W. and Guttmann, A. J. (1995). *J. Phys. A* **28**, p. 3591.

Essam, J. W., Guttmann, A. J., Jensen, I. and TanlaKishani, D. (1996). *J. Phys. A* **29**, p. 1619.

Essam, J. W. and TanlaKishani, D. (1994). *J. Phys. A* **27**, p. 3743.

Evans, M. R. (1996). *Europhys. Lett.* **36**, p. 13.

Evans, M. R. (2000). *Braz. J. Phys.* **30**, p. 42.

Evans, M. R. and Hanney, T. (2005). *J. Phys. A: Math. Gen.* **38**, pp. R195–R239.

Evans, M. R., Kafri, Y., Koduvely, H. M. and Mukamel, D. (1998). *Phys. Rev. E* **58**, p. 2764.

Ez-Zahraouy, H., Jetto, K. and Benyoussef, A. (2006). *Chi. J. Phys.* **44**, p. 486.

Faillettaz, J., Louchet, F. and Grasso, J. (2004). *Phys. Rev. Lett.* **93**, p. 208001.

Family, F. and Vicsek, T. (1985). *J. Phys. A* **18**, p. L75.

Fan, S. and Zhong, F. (2007). *Phys. Rev. E* **76**, p. 041141.

Feldman, D. E. (2005). *Phys. Rev. Lett.* **95**, p. 177201.

Ferrell, R. A., Menyhárd, N., Schmidt, H., Schwabl, F. and Szépfalusy, P. (1967). *Phys. Rev. Lett.* **18**, p. 891.

Fiore, C. E. and de Oliveira, M. J. (2004). *Phys. Rev. E* **70**, p. 046131.

Fisher, D. (1985). *Phys. Rev. B* **31**, p. 1396.

Fisher, D. and Huse, D. A. (1988). *Phys. Rev. B* **38**, p. 373.

Fisher, D. S. (1999). *Physica A* **263**, p. 222.

Fisher, M. E. (1967). *Rep. Prog. Phys.* **30**, p. 615.

Folk, R., Holovatch, Y. and Yavors'kii, T. (2003). *Physics Uspekhi* **46**, p. 169.

Forster, D., Nelson, D. R. and Stephen, M. J. (1977). *Phys. Rev. A* **16**, p. 732.

Fortuin, C. M. and Kasteleyn, P. W. (1972). *Physica* **57**, p. 536.

Fortunato, S. (2002). *Phys. Rev. B* **66**, p. 054107.

Fortunato, S. and Satz, H. (2001). *Nucl. Phys. B* **598**, p. 601.

Frachebourg, L. and Krapivsky, P. L. (1996). *Phys. Rev. E* **53**, p. R3009.

Fradkin, E. and Susskind, L. (1978). *Phys. Rev. D* **17**, p. 2637.

Fredrickson, G. H. and Andersen, H. C. (1984). *Phys. Rev. Lett.* **53**, p. 1244.
Frey, E., Täuber, U. C. and Schwabl, F. (1993). *Phys. Rev. E* **49**, p. 5058.
Friedan, D., Qiu, Z. and Shenker, S. (1984). *Phys. Rev. Lett.* **52**, p. 1575.
Fröjdh, P. and den Nijs, M. (1997). *Phys. Rev. Lett.* **78**, p. 1850.
Fröjdh, P., Howard, M. and Lauritsen, K. B. (1998). *J. Phys. A* **31**, p. 2311.
Fröjdh, P., Howard, M. and Lauritsen, K. B. (2001). *Int. J. Mod. Phys. B* **15**, p. 1761.
Fulco, U. L., Messias, D. N. and Lyra, M. L. (2001). *Phys. Rev. E* **63**, p. 066118.
Furtado, L. S. and Copelli, M. (2006). *Phys. Rev. E* **73**, p. 011907.
G. Grinstein, M. A. M. and Tu, Y. (1996). *Phys. Rev. Lett.* **76**, p. 4376.
Gálfi, L. and Rácz, Z. (1988). *Phys. Rev. A* **38**, p. 3151.
Garel, T. and Orland, H. (1997). *Phys. Rev. B* **55**, p. 226.
Garrido, P. L., Lebowitz, J. L., Maes, C. and Spohn, H. (1990). *Phys. Rev. A* **42**, p. 1954.
Garrido, P. L., Marro, J. and Gonzalez-Miranda, J. M. (1989). *Phys. Rev. A* **40**, p. 5802.
Garrido, P. L., Munoz, M. A. and de los Santos, F. (2000). *Phys. Rev. E* **61**, p. R4683.
Genovese, W. and Muñoz, M. (1999). *Phys. Rev. E* **60**, p. 69.
Ginelli, F., Hinrichsen, H., Livi, R., Mukamel, D. and Torcini, A. (2006). *J. Stat. Mech.* p. P08008.
Glauber, R. J. (1963). *J. Math. Phys.* **4**, p. 191.
Glumac, Z. and Uzelac, K. (1998). *Phys. Rev. E* **58**, p. 4372.
Godrèche, C. (2003). *J. Phys. A* **36**, p. 6313.
Godrèche, C. and Luck, J.-M. (2000a). *J. Phys. A: Math. Gen.* **33**, p. 9141.
Godrèche, C. and Luck, J.-M. (2000b). *J. Phys. A: Math. Gen.* **33**, p. 1151.
Godrèche, C., Luck, J.-M., Evans, M. R., D. Mukamel, S. S. and Speer, E. R. (1995). *J. Phys. A: Math. Gen.* **28**, p. 6039.
Goldenfeld, N. (1992). *Lectures on Phase Transitions and the Renormalization Group* (Addison-Wesley).
Goldschmidt, Y. Y. (1998). *Phys. Rev. Lett.* **81**, p. 2178.
Goldschmidt, Y. Y., Hinrichsen, H., Howard, M., and Täuber, U. C. (1999). *Phys. Rev. E* **59**, p. 6381.
Goldstein, S. and Speer, E. (1998). *Phys. Rev. E* **58**, p. 4226.
Gonzaléz-Miranda, J. M., Garrido, P. L., Marro, J. and Lebowitz, J. (1987). *Phys. Rev. Lett.* **59**, p. 1934.
Gopinathan, A. (2001). *J. Phys. A* **31**, p. 5499.
Gorishny, S. G., Larin, S. A. and Tkachov, F. (1984). *Phys. Lett.* **101A**, p. 120.
Grassberger, P. (1982a). *Z. Phys. B* **47**, p. 365.
Grassberger, P. (1982b). *Z. Phys. B* **47**, p. 365.
Grassberger, P. (1982c). *Math. Biosci.* **63**, p. 157.
Grassberger, P. (1986). in L. Pietronero and E. Tosatti (eds.), *Fractals in Physics* (Elsevier).
Grassberger, P. (1989a). *J. Phys. A* **22**, p. 3673.
Grassberger, P. (1989b). *J. Phys. A* **22**, p. L1103.
Grassberger, P. (1992a). *J. Phys. A* **25**, p. 5867.

Grassberger, P. (1992b). *J. Phys. A* **25**, p. 5867.
Grassberger, P. (1995a). *Physica A* **214**, p. 547.
Grassberger, P. (1995b). *J. Phys. A* **28**, p. L67.
Grassberger, P. (1995c). *J. Stat. Phys.* **79**, p. 13.
Grassberger, P. (1996). Directed percolation: results and open problems, *WUB 96-2*.
Grassberger, P. (2006). *J. Stat. Mech.* p. P01004.
Grassberger, P., Chaté, H. and Rousseau, G. (1997). *Phys. Rev. E* **55**, p. 2488.
Grassberger, P. and de la Torre, A. (1979). *Ann. Phys.* **122**, p. 373.
Grassberger, P. and Kantz, H. (1991). *J. Stat. Phys.* **63**, p. 685.
Grassberger, P., Krause, F. and von der Twer, T. (1984). *J. Phys. A* **17**, p. L105.
Grassberger, P. and Procaccia, I. (1982). *J. Chem. Phys.* **77**, p. 6281.
Grégorie, G. and Chaté, H. (2004). *Phys. Rev. Lett.* **92**, p. 025702.
Griffiths, R. B. (1969). *Phys. Rev. Lett.* **23**, p. 17.
Grimmett, G. (1999). *Percolation* (Springer-Verlag).
Grinstein, G., Jayaprakash, C. and Yu, H. (1985). *Phys. Rev. Lett.* **55**, p. 2527.
Grinstein, G., Lai, Z.-W. and Browne, D. A. (1989). *Phys. Rev. A* **40**, p. 4820.
Gropengiesser, U. (1994). *Physisca A* **207**, p. 492.
Grosskinsky, S., Schütz, G. M. and Spohn, H. (2003). *J. Stat. Phys.* **113**, p. 389.
Guida, R. and Zinn-Justin, J. (1998). *J.Phys. A* **31**, p. 8103.
H. W. J. Blote, A. H., J. R. Heringa and Zia, R. K. P. (1990). *Int. J. Mod. Phys. B* **5**, p. 685.
H. Yang, G. C. W. and Lu, T. M. (1994). *Phys. Rev. Lett.* **73**, p. 2348.
Halperin, B. I. and Hohenberg, P. C. (1967). *Phys. Rev. Lett.* **19**, p. 700.
Halpin-Healy, T. (1990). *Phys. Rev. A* **42**, p. 711.
Halpin-Healy, T. and Zhang, Y.-C. (1995). *Phys. Rep.* **254**, p. 215.
Hammal, O. A., Bonachela, J. A. and Munoz, M. A. (2007). *J. Stat. Mech.* p. P12007.
Harris, A. B. (1974a). *J. Phys. C* **7**, p. 1671.
Harris, T. E. (1974b). *Ann. Prob.* **2**, p. 969.
Hasenbuch, M. (2001). *J.Phys. A* **34**, pp. 8221–8236.
Havlin, S. and ben Avraham, D. (1987). *Adv. Phys.* **36**, p. 695.
Henkel, M. (1999). *Conformal Invariance and Critcal Phenomena* (Springer).
Henkel, M. (2002). *Nucl. Phys. B* **641**, p. 405.
Henkel, M. (2007). *J. Phys. Cond. Matter* **19**, p. 065101.
Henkel, M. and Baumann, F. (2007). *J. Stat. Mech.* p. P07015.
Henkel, M., Enns, T. and Pleimling, M. (2006). *J. Phys. A* **39**, p. L589.
Henkel, M. and Hinrichsen, H. (2001). *J. Phys. A* **34**, p. 1561.
Henkel, M., Orlandini, M. and Santos, J. (1997). *Ann. Phys. (NY)* **259**, p. 163.
Henkel, M. and Pleimling, M. (2003). *Phys. Rev. E* **68**, p. 065101(R).
Henkel, M. and Pleimling, M. (2005). *J. Phys. Cond. Matt* **S1899**, p. 17.
Henkel, M., Pleimling, M., Godreche, C. and Luck, J.-M. (2001). *Phys. Rev. Lett.* **87**, p. 265701.
Henkel, M. and Schütz, G. (2004). *J. Phys. A: Math. Gen.* **37**, p. 591.
Herring, C. (1950). *J. Appl. Phys.* **21**, p. 301.
Hieida, Y. (1998). *J. Phys. Soc. Jpn.* **67**, p. 369.

Hilhorst, H. J., Washenberger, M. and C, T. U. (2004). *J. Stat. Mech.* p. P10002.
Hinrichsen, H. (1997). *Phys. Rev. E* **55**, p. 219.
Hinrichsen, H. (2000a). *Adv. Phys.* **49**, p. 815.
Hinrichsen, H. (2000b). *J. Braz. Phys.* **30**, pp. 69–82.
Hinrichsen, H. (2000c). First-order transitions in fluctuating 1+1-dimensional nonequilibrium systems, Eprint: cond-mat/0006212.
Hinrichsen, H. (2001a). *Phys. Rev. E* **63**, pp. 036102–1.
Hinrichsen, H. (2001b). *Physica A* **291**, p. 275.
Hinrichsen, H. (2001c). *Phys. Rev. E* **63**, pp. 16109.
Hinrichsen, H. (2003). *Physica A* **320**, p. 249.
Hinrichsen, H. (2006). *J. Stat. Mech.* p. L06001.
Hinrichsen, H. (2007a). *J. Stat. Mech.* p. P07066.
Hinrichsen, H. (2007b). Ageing in homogeneous systems at criticality, Eprint: arXiv:0711.1106.
Hinrichsen, H. (2007c). Dynamical response function of the disordered kinetic ising model, Eprint: arXiv:0711.2421.
Hinrichsen, H. and Domany, E. (1997). *Phys. Rev. E* **56**, p. 94.
Hinrichsen, H., Domany, E. and Stauffer, D. (1998). *J. Stat. Phys.* **91**, pp. 807–814.
Hinrichsen, H. and Howard, M. (1999). *Eur. Phys. J. B* **7**, p. 635.
Hinrichsen, H. and Koduvely, H. M. (1998). *Eur. Phys. J. B* **5**, p. 257.
Hinrichsen, H., Livi, R., Mukamel, D. and Politi, A. (1997a). *Phys. Rev. Lett.* **79**, p. 2710.
Hinrichsen, H., Livi, R., Mukamel, D. and Politi, A. (2000). *Phys. Rev. E* **61**, p. R1032.
Hinrichsen, H. and Mukamel, D. (2003). *Phys. Rev. E* **67**, p. 016110.
Hinrichsen, H. and Ódor, G. (1998). *Phys. Rev. E* **58**, p. 311.
Hinrichsen, H. and Ódor, G. (1999a). *Phys. Rev. E* **60**, p. 3842.
Hinrichsen, H. and Ódor, G. (1999b). *Phys. Rev. Lett.* **82**, p. 1205.
Hinrichsen, H., Weitz, S. and Domany, E. (1997b). *J. Stat. Phys.* **88**, p. 617.
Hohenberg, P. C. and Halperin, B. I. (1977). *Rev. Mod. Phys.* **49**, p. 435.
Hooyberghs, J., Carlon, E. and Vanderzande, C. (2001). *Phys. Rev. E* **64**, p. 036124.
Hooyberghs, J., Iglói, F. and Vanderzande, C. (2003). *Phys. Rev. Lett.* **90**, p. 100601.
Hooyberghs, J., Iglói, F. and Vanderzande, C. (2004). *Phys. Rev. E* **69**, p. 066140.
Howard, M., Fröjdh, P. and Lauritsen, K. B. (2000). *Phys. Rev. E* **61**, p. 167.
Howard, M. J. (1996). *J. Phys. A* **29**, p. 3437.
Howard, M. J. and Barkema, G. T. (1996). *Phys. Rev. E* **53**, p. 5949.
Howard, M. J. and Täuber, U. C. (1997). *J. Phys. A* **30**, p. 7721.
Huse, D. A. (1989). *Phys. Rev. B* **40**, p. 304.
Huse, D. A. and Henley, C. L. (1985). *Phys. Rev. Lett.* **54**, p. 2708.
Hwa, T. (1992). *Phys. Rev. Lett.* **69**, p. 1552.
Hwang, W., Kwon, S., Park, H. and Park, H. (1998). *Phys. Rev. E* **57**, p. 6438.
Hwang, W. and Park, H. (1999). *Phys. Rev. E* **59**, p. 4683.
I. Ispolatov, S. R., P. L. Krapivsky (1995). *Phys. Rev. E* **52**, p. 2540.

Iglói, F. and Carlon, E. (1999). *Phys. Rev. B* **59**, p. 3783.
Iglói, F. and Monthus, C. (2005). *Physics Reports* **412**, p. 277.
Iglói, F., Peschel, I. and Turban, L. (1993). *Adv. Phys.* **42**, p. 683.
Inui, N., Tretyakov, A. Y. and Takayasu, H. (1995). *J. Phys. A* **28**, p. 1145.
Ising, E. (1925). *Z. Phys.* **31**, p. 253.
Itzykson, C. and Drouffe, J. M. (1989). *Statistical Field Theory* (Camb. U. Press, Cambridge).
J.-Dalmaroni, A. and Hinrichsen, H. (2003). *Phys. Rev. E* **68**, p. 036103.
Jabeen, Z. and Gupte, N. (2005). *Phys. Rev. E* **72**, p. 016202.
Jack, R., Mayer, P. and Sollich, P. (2006). *J. Stat. Mech* , p. P03006.
Jan, N. and de Arcangelis, L. (1994) (World Scientific, Singapore), p. 1.
Janke, W. and Schakel, A. M. J. (2006). *Braz. J. Phys.* **36**, p. 708.
Janowsky, S. A. (1995). *Phys. Rev. E* **51**, p. 1858.
Janssen, H. K. (1976). *Z. Phys. B* **23**, p. 377.
Janssen, H. K. (1981). *Z. Phys. B* **42**, p. 151.
Janssen, H. K. (1985). *Phys. B* **58**, p. 311.
Janssen, H. K. (1992). *From Phase Transition to Chaos* (World Scientific, Singapore).
Janssen, H. K. (1997a). *Phys. Rev. E* **55**, p. 6253.
Janssen, H. K. (1997b). *Phys. Rev. Lett.* **78**, p. 2890.
Janssen, H. K. (2001). *J. Stat. Phys.* **103**, p. 801.
Janssen, H. K. (2003). Survival and percolation probabilities in the field theory of growth models, Eprint: cond-mat/0304631.
Janssen, H. K. (2005). *J. Phys.: Cond. Mat.* **17**, p. S1973.
Janssen, H. K., Müller, M. and Stenull, O. (2004a). *Phys. Rev. E* **70**, p. 026114.
Janssen, H. K., Oerding, K. and Sengespeick, E. (1995). *J. Phys. A* **28**, p. 6073.
Janssen, H. K., Oerding, K., van Wijland, F. and Hilhorst, H. (1999). *Eur. Phys. J. B* **7**, p. 137.
Janssen, H. K., Schaub, B. and Schmittman, B. (1989). *Z. Phys.* **73**, p. 539.
Janssen, H. K., Schaub, B. and Schmittmann, B. (1988). *Z. Phys. B* **72**, p. 111.
Janssen, H. K. and Schmittman, B. (1986). *Z. Phys. B* **64**, p. 503.
Janssen, H. K. and Schmittmann, B. (1986). *Z. Phys. B* **63**, p. 517.
Janssen, H. K., van Wijland, F., Deloubriere, O. and Täuber, U. C. (2004b). *Phys. Rev. E* **70**, p. 056114.
Jaster, A., Mainville, J., Schuelke, L. and Zheng, B. (1999). *J. Phys. A* **32**, p. 1395.
Jeffries, J. H., Zuo, J. K. and Craig, M. M. (1996). *Phys. Rev. Lett.* **76**, p. 4931.
Jensen, H. J. (1998). *Self-Organized Criticality* (Cambridge Univ. Press, Cambridge England).
Jensen, I. (1991). *Phys. Rev. A* **43**, p. 3187.
Jensen, I. (1992). *Phys. Rev. E* **45**, p. R563.
Jensen, I. (1993a). *Phys. Rev. Lett.* **70**, p. 1465.
Jensen, I. (1993b). *Phys. Rev. E* **47**, p. R1.
Jensen, I. (1993c). *J. Phys. A* **26**, p. 3921.
Jensen, I. (1994). *Phys. Rev. E* **50**, p. 3623.
Jensen, I. (1996a). *Phys. Rev. Lett.* **77**, p. 4988.

Jensen, I. (1996b). *Phys. Rev. Lett.* **77**, p. 4988.
Jensen, I. (1997). *J. Phys. A* **30**, p. 8471.
Jensen, I. (1999a). *J. Phys. A* **32**, p. 5233.
Jensen, I. (1999b). *J. Phys. A* **32**, p. 6055.
Jensen, I. (2005). *J. Phys. A* **38**, p. 1441.
Jensen, I. and Dickman, R. (1993a). *Phys. Rev. E* **48**, p. 1710.
Jensen, I. and Dickman, R. (1993b). *Phys. Rev. E* **48**, p. 1710.
Jensen, I. and Dickman, R. (1993c). *J. Stat. Phys.* **71**, p. 89.
Jensen, I. and Dickman, R. (1993d). *J. Phys. A* **26**, p. L151.
Jensen, I. and Guttmann, A. J. (1995). *J. Phys. A* **28**, p. 4813.
Jensen, I. and Guttmann, A. J. (1996). *Nucl. Phys. B* **47**, p. 835.
Jeong, H., Kahng, B. and Kim, D. (1999). *Phys. Rev. E* **59**, p. 1570.
Juhász, R., Santen, L. and Iglói, F. (2005). *Phys. Rev. Lett.* **94**, p. 010601.
K. A, M. and Peak, D. (2006). *Annals of Botany* **99**, p. 219.
K. Park, S. K. and mook Kim, I. (2005). *Phys. Rev. E* **71**, p. 066129.
Kadanoff, L. P. *et al.* (1967). *Rev. Mod. Phys.* **39**, p. 395.
Kaiser, C. and Turban, L. (1994). *J. Phys. A* **27**, p. L579.
Kaiser, C. and Turban, L. (1995). *J. Phys. A* **28**, p. 351.
Kamphorst, J., da Silva, L. and Dickman, R. (1999). *Phys. Rev. E* **60**, p. 5126.
Kang, K. and Redner, S. (1985). *Phys. Rev. A* **32**, p. 435.
Kardar, M. (1987). *Nucl. Phys. B* **290**, p. 582.
Kardar, M., Parisi, G. and Zhang, Y. (1986). *Phys. Rev. Lett.* **56**, p. 889.
Katz, S., Lebowitz, J. L. and Spohn, H. (1984). *J. Stat. Phys.* **34**, p. 497.
Kauffman, S. A. (1969). *J. Theor. Biol.* **22**, p. 437.
Kawasaki, K. (1966). *Phys. Rev.* **145**, p. 224.
Kawasaki, K. (1972). in C. D. C and M. S. Green (eds.), *Phase Transitions and Critical Phenomena*, Vol. 2 (New York: Academic), p. 443.
Keldysh, L. V. (1964). *Zh. Eksp. Theor. Teor. Fiz.* **47**, p. 1515.
Kertész, J. and Wolf, D. E. (1989). *Phys. Rev. Lett.* **62**, p. 2571.
Kim, H., Park, K. and Kim, I. (2001). *Phys. Rev. E* **65**, p. 017104.
Kim, J. M. and Kosterlitz, J. M. (1989). *Phys. Rev. Lett.* **62**, p. 2289.
Kim, M. H. and Park, H. (1994). *Phys. Rev. Lett.* **73**, p. 2579.
Kinzel, W. (1983). in G. Deutscher, R. Zallen and J. Adler (eds.), *Percolation Structures and Processes*, Vol. 5 (Hilger, Bristol).
Kinzel, W. (1985). *Z. Phys. B* **58**, p. 229.
Kiyono, K., Struzik, Z. R., Aoyagi, N., Togo, F. and Yamamoto, Y. (2005). *Phys. Rev. Lett.* **95**, p. 058101.
Kiyono, K., Struzik, Z. R. and Yamamoto, Y. (2006). *Phys. Rev. Lett.* **96**, p. 068701.
Knapek, C. A., Ivlev, A. V., Klumov, B. and *et al.* (2007). *Phys. Rev. Lett.* **98**, p. 015001.
Kockelkoren, J. and Chaté, H. (2003a). *Phys. Rev. Lett.* **90**, p. 125701.
Kockelkoren, J. and Chaté, H. (2003b). Absorbing phase transitions with coupling to a static field and a conservation law, Eprint: cond-mat/0306039.
Koiller, B. and Robbins, M. (2000). *Phys. Rev. B* **62**, p. 5771.

Kolakowska, A., Novotny, M. A. and Verma, P. S. (2004). *Phys. Rev. E* **70**, p. 051602.

Kolton, A. B., Rosso, A., Giamarchi, T. and Krauth, W. (2006). *Phys. Rev. Lett.* **97**, p. 057001.

Konkoli, Z., Johannesson, H. and Lee, B. P. (1999). *Phys. Rev. Lett.* **59**, p. R3787.

Kosterlitz, J. M. and Thouless, D. J. (1973). *J. Phys. C* **6**, p. 1181.

Krapivsky, P. L. (1992). *Phys. Rev. A* **45**, p. 1067.

Krapivsky, P. L. and Ben-Naim, E. (1997). *Phys. Rev. E* **56**, p. 3788.

Krapivsky, P. L., Ben-Naim, E. and Redner, S. (1994). *Phys. Rev. E* **50**, p. 2474.

Kree, R., Schaub, B. and Schmitmann, B. (1989). *Phys. Rev. A* **39**, p. 2214.

Kroon, R., Fleurent, H. and Sprik, R. (1993). *Phys. Rev. E* **47**, p. 2462.

Krug, J. (1991). *Phys. Rev. Lett.* **67**, p. 1882.

Krug, J. (1997). *Adv. Phys.* **46**, p. 139.

Krug, J. and Ferrari, P. A. (1996). *J. Phys. A* **29**, p. L465.

Kun, F. and Hermann, H. J. (1999). *Phys. Rev. E* **59**, p. 2623.

Kuramoto, Y. and Tsuzuki, T. (1977). *Prog. Theor. Phys.* **55**, p. 356.

Kwon, S., Lee, J. and Park, H. (2000). *Phys. Rev. Lett.* **85**, p. 1682.

Kwon, S. and Park, H. (1995). *Phys. Rev. E* **52**, p. 5955.

Landau, L. D. and Lifshitz, E. M. (1981). *Statistical Mechanics* (Pergamon, London).

Lässig, M. (1995). *Nucl. Phys. B* **559**.

Lässig, M. and Kinzelbach, H. (1997). *Phys. Rev. Lett.* **78**, p. 906.

Lauritsen, K. B., Fröjdh, P. and Howard, M. (1998). *Phys. Rev. Lett.* **81**.

Lauritsen, K. B., Sneppen, K., Markovsová, M. and Jensen, M. H. (1997). *Physica A* **247**, p. 1.

Lawrie, I. and Sarbach, S. (1984). in C. Domb and J. Lebowitz (eds.), *Phase Transitions and Critical Phenomena*, Vol. 9 (Academic Press, London).

Le Doussal, P. and Wiese, K. J. (2003). *Phys. Rev. B* **68**, p. 174202.

Le Doussal, P. and Wiese, K. J. (2005). *Phys. Rev. E* **72**, p. 035101(R).

Lee, B. P. (1994). *J. Phys. A* **27**, p. 2633.

Lee, B. P. and Cardy, J. (1995). *J. Stat. Phys.* p. 971.

Lee, J. W. (2000). *Physica A* **292**, p. 231.

Lehner, C., Rajewsky, N., Wolf, D. and Kertész, J. (1990). *Physica A* **164**, p. 81.

Lei, X. W. and Zheng, B. (2007). *Phys. Rev. E* **75**, p. 040104.

Leschhorn, H. (1996). *Phys. Rev. E* **54**, p. 1313.

Leung, K. T. and Cardy, J. L. (1986). *J. Stat. Phys.* **44**, p. 567.

Ligget, T. (1985). *Interacting Particle Systems* (Springer-Verlag, Berlin).

Lipowski, A. (1996). *J. Phys. A* **29**, p. L355.

Lipowski, A. (1999). *Phys. Rev. E* **60**, p. R6255.

Lipowski, A. and Droz, M. (2001). *Phys. Rev. E* **64**, p. 031107.

Lipowski, A. and Droz, M. (2002a). *Phys. Rev. E* **65**, p. 056114.

Lipowski, A. and Droz, M. (2002b). *Phys. Rev. E* **66**, p. 016106.

Lipowski, A. and Lopata, M. (1999). *Phys. Rev. E* **60**, p. 1516.

Lipowsky, R. (1982). *Phys. Rev. Lett.* **49**, p. 1575.

Lippiello, E. and Zannetti, M. (2000). *Phys. Rev. E* **61**, p. 3369.

López, C. and Muñoz, M. A. (1997). *Phys. Rev. E* **56**, p. 4864.

López, J. M. and Rodríguez, M. A. (1996). *Phys. Rev. E* **54**, p. R2189.
López, J. M. and Rodríguez, M. A. (1997). *J. Phys. I* **7**, p. 1191.
López, J. M. and Schmittbuhl, J. (1998). *Phys. Rev. E* **57**, p. 6405.
Lorenz, E. and Janke, W. (2007). *Eur. Phys. Lett.* **77**, p. 10003.
Lu, E. and Hamilton, R. (1991). *Astrophys. J.* **380**, p. L89.
Lübeck, S. (2001). *Phys. Rev. E* **64**, p. 016123.
Lübeck, S. (2002). *Phys. Rev. E* **65**, p. 046150.
Lübeck, S. (2004). *Int. J. of Mod. Phys. B* **18**, p. 3977.
Lübeck, S. (2006a). *J. Stat. Mech.* p. P09009.
Lübeck, S. (2006b). *J. Stat. Phys.* **123**, p. 193.
Lübeck, S. and Heger, P. (2003a). *Phys. Rev. E* **68**, p. 056102.
Lübeck, S. and Heger, P. (2003b). *Phys. Rev. Lett.* **90**, p. 230601.
Lübeck, S. and Hucht, A. (2002). *J. Phys. A* **35**, p. 4853.
Lübeck, S. and Willmann, R. D. (2005). *Nuclear Physics B* **718**, p. 348.
Lushnikov, A. A. (1987). *Phys. Lett. A* **120**, p. 135.
L'vov, V. S., Lebedev, V. V., Paton, M. and Procaccia, I. (1993). *Nonlinearity* **6**, p. 25.
Ma, S. K. (1976). *Modern Theory of Critical Phenomena* (Addison-Wesley).
Maia, M. D. and Dickman, R. (2007). *J. Cond. Matt.* **19**, p. 065143.
Majaniemi, S., A-Nissila, T. and Krug, J. (1996). *Phys. Rev. B* **53**, p. 8071.
Majumdar, S. N., Bray, A. J., Cornell, S. J. and Sire, C. (1996). *Phys. Rev. Lett.* **77**, p. 3704.
Majumdar, S. N., Dean, D. S. and Grassberger, P. (2001). *Phys. Rev. Lett.* **86**, p. 2301.
Majumdar, S. N., Huse, D. and Lubachevsky, B. (1994). *Phys. Rev. Lett.* **73**, p. 182.
Majumdar, S. N., Krishnamurthy, S. and Barma, M. (1998). *Phys. Rev. Lett.* **81**, p. 3691.
Majumdar, S. N., Krishnamurthy, S. and Barma, M. (2000a). *Phys. Rev. E* **61**, p. 6337.
Majumdar, S. N., Krishnamurthy, S. and Barma, M. (2000b). *J. Stat. Phys.* **99**, p. 1.
Makeev, M., Cuerno, R. and Barabási, A.-L. (2002). *Nucl. Inst. and Meth. B* **197**, pp. 185–227.
Mandache, C. and ben Avraham, D. (2000). *J. Chem. Phys.* **112**, p. 7735.
Manna, S. S. (1991). *J. Phys. A* **24**, p. L363.
Margaritis, A., Ódor, G. and Patkós, A. (1988). *Z. Phys. C.* **39**, p. 109.
Marinari, E., Pagnani, A. and Parisi, G. (2000). *J. Phys. A* **33**, p. 8181.
Marinari, E., Pagnani, A., Parisi, G. and Rácz, Z. (2002). *Phys. Rev. E* **65**, p. 026136.
Mariz, A. M., Herrmann, H. J. and de Arcangelis, L. (1990). *J. Stat. Phys.* **59**, p. 1043.
Marques, M. C. (1989). *J. Phys. A* **22**, p. 4493.
Marques, M. C. (1990). *Phys. Lett. A* **145**, p. 379.
Marques, M. C. and Ferreira, A. L. (1994). *J. Phys. A* **27**, p. 3389.
Marques, M. C. and Mendes, J. F. (1999). *Eur. Phys. B* **12**, p. 123.

Marques, M. C., Santos, M. A. and Mendes, J. F. (2001). *Phys. Rev. E* **65**, p. 016111.

Marro, J. and Dickman, R. (1999). *Nonequilibrium Phase Transitions in Lattice Models* (Cambridge University Press, Cambridge).

Marro, J., Lebowitz, J. and Kalos, M. H. (1979). *Phys. Rev. Lett.* **43**, p. 282.

Marsili, M. (1994). *J. Stat. Phys.* **77**, p. 733.

Martin, P. C., Siggia, E. D. and Rose, H. A. (1973). *Phys. Rev. A* **8**, p. 423.

Martins, M. L., Ferreira, S. and Vilela, M. J. (2007). *Phys. of Life Rev.* **4**, p. 128.

Mayer, P., Léonard, S., Berthier, L., Garrahan, J. P. and Sollich, P. (2006). *Phys. Rev. Lett.* **96**, p. 030602.

Mayer, P. and Sollich, P. (2007). *J. Phys. A* **40**, p. 5823.

Mazenko, G. (2004). *Phys. Rev. E* **69**, p. 016114.

McDonald, C. T., Gibbs, J. H. and Pipkin, A. C. (1968). *Biopolymers* **6**, p. 1.

McKane, A. J. (1977). *J. Phys. G* **3**, p. 1165.

Meakin, P., Ramanlal, P., Sander, L. and Ball, R. (1986). *Phys. Rev. A* **34**, p. 5091.

Mendes, J. F. F., Dickman, R., Henkel, M. and Marques, M. C. (1994). *J. Phys. A* **27**, p. 3019.

Mendes, J. F. F., Dickman, R. and Hermann, H. (1996). *Phys. Rev. E* **54**, pp. R3071–R3074.

Menon, G. I. and Ray, P. (2001). *J. Phys. A* **34**, p. L735.

Menyhárd, N. (1994). *J. Phys. A* **27**, p. 6139.

Menyhárd, N. and Ódor, G. (1995). *J. Phys. A.* **28**, p. 4505.

Menyhárd, N. and Ódor, G. (1996). *J. Phys. A.* **29**, p. 7739.

Menyhárd, N. and Ódor, G. (1997). *J. Phys. A* **30**, p. 8515.

Menyhárd, N. and Ódor, G. (1998). *J. Phys. A* **31**, p. 6771.

Menyhárd, N. and Ódor, G. (2000). *Braz. J. of Phys.* **30**, p. 113.

Menyhárd, N. and Ódor, G. (2002). *Phys. Rev. E* **66**, p. 016127.

Menyhárd, N. and Ódor, G. (2003). *Phys. Rev. E* **68**, p. 056106.

Menyhárd, N. and Ódor, G. (2006). *Phys. Rev. E* **73**, p. 036130.

Menyhárd, N. and Ódor, G. (2007). *Phys. Rev. E* **76**, p. 021103.

Mermin, N. D. and Wagner, H. (1996). *Phys. Rev. Lett.* **17**, p. 1133.

Mézard, M. and Parisi, G. (1991). *J. Phys. I* **1**, p. 809.

Mohanty, P. and Dhar, D. (2002). *Phys. Rev. Lett.* **89**, p. 104303.

Mollison, D. (1977). *J. R. Stat. Soc. B* **39**, p. 283.

Monson, E. and Kopelman, R. (2004). *Phys. Rev. E* **69**, p. 021103.

Monthus, C. and Garel, T. (2004). *Phys. Rev. E* **69**, p. 061112.

Monthus, C. and Garel, T. (2005). *J. Stat. Mech.* p. P12011.

Moreira, A. G. and Dickman, R. (1996). *Phys. Rev. E* **54**, p. R3090.

Morel, S. *et al.* (1998). *Phys. Rev. E* **58**, p. 6999.

Moser, K., Wolf, D. E. and Kertész, J. (1991). *Physica A* **178**, p. 215.

Mullins, W. W. (1957). *J. Appl. Phys.* **28**, p. 333.

Muñoz, M. A. (2004). Nonequilibrium phase transitions and multiplicative noise, in E. Korutcheva and R. Cuerno (eds.), *Advances in Condensed Matter and Statistical Mechanic* (New York: Nova Science Publishers).

Muñoz, M. A., Dickman, R., Grinstein, G. and Livi, R. (1996). *Phys. Rev. Lett.*

**76**, p. 451.
Muñoz, M. A., Dickman, R., P-Satorras, R., Vespignani, A. and Zapperi, S. (2001). in J. Marro and P. L. Garrido (eds.), *Proc. of the 6th Granada Seminar on Computation Physics*, Vol. 574 (AIP Conference Proceedings).
Muñoz, M. A., Dickman, R., Vespagnani, A. and Zapperi, S. (1999). *Phys. Rev. E* **59**, p. 6175.
Muñoz, M. A., Grinstein, G. and Dickman, R. (1998). *J. Stat. Phys.* **91**, p. 541.
Muñoz, M. A., Grinstein, G. and Tu, Y. (1997). *Phys. Rev. E* **56**, p. 5101.
Mussawisade, K., Santos, J. E. and Schütz, G. M. (1998). *J. Phys. A* **31**, p. 4381.
Nagy, M., Daruka, I. and Vicsek, T. (2006). *Physica A* **373**, p. 445.
Narayan, O. and Fisher, D. S. (1993). *Phys. Rev. B* **48**, p. 7030.
Nattermann, T., Stepanow, S., Tang, L.-H. and Leschhorn, H. (1992). *J. Phys. II* **2**, p. 1483.
Neugebauer, C. J. and Taraskin, S. N. (2007). *Phys. Rev. E* **76**, p. 011119.
Noest, A. J. (1986). *Phys. Rev. Lett.* **57**, p. 90.
Noest, A. J. (1988). *Phys. Rev. B* **38**, p. 2715.
Noh, J. D., Park, H. and den Nijs, M. (2000). *Phys. Rev. Lett.* **84**, p. 3891.
Noh, J. D., Park, H., Kim, D. and den Nijs, M. (2001). *Phys. Rev. E* **64**, p. 046131.
O'Donoghue, S. J. and Bray, A. J. (2001). *Phys. Rev. E* **64**, p. 041105.
Ódor, G. (1995). *Phys. Rev. E* **51**, p. 6261.
Ódor, G. (2000). *Phys. Rev. E* **62**, p. R3027.
Ódor, G. (2001a). *Phys. Rev. E* **63**, p. 067104.
Ódor, G. (2001b). *Phys. Rev. E* **63**, p. 056108.
Ódor, G. (2001c). *Phys. Rev. E* **63**, p. 021113.
Ódor, G. (2001d). *Phys. Rev. E* **63**, p. 056108.
Ódor, G. (2002). *Phys. Rev. E* **65**, p. 026121.
Ódor, G. (2003a). *Phys. Rev. E* **67**, p. 056114.
Ódor, G. (2003b). *Braz. J. of Phys.* **33**, p. 431.
Ódor, G. (2003c). *Phys. Rev. E* **67**, p. 016111.
Ódor, G. (2004a). *Rev. Mod. Phys.* **76**, p. 663.
Ódor, G. (2004b). *Phys. Rev. E* **70**, p. 026119.
Ódor, G. (2004c). *Phys. Rev. E* **70**, p. 036122.
Ódor, G. (2004d). *Phys. Rev. E* **70**, p. 066122.
Ódor, G. (2006a). *J. Stat. Mech.* p. L11002.
Ódor, G. (2006b). *Phys. Rev. E* **73**, p. 047103.
Ódor, G. (2008). *Int. J. of Mod. Phys. C* **19**.
Odor, G., Mendes, J. F., Santos, M. A. and Marques, M. C. (1998). *Phys. Rev. E* **58**, p. 7020.
Ódor, G., Boccara, N. and Szabó, G. (1993). *Phys. Rev. E* **48**, p. 3168.
Ódor, G., Krikelis, A., Vesztergombi, G. and Rohrbach, F. (1999). Proceedings of the 7-th euromicro workshop on parallel and distributed process funchal (portugal), Eprint: physics/9909054.
Ódor, G. and Menyhárd, N. (1998). *Phys. Rev. E* **57**, p. 5168.
Ódor, G. and Menyhárd, N. (2000). *Phys. Rev. E* **61**, p. 6404.
Ódor, G. and Menyhárd, N. (2002). *Physica D* **168**, p. 305.

Ódor, G., Santos, M. A. and Marques, M. (2002). *Phys. Rev. E* **65**, p. 056113.
Ódor, G. and Szolnoki, A. (1996). *Phys. Rev. E* **53**, p. 2231.
Ódor, G. and Szolnoki, A. (2005). *Phys. Rev. E* **71**, p. 066128.
Oerding, K. (1996a). *J. Phys. A* **29**, p. 7051.
Oerding, K. (1996b). *J. Phys. A* **29**, p. 7051.
Oerding, K., Cornell, S. J. and Bray, A. J. (1997). *Phys. Rev. E* **56**, p. R25.
Oerding, K. and van Wijland, F. (1998). *J. Phys. A* **31**, p. 7011.
Oerding, K., van Wijland, F., Leroy, J. P. and Hilhorst, H. J. (2000). *J. Stat. Phys.* **99**, p. 1365.
Ohtsuki, T. and Keyes, T. (1986). *Phys. Rev. A* **35**, p. 2697.
Ohtsuki, T. and Keyes, T. (1987). *Phys. Rev. A* **36**, p. 4434.
Oliveira, M. J. (1992). *J. Stat. Phys.* **66**, p. 273.
Oliveira, M. J., Mendes, J. F. F. and Santos, M. A. (1993). *J. Phys. A* **26**, p. 2317.
Onsager, L. (1944). *Phys. Rev.* **65**, p. 117.
Ovchinnikov, A. A. and Zel'dovich, Y. B. (1978). *Chem. Phys.* **28**, p. 215.
Park, H., Kim, M. H. and Park, H. (1995). *Phys. Rev. E* **52**, p. 5664.
Park, H. and Park, H. (1995). *Physica A* **221**, p. 97.
Park, H. S. and Park, H. (2001). *J. of Kor. Phys. Soc.* **38**, p. 494.
Park, J.-M. and Deem, M. W. (1998). *Phys. Rev. E* **57**, p. 3618.
Park, K., Hinrichsen, H. and Kim, I. (2002). *Phys. Rev. E* **66**, p. 025101.
Park, K., Hinrichsen, H. and mook Kim, I. (2001). *Phys. Rev. E* **63**, p. 065103.
Park, K. and Kim, I. (2002). *Phys. Rev. E* **66**, p. 027106.
Park, S., Kim, D. and Park, J. (2000). *Phys. Rev. E* **62**, p. 7642.
Park, S. and Park, H. (2005a). *Phys. Rev. Lett.* **94**, p. 065701.
Park, S. and Park, H. (2005b). *Phys. Rev. E.* **71**, p. 016137.
Park, S. and Park, H. (2006). *Phys. Rev. E* **73**, p. 025105(R).
Park, S. and Park, H. (2007). *Phys. Rev. E* **76**, p. 051123.
Pastor-Satorras, R. and Vespagnani, A. (2000). *Phys. Rev. E* **62**, p. 5875.
Paul, R., Puri, S. and Rieger, H. (2004). *Eur. Phys. Lett.* **68**, p. 881.
Pázmándi, K. B. F. and Batrouni, G. (2000). *Phys. Rev. Lett.* **80**, p. 4477.
Pelissetto, A. and Vicari, E. (2000a). *Phys. Rep.* **368**, p. 549.
Pelissetto, A. and Vicari, E. (2000b). *Phys. Rev. B* **62**, p. 6393.
Peliti, L. (1985). *J. Phys.* **46**, p. 1496.
Peliti, L. (1986). *J. Phys. A* **19**, p. L365.
Peters, O., Hertlein, C. and Christensen, K. (2002). *Phys. Rev. Lett.* **88**, p. 018701.
Picone, A. and Henkel, M. (2002). *J. Phys. A* **35**, p. 5575.
Picone, A. and Henkel, M. (2004). *Nucl. Phys.* **B688**, p. 217.
Pleimling, M. and Gambassi, A. (2005). *Phys. Rev. B* **71**, p. 180401(R).
Pleimling, M. and Iglói, F. (2007). *Europhys. Lett.* **79**, p. 56002.
Politzer, P. A. (2000). *Phys. Rev. Lett.* **84**, p. 1192.
Potts, R. B. (1952). *Proc. Camb. Phil. Soc.* **48**, p. 106.
Praestgaard, E., Schmittmann, B. and Zia, R. K. P. (2000). *Eur. Phys. J. B* **18**, p. 675.

Praestraad, E. *et al.* (1994). *Europhys. Lett.* **25**, p. 447.
Privman, V. (1996). *Nonequilibrium Statistical Mechanics in one Dimension* (Cambridge University Press).
Privman, V., Cadilhe, A. M. R. and Glasser, M. L. (1995). *J. Stat. Phys.* **81**, p. 881.
Prudnikov, V., Belim, S., Osintsev, E. and Fedorenko, A. (1998). *Phys. Sol. State* **40**, p. 1383.
Pruessner, G. (2007). Unconventional finite size scaling in the directed percolation universality class, Eprint: arXiv:0706.1144.
Rácz, Z. (1976). *Phys. Rev. B* **13**, p. 263.
Rácz, Z. (1985). *Phys. Rev. Lett.* **55**, p. 1707.
Rácz, Z. (1996). Kinetic ising models with competing dynamics: Mappings, correlations, steady states and phase transitions, in V. Privman (ed.), *Nonequilibrium Statistical Mechanics in one Dimension* (Cambridge University Press).
Rácz, Z. (2002). Nonequilibrium phase transitions, in *Lecture Notes, Les Houches*, eprint: cond-mat/0210435.
Rácz, Z. and Zia, R. K. P. (1994). *Phys. Rev. E* **49**, p. 139.
Rajesh, R. and Majumdar, S. N. (2001). *Phys. Rev. E* **63**, p. 036114.
Rajewsky, N., Santen, L., Schadschneider, A. and Schreckenberg, M. (1998). *J. Stat. Phys.* **92**, p. 151.
Ramasco, J. J., Henkel, M., Santos, M. A. and da Silva Santos C A (2004). *J. Phys. A: Math. Gen.* **37**, p. 10497.
Ramaswamy, R. and Barma, M. (1987). *J. Phys. A* **20**, p. 2973.
Reichenbach, T., Frey, E. and Franosch, T. (2007). *New J. Phys.* **9**, p. 159.
Rey, P.-A. and Cardy, J. (1998). *J. Phys. A* **32**, p. 1585.
Richardson, M. J. E. and Cardy, J. (1999). *J. Phys. A: Math. Gen.* **32**, p. 4035.
Riedel, E. K. and Wegner, F. J. (1969). *Z. Phys.* **225**, p. 195.
Rossi, M., Pastor-Satorras, R. and Vespagnani, A. (2000). *Phys. Rev. Lett.* **85**, p. 1803.
Sales, J. A., Martins, M. L. and Moreira, J. G. (1997). *Physica A* **245**, p. 461.
Samukhin, A. N., Dorogovtsev, S. N. and Mendes, J. F. F. (2007) Eprint: arXiv:0706.1176.
Santos, M. A. and Teixeira, S. (1995). *J. Stat. Phys.* **78**, p. 963.
Sarma, S. D. *et al.* (1996). *Phys. Rev. E* **53**, p. 359.
Scalettar, F. P. R. T. and Zimányi, G. T. (1997). *Phys. Rev. Lett.* **79**, p. 5130.
Schehr, G. and Rieger, H. (2005). *Phys. Rev. B* **71**, p. 184202.
Scheidegger, A. E. (1967). *Bull. Assoc. Sci. Hydro.* **12**, p. 15.
Scheucher, M. and Spohn, H. (1988). *J. Stat. Phys.* **53**, p. 279.
Schmittman, B. (1993). *Europhys. Lett.* **24**, p. 109.
Schmittman, B. and Zia, R. K. P. (1991). *Phys. Rev. Lett.* **66**, p. 357.
Schmittman, B. and Zia, R. K. P. (1996). Statistical mechanics of driven diffusive systems, in C. Domb and J. L. Lebowitz (eds.), *Phase Transitions and Critical Phenomena*, Vol. 17 (Academic Press, London).
Schroeder, M. *et al.* (1993). *Europhys. Lett.* **24**, p. 563.

Schütz, G. (2001). Phase Transitions and Critical Phenomena (London: Academic).

Schütz, G. M. and Mussawisade, K. (1998). *Phys. Rev. E* **57**, p. 2563.

Sivashinsky, G. I. (1979). *Acta Astronaut.* **6**, p. 569.

Sneppen, K., Krug, J., Jensen, M. H., Jayaprakash, C. and Bohr, T. (1992). *Phys. Rev. A* **46**, p. R7351.

Sornette, D. (1989). *Europhys. Lett.* **9**, p. 197.

Sornette, D. (1998). *Phys. Rep.* **297**, pp. 239–270.

Stanley, H. E. (1968). *Phys. Rev.* **176**, p. 718.

Stanley, H. E. (1971). *Introduction to Phase Transitions and Critical Phenomena* (Oxford University Press, Oxford).

Stanley, H. E., Staufer, D., Kertész, J. and Herrmann, H. (1986). *Phys. Rev. Lett.* **59**, p. 2326.

Stauffer, D. (1996). *Int. J. Mod. Phys. C* **7**, p. 753.

Stauffer, D. and Aharony, A. (1994). *Introduction to Percolation Theory* (Taylor & Francis, London).

Struik, L. C. E. (1978). *Physical Ageing in Amorphus Polymers and Other Materials* (Elsevier (Amsterdam)).

Sudbury, A. (1990). *Ann. Prob.* **18**, p. 581.

Suzuki, M. (1971). *Prog. Theor. Phys.* **46**, p. 1337.

Suzuki, M. (1974). *Prog. Theor. Phys.* **51**, p. 1992.

Szabó, G. and Czárán, T. (2001). *Phys. Rev. E* **50**, p. 061904.

Szabó, G. and Ódor, G. (1994). *Phys. Rev. E* **49**, p. 2764.

Szolnoki, A. (2000). *Phys. Rev. E* **62**, p. 7466.

T. Tomé, M. J. O. and Santos, M. A. (1991). *J. Phys. A* **24**, p. 3677.

Takayasu, H. (1989). *Phys. Rev. Lett.* **63**, p. 2563.

Takayasu, H., Nishikawa, I. and Tasaki, H. (1988). *Phys. Rev. A* **37**, p. 3110.

Takayasu, H. and Tretyakov, A. Y. (1992). *Phys. Rev. Lett.* **68**, p. 3060.

Takayasu, M. and Takayasu, H. (1997). in V. Privman (ed.), *Nonequilibrium Statistical Mechanics in One Dimension.*

Takeuchi, K. A., Kuroda, M., Chaté, H. and Sano, M. (2007). Unconventional finite size scaling in the directed percolation universality class, Eprint: arXiv:0706.1144.

Tamayo, P., Alexander, F. J. and Gupta, R. (1995). *Phys. Rev. E* **50**, p. 3474.

Tang, L. and Leschhorn, H. (1992). *Phys. Rev. A* **45**, p. R8309.

Tang, L. H., Forrest, B. M. and Wolf, D. E. (1992). *Phys. Rev. A* **45**, p. 7162.

Tang, S. and Landau, D. P. (1987). *Phys. Rev. B* **36**, p. 567.

Täuber, U. C. (to be published). *Critical Dynamics: A Field Theory Approach to Equilibrium and Nonequilibrium Sealing Behavior* (Cambridge University Press).

Täuber, U. C. (2000) Unpublished, private communication.

Täuber, U. C., Howard, M. and Vollmayr-Lee, B. P. (2005). *J. Phys. A* **38**, pp. R79–R131.

Täuber, U. C., Howard, M. J. and Hinrichsen, H. (1998). *Phys. Rev. Lett.* **80**, p. 2165.

Täuber, U. C. and Rácz, Z. (1997). *Phys. Rev. E* **55**, p. 4120.

Täuber, U. C., Santos, J. E. and Rácz, Z. (1999). *Eur. Phys. J. B* **7**, p. 309.

Tomé, T. and de Oliviera, M. J. (1989). *Phys. Rev. A* **40**, p. 6643.

Toom, A. (1994a). *J. Stat. Phys.* **74**, p. 91.

Toom, A. (1994b). *J. Stat. Phys.* **74**, p. 111.

Toussaint, D. and Wilczek, F. (1983). *J. Chem. Phys.* **78**, p. 2642.

Tripathy, G. and Barma, M. (1997). *Phys. Rev. Lett.* **78**, p. 3039.

Tu, Y. (1994). *Phys. Rev. Lett.* **73**, p. 3109.

Tu, Y., Grinstein, G. and Muñoz, M. A. (1997). *Phys. Rev. Lett.* **78**, p. 274.

Tu, Y. and Toner, J. (1995). *Phys. Rev. Lett.* **75**, p. 4326.

Vallés, J. L. and Marro, J. (1987). *J. Stat. Phys.* **49**, p. 89.

van Beijeren, H., Kutner, R. and Spohn, H. (1985). *Phys. Rev. Lett.* **54**, p. 2026.

Vernon, D. and Howard, M. (2001). *Phys. Rev. E* **63**, p. 041116.

Vespignani, A. and Zapperi, S. (1997). *Phys. Rev. Lett.* **78**, p. 4793.

Vesztergombi, G., Odor, G., Rohrbach, F. and Varga, G. (1997). Eprint: CERN-OPEN-97-034.

Vicsek, T., Czirók, A., Ben-Jacob, E., Cohen, I. and Shochet, O. (1995). *Phys. Rev. Lett.* **75**, p. 1226.

Villain, J. (1991). *J. Phys. I* **1**, p. 19.

Voigt, C. A. and Ziff, R. M. (1997). *Phys. Rev. E* **56**, p. R6241.

Vojta, T. (1998). *J. Phys. A* **31**, p. 6595.

Vojta, T. (2003). *Phys. Rev. Lett.* **90**, p. 107202.

Vojta, T. (2006). *J. Phys. A* **39**, pp. R143–R205.

Vojta, T. and Lee, M. (2006). *Phys. Rev. Lett.* **96**, p. 035701.

Wang, F. and Suzuki, M. (1996). *Physica A* **223**, p. 34.

Wang, J. S. and Lebowitz, J. L. (1988). *J. Stat. Phys.* **51**, p. 893.

Washenberger, M., Mobilia, M. and Tauber, U. (2007). *J. Phys. Condens. Matter* **19**, p. 065139.

Webman, I., ben Avraham, D., Cohen, A. and Havlin, S. (1998). *Phil. Mag. B* **77**, p. 1401.

Weger, M. and Goldberg, J. B. (1973). *Solid State Physics* (Academic, New York), p. 2.

Wegner, F. J. and Houghton, A. (1972). *Phys. Rev. A* **8**, p. 401.

Werner, G. (2007). *Biosystems* **90**, p. 496.

Wijland, F. (2001). *Phys. Rev. E* **63**, p. 022101.

Wijland, F. (2002). *Phys. Rev. Lett.* **89**, p. 190602.

Wijland, F., Oerding, K. and Hilhorst, H. J. (1998). *Physica A* **251**, p. 179.

Wilson, K. G. and Kogut, J. B. (1974). *Phys. Rept.* **12**, p. 75.

Windus, A. and Jensen, H. J. (2007). *J. Phys. A* **40**, p. 2287.

Wiseman, S. and Domany, E. (1995). *Phys. Rev. E* **52**, p. 3469.

Wiseman, S. and Domany, E. (1998). *Phys. Rev. Lett.* **81**, p. 22.

Wolf, D. E. (1991). *hys. Rev. Lett.* **67**, p. 1783.

Wolf, D. E. and Kertész, J. (1987). *Europhys. Lett.* **4**, p. 651.

Wolfram, S. (1983). *Rev. Mod. Phys.* **55**, p. 601.

Wu, F. Y. (1982). *Rev. Mod. Phys.* **54**, p. 235.

Yakhot, V. (1981). *Phys. Rev. A* **24**, p. 642.

Yanguas-Gil, A., Cotrino, J., Barrancod, A. and Gonzalez-Elipe, A. R. (2006). *Phys. Rev. Lett.* **96**, p. 236101.

Hua, D.-Y. (2004). *Phys. Rev. E* **70**, p. 066101.

Ying, H. P., Zheng, B., Yu, Y. and Trimper, S. (2001). *Phys. Rev. E* **63**, p. 035101.

Zaleski, S. (1989). *Physica D* **34**, p. 427.

Zel'dovich, Y. B. and Ovchinnikov, A. A. (1977). *Sov. Phys.-JETP* **47**, p. 829.

Zheng, B. (1998). *Int. J. Mod. Phys. B* **12**, p. 1419.

Zheng, B. (2000). *Phys. Lett. A* **277**, p. 257.

Zheng, B. (2001). *Phys. Lett. A* **282**, p. 132.

Zheng, B., Ren, F. and Ren, H. (2003). *Phys. Rev. E* **68**, p. 046120.

Zhong, D. and Avraham, D. (1995). *Phys. Lett. A* **209**, p. 333.

Zhuo, J., Redner, S. and Park, H. (1993). *J. Phys. A* **26**, p. 4197.

Ziff, R., Gulari, E. and Barshad, Y. (1986). *Phys. Rev. Lett.* **56**, p. 2553.

Zinn-Justin, J. (2002). *Quantum Field Theory and Critical Phenomena* (Oxford University Press).

Zumofen, G. and Klafter, J. (1994). *Phys. Rev. E* **50**, p. 5119.

Zwerger, W. (1981). *Phys. Lett. A* **84**, p. 269.

# Index